STUDENT'S S

MODELING, FUNCTIONS, AND GRAPHS
Algebra for College Students

Second Edition

Katherine Yoshiwara
Los Angeles Pierce College

Bruce Yoshiwara
Los Angeles Pierce College

Irving Drooyan
Los Angeles Pierce College

Prepared by Laurel Technical Services

PWS Publishing Company

I(T)P An International Thomson Publishing Company
Boston • Albany • Bonn • Cincinnati • Detroit • London • Madrid • Melbourne • Mexico City • New York • Paris • San Francisco • Singapore • Tokyo • Toronto • Washington

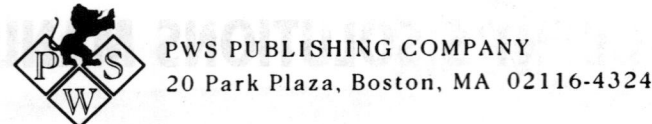

PWS PUBLISHING COMPANY
20 Park Plaza, Boston, MA 02116-4324

Copyright © 1996 by PWS Publishing Company,
  a division of International Thomson Publishing Inc.

All rights reserved. Instructors of classes adopting ***Modeling, Functions, and Graphs: Algebra for College Students, Second Edition*** by Katherine Yoshiwara, Bruce Yoshiwara, and Irving Drooyan as a required text may reproduce materials for classroom use. Otherwise, no part of this book may be reproduced, stored in a retrieval system, or transcribed, in any form or by any means–electronic, mechanical, photocopying, recording, or otherwise–without the prior written permission of the publisher, PWS Publishing Company.

I(T)P™
International Thomson Publishing
The trademark ITP is used under license.

ISBN: 0-534-95042-6

Printed and bound in the United States by Malloy.
  99 — 10 9 8 7 6

Preface

This manual contains solutions to all of the odd-numbered exercises in the text, as well as solutions to all of the exercises in the review sections. We refer to the art in the back of the textbook, where applicable, rather than reproducing it here.

Table of Contents

Chapter 1	Fundamentals	1
Chapter 2	Linear Models	39
Chapter 3	Applications of Linear Models	67
Chapter 4	Quadratic Models	107
Chapter 5	Functions and Their Graphs	155
Chapter 6	Powers and Roots	177
Chapter 7	Exponential and Logarithmic Functions	245
Chapter 8	Polynomial and Rational Functions	267
Chapter 9	Sequences and Series	295
Chapter 10	Additional Topics	323
Appendix 1	Subsets and Properties of the Real Numbers	363
Appendix 2	Review of Products and Factoring	363
Appendix 3	Complex Numbers	369

Chapter 1

Section 1.2

1. **a.** Jim is 27 years older than Ana, so $j = 27 + a$.

   b. Evaluating the above expression for $a = 22$ we get $j = 27 + 22$. So $j = 49$.

3. **a.** Rate multiplied by time equals distance. Helen must travel a distance of 1260 miles so $rh = 1260$. We solve for h by dividing both sides by r. This gives $h = \dfrac{1260}{r}$.

   b. Evaluating the equation above at $r = 45$, gives
   $h = \dfrac{1260}{45} = 28$ hours.

5. **a.** The number of feet of fabric is three times the number of yards of fabric, so $f = 3y$. We solve for y by dividing both sides by 3, therefore $y = \dfrac{f}{3}$.

   b. Fabric costs \$5.79 per yard so the cost of y yards is $5.79y$ dollars. Substituting $\dfrac{f}{3}$ in for y gives the cost of f feet of fabric
   $= 5.79\left(\dfrac{f}{3}\right)$ dollars.

   c. Evaluate the expression from b. at $f = 10$. This gives:
   cost $= 5.79\left(\dfrac{10}{3}\right) = \$19.30$

7. **a.** Area is π times radius squared, so $A = \pi r^2$.

   b. Evaluate the equation in a. at $r = 5$;
   $A = \pi(5)^2 = 25\pi = 78.5398 \text{ cm}^2$.

9. **a.** The total number of assignments equals the number Farshid turned in (20) plus the number he missed. So $n = 20 + m$.

   b. His average homework score is the total number of points (198) divided by the total number of assignments. So, $a = \dfrac{198}{n}$. Now substitute $20 + m$ in for n to get: $a = \dfrac{198}{20 + m}$.

   c. Evaluate the equation above at $m = 5$; $a = \dfrac{198}{20 + 5} = 7.92$.

11. **a.** The amount of sales tax is equal to the price of an item times the sales tax rate. So $t = p(7.9\%) = p(0.079)$.

    b. The total bill is the price plus the amount of sales tax. Using the expression from a., $b = p + p(0.079)$.

    c. Evaluate the expression in b. at $p = 490$;
    $b = 490 + 490(0.079) = \$528.71$.

13. **a.** Let C be the total cost of talking for m minutes. Then C equals \$1.97 plus \$0.39 times each minute. So $C = 1.97 + 0.39m$.

    b. Evaluate the equation from a. at $m = 27$;
    $C = 1.97 + 0.39(27) = \$12.50$.

Chapter 1, Section 1.2

15. a. Let c represent the amount of rice Juan consumes in w weeks. Since Juan consumes 0.4 lbs per week, the amount he consumes in w weeks is 0.4 times w. So $c = 0.4w$.

   b. Let r = the amount of rice left in the bag after w weeks. Since the bag originally held 50 lbs, if Juan has consumed c lbs after w weeks, then $r = 50 - c$. However, we can substitute $0.4w$ in for c, therefore the amount left is given by the formula: $r = 50 - 0.4w$.

   c. Evaluate the formula in b. at $w = 6$; $r = 50 - 0.4(6) = 47.6$ lbs.

17. a. If Leon's truck uses u gallons of gas to drive m miles then the gas mileage that Leon's truck gets is $\frac{m}{u}$ miles per gallon. We know that Leon's truck gets 20 miles per gallon. So $\frac{m}{u} = 20$. To solve for u, multiply both sides by $\frac{u}{20}$, therefore $u = \frac{m}{20}$.

   b. The truck holds 14.6 gallons, so if Leon has used u gallons of gas then there are $g = 14.6 - u$ gallons remaining in the tank. Substituting in $\frac{m}{20}$ for u (from the expression in a.), we get $g = 14.6 - \frac{m}{20}$.

   c. Evaluate the expression in b. at $m = 110$. Then:
$$g = 14.6 - \frac{110}{20} = 9.1 \text{ gallons.}$$

19. $\quad \dfrac{3(6-8)}{-2} - \dfrac{6}{-2}$ Evaluate expression in parentheses

$\quad = \dfrac{3(-2)}{-2} - \dfrac{6}{-2}$ Evaluate products and quotients

$\quad = 3 - (-3)$ Evaluate sums and differences
$\quad = 6$

21. $\quad 6[3 - 2(4+1)]$ Evaluate expression in inner parentheses
$\quad = 6[3 - 2(5)]$ Evaluate product in parentheses
$\quad = 6[3 - 10]$ Evaluate difference in parentheses
$\quad = 6[-7]$ Evaluate final product
$\quad = -42$

23. $\quad (4-3)[2 + 3(2-1)]$ Evaluate expressions in inner parentheses
$\quad = (1)[2 + 3(1)]$ Evaluate product in parentheses
$\quad = (1)[2 + 3]$ Evaluate sum in parentheses
$\quad = (1)[5]$ Evaluate final product
$\quad = 5$

Chapter 1, Section 1.2

25. $\dfrac{64}{8[4-2(3+1)]}$ Evaluate sum in inner parentheses

$= \dfrac{64}{8[4-2(4)]}$ Evaluate product in inner parentheses

$= \dfrac{64}{8[4-8]}$ Evaluate difference in parentheses

$= \dfrac{64}{8[-4]}$ Evaluate product in denominator

$= \dfrac{64}{-32}$ Evaluate final quotient

$= -2$

27. $\dfrac{5[3+(8-1)]}{-25}$ Evaluate difference in parentheses

$= \dfrac{5[3+7]}{-25}$ Evaluate sum in parentheses

$= \dfrac{5[10]}{-25}$ Evaluate product

$= \dfrac{50}{-25}$ Evaluate final quotient

$= -2$

29. $[-3(8-2)+3]\left[\dfrac{24}{6}\right]$ Evaluate difference in parentheses

$= [-3(6)+3]\left[\dfrac{24}{6}\right]$ Evaluate quotient in parentheses

$= [-3(6)+3][4]$ Evaluate product in parentheses

$= [-18+3]4$ Evaluate sum in parentheses

$= [-15]4$ Evaluate final product

$= -60$

31. $-5^2 = -(5 \cdot 5)$ First evaluate product in parentheses

$= -(25)$ Evaluate final product

$= -25$

33. $(-3)^4 = (-3) \cdot (-3) \cdot (-3) \cdot (-3)$ Evaluate product

$= 81$

35. $-4^3 = -(4 \cdot 4 \cdot 4)$ Evaluate product in parentheses

$= -(64)$ Evaluate final product

$= -64$

Chapter 1, Section 1.2

37. $(-2)^5 = (-2)\cdot(-2)\cdot(-2)\cdot(-2)\cdot(-2)$ Evaluate product
$= -32$

39. $\dfrac{4\cdot 2^3}{16} + 3\cdot 4^2$ Evaluate exponents first

$= \dfrac{4\cdot 8}{16} + 3\cdot 16$ Evaluate products

$= \dfrac{32}{16} + 48$ Evaluate quotient

$= 2 + 48$ Evaluate sum

$= 50$

41. $\dfrac{3^2-5}{6-2^2} - \dfrac{6^2}{3^2}$ Evaluate exponents first

$= \dfrac{9-5}{6-4} - \dfrac{36}{9}$ Evaluate differences within fraction

$= \dfrac{4}{2} - \dfrac{36}{9}$ Evaluate quotients

$= 2 - 4$ Evaluate difference

$= -2$

43. $\dfrac{(-5)^2-3^2}{4-6} + \dfrac{(-3)^2}{2+1}$ Evaluate exponents first

$= \dfrac{25-9}{4-6} + \dfrac{9}{2+1}$ Evaluate sums and differences within fractions

$= \dfrac{16}{-2} + \dfrac{9}{3}$ Evaluate quotients

$= -8 + 3$ Evaluate sum

$= -5$

45. The keying sequence is:
(−) 8398 + (26 × 17) ENTER which evaluates to −19.

47. The keying sequence is:
(112.78 + 2599.124) + 27.56 ENTER which evaluates to 98.4.

49. The keying sequence is:
2nd √ (24 × 54) ENTER which evaluates to 36.

51. The keying sequence is:
(116 − 35) + (215 − 242) ENTER which evaluates to −3.

53. The keying sequence is:
2nd √ (27 × + 36 ×) ENTER which evaluates to 45.

55. The keying sequence is:
((−) 27 − 2nd √ (27 × − 4 × 4 × 35)) + (2 × 4) ENTER which evaluates to −5.

4

Chapter 1, Section 1.3

57. a. $2 + \frac{3}{4}$

b. $\frac{2+3}{4}$

59. a. $-(23^2)$

b. $(-23)^2$

61. a. $\sqrt{9+16}$

b. $\sqrt{9} + 16$

63. $\frac{5(F-32)}{9}$; $F = 212$.

$\frac{5(212-32)}{9} = \frac{5(180)}{9}$

$= \frac{900}{9} = 100$

65. $P + Prt$; $P = 1000$, $r = 0.04$, $t = 2$.
$1000 + 1000 \cdot (0.04) \cdot 2$
$= 1000 + 80 = 1080$

67. $\frac{1}{2}gt^2 - 12t$; $g = 32$, $t = \frac{3}{4}$.

$\frac{1}{2} \cdot 32 \cdot \left(\frac{3}{4}\right)^2 - 12\left(\frac{3}{4}\right)$

$= \frac{1}{2} \cdot 32 \cdot \frac{9}{16} - \frac{36}{4} = \frac{288}{32} - 9$

$= 9 - 9 = 0$

69. $\frac{32(V-v)^2}{g}$; $V = 12.78$, $v = 4.26$, $g = 32$.

$\frac{32(12.78-4.26)^2}{32}$

$= (12.78 - 4.26)^2 = (8.52)^2$

$= 72.5904$

Section 1.3

1. $3x + 5 = 26$ Subtract 5 from both sides

$3x = 21$ Divide both sides by 3

$x = 7$

3. $3(z+2) = 37$ Divide both sides by 3

$z + 2 = \frac{37}{3}$ Subtract 2 from both sides

$z = \frac{31}{3}$

$z = 10.3\overline{3}$

Chapter 1, Section 1.3

5. $3y - 2(y - 4) = 12 - 5y$ Distributive property
    $3y - 2y + 8 = 12 - 5y$ Collect like terms
    $y + 8 = 12 - 5y$ Add $5y$ to both sides
    $6y + 8 = 12$ Subtract 8 from both sides
    $6y = 4$ Divide both sides by 6
    $y = \frac{4}{6}$
    $y = 0.6\overline{6}$

7. $0.8w - 2.6 = 1.4w + 0.3$ Subtract $0.8w$ from both sides
    $-2.6 = 0.6w + 0.3$ Subtract 0.3 from both sides
    $-2.9 = 0.6w$ Divide both sides by 0.6
    $-4.8\overline{3} = w$

9. $0.25t + 0.10(t - 4) = 11.60$ Distributive property
    $0.25t + 0.10t - 0.40 = 11.60$ Combine like terms
    $0.35t - 0.40 = 11.60$ Add 0.40 to both sides
    $0.35t = 12.00$ Divide both sides by 0.35
    $t = 34.2857$

11. Celine's demand is given by the expression $200 - 5p$, and their supply is given by the expression $56 + 3p$. Setting supply equal to demand gives the equation
    $56 + 3p = 200 - 5p$ Add $5p$ to both sides
    $56 + 8p = 200$ Subtract 56 from both sides
    $8p = 144$ Divide both sides by 8
    $p = \$18.00$ Include appropriate units

13. **a.** We are asked to find the amount of time that it takes Roger's wife to find him. Let t = the time in hours that has gone by since Roger's wife left the house.

    b. Roger's wife travels at 45 miles per hour so she will have traveled $45t$ miles after t hours.

    c. Roger leaves six hours ahead of his wife and cycles at 16 miles per hour so the distance that he has traveled is $16(t + 6)$.

Chapter 1, Section 1.3

   d. Roger's wife has caught up to Roger when their distances traveled are equal. So we solve the equation:

     $45t = 16(6+t)$ Distributive property

     $45t = 96 + 16t$ Subtract $16t$ from both sides

     $29t = 96$ Divide both sides by 29

       $t = 3.3103$ hours   Include appropriate units

15. a. We are asked to find the number of copies that would need to be made in order for the total cost of both machines to be the same. Let c be the number of copies made.

   b. The total cost of the first machine listed is $20,000 plus $0.02 per copy, or $20{,}000 + 0.02c$. The total cost of the second machine is $17,500 plus $0.025 per copy or $17{,}500 + 0.025c$.

   c. We want these total costs to be the same so we set
$20{,}000 + 0.02c = 17{,}500 + 0.025c$ and solve for c.

     $20{,}000 + 0.02c = 17{,}500 + 0.025c$ Subtract $0.02c$ from both sides

         $20{,}000 = 17{,}500 + 0.005c$ Subtract 17,500 from both sides

              $2500 = 0.005c$ Divide both sides by 0.005

          $500{,}000 = c$

17. a. Population now = 135,000; rate of increase = 8% per year. Expect that in one year population will increase by 0.08(135,000). So in one year the population will be $135{,}000 + 0.08(135{,}000) = 145{,}800$ people.

   b. Let x represent the population last year. Then, as in a., an 8% increase would make the population $x + (0.08)x$ this year. We know the population this year is 135,000 so we solve for x in:

     $x + (0.08)x = 135{,}000$ Combine like terms

        $(1.08)x = 135{,}000$ Divide both sides by 1.08

            $x = 125{,}000$

19. Virginia's salary was $24,000 per year. A 7% pay cut would reduce her salary to $24{,}000 - 24{,}000(0.07) = \$22{,}320$ per year. If she wants to make $24,000 again, then she must receive a pay increase r so that $22{,}320 + 22{,}320r = 24{,}000$. Solve for r.

     $22{,}320 + 22{,}320r = 24{,}000$ Subtract 22,320 from both sides

           $22{,}320r = 1680$ Divide both sides by 22,320

                $r = 0.07527$ Express r as a percentage

                $r = 7.527\%$

Chapter 1, Section 1.3

21. Let F = Delbert's final exam score. Then we have the weighted average: $\dfrac{77 \cdot (0.70) + F \cdot (0.30)}{1.00}$ = term average. Delbert wants a term average of 80, so solve for F in the following equation:

$77 \cdot (0.70) + F \cdot (0.30) = 80$ Multiply $77 \cdot (0.70)$

$\qquad 53.9 + F \cdot (0.30) = 80$ Subtract 53.9 from both sides

$\qquad\qquad F \cdot (0.30) = 26.1$ Divide both sides by 0.30

$\qquad\qquad\qquad F = 87$

23. a. Let A represent the amount in pounds of fertilizer one (6% potash) that we use. We are asked to find A so that a mixture of A pounds of fertilizer one and $(10 - A)$ pounds of fertilizer two (15% potash) has 8% potash.

    b.

Pounds of fertilizer	% Potash	Pounds of potash
A	6%	$(0.06)A$
$10 - A$	15%	$(0.15) \cdot (10 - A)$
10	8%	$(0.08)10 = 0.8$

    c. According to the table above, there are $(0.06)A$ pounds of potash in A pounds of fertilizer one, and $(0.15) \cdot (10 - A)$ pounds of potash in $(10 - A)$ pounds of fertilizer two.

    d. The amount of potash in the mixture is $(0.06)A + (0.15) \cdot (10 - A)$. We want this amount to be 0.8 pounds so we solve for A in the following equation.

$(0.06)A + (0.15) \cdot (10 - A) = 0.8$ Distributive property

$\quad (0.06)A + 1.5 - (0.15)A = 0.8$ Combine like terms

$\qquad\qquad\quad 1.5 - (0.09)A = 0.8$ Add $(0.09)A$ to both sides

$\qquad\qquad\quad 1.5 = 0.8 + (0.09)A$ Subtract 0.8 from both sides

$\qquad\qquad\qquad 0.7 = (0.09)A$ Divide both sides by (0.09)

$\qquad\qquad\qquad 7.78 = A$

So we should use 7.78 pounds of the first fertilizer and 2.22 pounds of the second fertilizer.

Chapter 1, Section 1.3

25. a. Let s represent the annual salary of Lacy's clerks. We are asked to find s so that the average salary at the store is under $19,000.

b. The total amount that Lacy's pays out:
to managers is $4(28,000) = 112,000$
to department heads is $12(22,000) = 264,000$
to clerks is $30(s) = 30s$

c. The total amount Lacy's pays in salaries is $112,000 + 264,000 + 30s$. If Lacy's has an average salary of under $19,000, then a different expression for the total amount Lacy's pays in salaries is: $(19,000) \cdot (4 + 12 + 30) = (19,000) \cdot (46) = 874,000$.

d. Solve for s in: $112,000 + 264,000 + 30s = 874,000$

$112,000 + 264,000 + 30s = 874,000$ Combine constants on the left

$376,000 + 30s = 874,000$ Subtract 376,000 from both sides

$30s = 498,000$ Divide both sides by 30

$s = 16,600$

Lacy's needs to pay clerks less than $16,600 annually to keep the average below $19,000.

27. $v = k + gt$ Subtract k from both sides

$v - k = gt$ Divide both sides by g

$\dfrac{v-k}{g} = t$

29. $S = 2w(w + 2h)$ Distributive property

$S = 2w^2 + 4wh$ Subtract $2w^2$ from both sides

$S - 2w^2 = 4wh$ Divide both sides by $4w$

$\dfrac{S - 2w^2}{4w} = h$

31. $P = a + (n-1)d$ Distributive property

$P = a + nd - d$ Subtract $(a - d)$ from both sides

$P - a + d = nd$ Divide both sides by d

$\dfrac{P - a + d}{d} = n$

33. $A = \pi rh + \pi r^2$ Subtract πr^2 from both sides

$A - \pi r^2 = \pi rh$ Divide both sides by πr

$\dfrac{A - \pi r^2}{\pi r} = h$

Chapter 1, Section 1.4

Section 1.4

1. $9x^2 = 25$ Divide both sides by 9
   $x^2 = \frac{25}{9}$ Take square roots of both sides
   $x = \pm\frac{5}{3}$

3. $2x^2 = 14$ Divide both sides by 2
   $x^2 = 7$ Take square roots of both sides
   $x = \pm\sqrt{7}$

5. $4x^2 - 24 = 0$ Add 24 to both sides
   $4x^2 = 24$ Divide both sides by 4
   $x^2 = 6$ Take square roots of both sides
   $x = \pm\sqrt{6}$

7. $\frac{2x^2}{3} = 4$ Multiply both sides by 3
   $2x^2 = 12$ Divide both sides by 2
   $x^2 = 6$ Take square roots of both sides
   $x = \pm\sqrt{6}$

9. $A = P(1+r)^2$. We have $P = \$1600$ and we want to find an r so that $A = \$2000$.
   Solve for r in $2000 = 1600(1+r)^2$:
   $2000 = 1600(1+r)^2$ Divide both sides by 1600
   $1.25 = (1+r)^2$ Take square roots of both sides
   $\pm 1.118 = 1+r$
   If $1 + r = -1.118$, then $r = -2.118$, but we can't have a negative interest rate, so the only viable solution is $1.118 = 1 + r$ or $0.118 = r$. That is, we need a 11.8% interest rate.

11. $\sqrt{x} - 5 = 3$ Add 5 to both sides
    $\sqrt{x} = 8$ Square both sides
    $x = 64$

Chapter 1, Section 1.4

13. $\sqrt{y+6} = 2$ Square both sides
      $y + 6 = 4$ Subtract 6 from both sides
          $y = -2$

15. $4\sqrt{z} - 8 = -2$ Add 8 to both sides
       $4\sqrt{z} = 6$ Divide both sides by 4
        $\sqrt{z} = \frac{3}{2}$ Square both sides
          $z = \frac{9}{4}$

17. $5 + 2\sqrt{6 - 2w} = 13$ Subtract 5 from both sides
      $2\sqrt{6 - 2w} = 8$ Divide both sides by 2
       $\sqrt{6 - 2w} = 4$ Square both sides
         $6 - 2w = 16$ Subtract 6 from both sides
            $-2w = 10$ Divide both side by -2
               $w = -5$

19. $T = 2\pi\sqrt{\frac{L}{32}}$. We have timed $T = 10.54$. Solve for L in $10.54 = 2\pi\sqrt{\frac{L}{32}}$.

      $10.54 = 2\pi\sqrt{\frac{L}{32}}$ Divide both sides by 2π
      $1.6775 = \sqrt{\frac{L}{32}}$ Square both sides
      $2.814 = \frac{L}{32}$ Multiply both sides by 32
   $90.0475 \text{ feet} = L$

21. $\frac{6}{w+2} = 4$ Multiply both sides by $w + 2$
     $6 = 4(w + 2)$ Distributive property
     $6 = 4w + 8$ Subtract 8 from both sides
    $-2 = 4w$ Divide both sides by 4
    $\frac{-1}{2} = w$

11

Chapter 1, Section 1.4

23. $9 = \dfrac{h-5}{h-2}$ Multiply both sides by $h-2$
$9(h-2) = h-5$ Distributive property
$9h - 18 = h - 5$ Subtract h from both sides
$8h - 18 = -5$ Add 18 to both sides
$8h = 13$ Divide both sides by 8
$h = \dfrac{13}{8}$

25. $\dfrac{15}{s^2} = 8$ Multiply both sides by s^2
$15 = 8s^2$ Divide both sides by 8
$\dfrac{15}{8} = s^2$ Take square roots of both sides
$\pm\sqrt{\dfrac{15}{8}} = s$

27. $4.3 = \sqrt{\dfrac{18}{y}}$ Square both sides
$18.49 = \dfrac{18}{y}$ Multiply both sides by y
$18.49y = 18$ Divide both sides by 18.49
$y = 0.9735$

29. $S = \dfrac{182.6wh^2}{l}$; $S = 1600$ pounds, $w = 4$ inches, $h = 9$ inches. Solve for l in:

$1600 = \dfrac{182.6 \times 4 \times (9)^2}{l}$ Simplify on the right
$1600 = \dfrac{59{,}162.4}{l}$ Multiply both sides by l
$1600l = 59{,}162.4$ Divide both sides by 1600
$l = 36.9765$ feet

31. $\dfrac{3}{4} = \dfrac{y+2}{12-y}$ Fundamental property of proportions
$3(12 - y) = 4(y + 2)$ Distributive property
$36 - 3y = 4y + 8$ Add $3y$ to both sides
$36 = 7y + 8$ Subtract 8 from both sides
$28 = 7y$ Divide both sides by 7
$4 = y$

Chapter 1, Section 1.4

33.
$$\frac{50}{r} = \frac{75}{r+20}$$ Fundamental property of proportions
$50(r+20) = 75r$ Distributive property
$50r + 1000 = 75r$ Subtract $50r$ from both sides
$1000 = 25r$ Divide both sides by 25
$40 = r$

35. The ratio that remains constant is the ratio of taxes to house value, $\frac{2700}{120,000} = \frac{9}{400}$.
To find the taxes on a house assessed at \$275,000, solve for x in $\frac{9}{400} = \frac{x}{275,000}$.

$\frac{9}{400} = \frac{x}{275,000}$ Fundamental property of proportions
$2,475,000 = 400x$ Divide both sides by 400
$6187.50 = x$
The taxes are \$6,187.50.

37. The ratio that remains constant is the amount of money that you get compared to the amount that Myrtle gets, $\frac{3}{5}$. Since \$100,000 was left, if you get x dollars, then Myrtle gets $100,000 - x$ dollars. So to find out how much you get, solve for x in:

$\frac{3}{5} = \frac{x}{100,000-x}$ Fundamental property of proportions
$3(100,000 - x) = 5x$ Distributive property
$300,000 - 3x = 5x$ Add $3x$ to both sides
$300,000 = 8x$ Divide both sides by 8
$37,500 = x$
You get \$37,500.

39. The ratio that remains constant is the ratio of the number of inches to the number of miles represented, $\frac{\left(\frac{3}{8}\right)}{10} = \frac{3}{80}$. If x represents the actual length of Isle Royale, then we find x by solving:

$\frac{3}{80} = \frac{\left(\frac{27}{16}\right)}{x}$ Simplify the right side
$\frac{3}{80} = \frac{27}{16x}$ Fundamental property of proportions
$48x = 2160$ Divide both sides by 48
$x = 45$
The actual length is 45 miles.

Chapter 1, Section 1.4

41. The ratio that remains constant is the number of tagged perch to the total number of perch, $\frac{18}{80} = \frac{9}{40}$. Since 200 were originally tagged, to estimate the original number we solve for x in:

$\frac{9}{40} = \frac{200}{x}$ Fundamental property of proportions

$9x = 8000$ Divide both sides by 9

$x = 889$

The original number of perch was 889.

43. $S = \frac{a}{1-r}$ Multiply both sides by $1-r$

$S(1-r) = a$ Distributive property

$S - Sr = a$ Subtract S from both sides

$-Sr = a - S$ Divide both sides by $-S$

$r = \frac{a-S}{-S}$ Multiply numerator and denominator by -1

$r = \frac{S-a}{S}$

45. $H = \frac{2xy}{x+y}$ Multiply both sides by $x+y$

$H(x+y) = 2xy$ Distributive property

$Hx + Hy = 2xy$ Subtract Hx from both sides

$Hy = 2xy - Hx$ Distributive property

$Hy = (2y - H)x$ Divide both sides by $2y - H$

$\frac{Hy}{2y-H} = x$

47. $F = \frac{mv^2}{r}$ Multiply both sides by r

$Fr = mv^2$ Divide both sides by m

$\frac{Fr}{m} = v^2$ Take square roots of both sides

$\pm\sqrt{\frac{Fr}{m}} = v$

49. $F = \dfrac{Gm_1m_2}{d^2}$ Multiply both sides by d^2

$Fd^2 = Gm_1m_2$ Divide both sides by F

$d^2 = \dfrac{Gm_1m_2}{F}$ Take square roots of both sides

$d = \pm\sqrt{\dfrac{Gm_1m_2}{F}}$

51. $T = 2\pi\sqrt{\dfrac{L}{g}}$ Square both sides

$T^2 = 4\pi^2 \dfrac{L}{g}$ Multiply both sides by g

$gT^2 = 4\pi^2 L$ Divide both sides by $4\pi^2$

$\dfrac{gT^2}{4\pi^2} = L$

53. $r = \sqrt{t^2 - s^2}$ Square both sides

$r^2 = t^2 - s^2$ Add s^2 to both sides

$r^2 + s^2 = t^2$ Subtract r^2 from both sides

$s^2 = t^2 - r^2$ Take square roots of both sides

$s = \pm\sqrt{t^2 - r^2}$

Section 1.5

1. Let the angles of the triangle be A, B, and C. Let angle A be the smallest, $B = 10 + A$, and $C = 29 + A$. We use that the sum of the angles of any triangle is $180°$.

$A + B + C = 180$ Substitution

$A + (10 + A) + (29 + A) = 180$ Combine like terms

$3A + 39 = 180$ Subtract 39 from both sides

$3A = 141$ Divide both sides by 3

$A = 47$

The angles are $47°$, $57°$, and $76°$.

Chapter 1, Section 1.5

3. Since the triangle is a right triangle, one angle is 90°. Let A be the smallest acute angle. Since the sum of the angles of any triangle is 180°, we have:
   $90 + A + 2A = 180$ Combine like terms
   $90 + 3A = 180$ Subtract 90 from both sides
   $3A = 90$ Divide both sides by 3
   $A = 30$
   The angles are 30°, 60°, and 90°.

5. Let A represent the measure of the equal angles. The vertex angle is $2A - 20$. Since the sum of the angles of any triangle is 180°, we have:
   $A + A + (2A - 20) = 180$ Combine like terms
   $4A - 20 = 180$ Add 20 to both sides
   $4A = 200$ Divide both sides by 4
   $A = 50$
   The angles are 50°, 50°, and 80°.

7. Let l be the length of the equal sides. Since the perimeter is 42 cm, we have:
   $l + l + 12 = 42$ Combine like terms
   $2l + 12 = 42$ Subtract 12 from both sides
   $2l = 30$ Divide both sides by 2
   $l = 15$
   The equal sides are 15 cm long.

9. Let x be the height of the TV. By the Pythagorean Theorem:
   $28^2 + x^2 = 35^2$ Simplify both sides
   $784 + x^2 = 1225$ Subtract 784 from both sides
   $x^2 = 441$ Take square roots of both sides
   $x = 21$
   The height is 21 inches.

11. Let x be the distance from the tip of the shadow to the top of the tree. By the Pythagorean Theorem:
    $30^2 + 30^2 = x^2$ Simplify the left side
    $900 + 900 = x^2$ Combine like terms on the left
    $1800 = x^2$ Take square roots of both sides
    $42.4264 = x$
    The distance is 42.4264 meters.

Chapter 1, Section 1.5

13. Let x be the length of a side of the inscribed square. By the Pythagorean Theorem:

$\left(\frac{x}{2}\right)^2 + \left(\frac{x}{2}\right)^2 = 8^2$ Simplify both sides

$\frac{x^2}{4} + \frac{x^2}{4} = 64$ Combine like terms on the left

$\frac{x^2}{2} = 64$ Multiply both sides by 2

$x^2 = 128$ Take square roots of both sides

$x = 11.3137$

The length is 11.3137 inches.

15. $3x - 2 > 1 + 2x$ Subtract $2x$ from both sides
$x - 2 > 1$ Add 2 to both sides
$x > 3$

17. $\frac{-2x-6}{-3} > 2$ Multiply both sides by -3
$-2x - 6 < -6$ Add 6 to both sides
$-2x < 0$ Divide both sides by -2
$x > 0$

19. $\frac{2x-3}{3} \leq \frac{3x}{-2}$ Multiply both sides by 3
$2x - 3 \leq \frac{9x}{-2}$ Multiply both sides by -2
$-4x + 6 \geq 9x$ Add $4x$ to both sides
$6 \geq 13x$ Divide both sides by 13
$\frac{6}{13} \geq x$

21. Let x be the length of the third side. By the triangle inequality $10 + 6 > x$ or $16 > x$. Using another pair of sides, $6 + x > 10$ or $x > 4$. Therefore, $16 > x > 4$, so the third side is greater than 4 feet but less than 16 feet long.

23. Let x represent the height of the lamppost. By similar triangles, we have:

$\frac{x}{12+9} = \frac{6}{9}$ Simplify both sides

$\frac{x}{21} = \frac{2}{3}$ Fundamental property of proportions

$3x = 42$ Divide both sides by 3

$x = 14$

The lamppost is 14 feet tall.

Chapter 1, Section 1.5

25. Let r represent the radius of the circle of the exposed surface of water. By similar triangles:

$\frac{4}{12} = \frac{r}{7}$ Fundamental property of proportions

$28 = 12r$ Divide both sides by 12

$\frac{7}{3} = r$

The area of this circle is $\pi r^2 = \pi \left(\frac{7}{3}\right)^2 = 17.1042 \text{ ft}^2$.

27. Let x represent the distance EB. By similar triangles:

$\frac{20}{13} = \frac{x}{58}$ Fundamental property of proportions

$1160 = 13x$ Divide both sides by 13

$89.2308 = x$

The distance EB is 89.2308 feet.

29. a. Volume of a sphere is $\frac{4}{3}\pi r^3$ where r is the radius. So the volume is

$\frac{4}{3}\pi(1.2)^3 = 7.2382 \text{ m}^3$.

b. Surface area of a sphere of radius r is $4\pi r^2$, so the surface area is

$4\pi(0.7)^2 = 6.1575 \text{ cm}^2$.

31. a. The volume of a cylinder with radius r and height h is $\pi r^2 h$. So the volume is

$\pi(6)^2 23.2 = 2623.8582 \text{ m}^3$.

b. The surface area of a cylinder is $2\pi(rh + r^2)$, so the surface area is

$2\pi\left([15.3]4.5 + [15.3]^2\right) = 1903.4282 \text{ in.}^2$

33. a. The surface area of a box with height h, length l, and width w is $2hl + 2hw + 2wl$. Since $l = 20$ and $w = 16$, we have surface area $= 2h(20) + 2h(16) + 2(16)(20) = 72h + 640$.

b. Solve for h in the equation:

$1216 = 72h + 640$ Subtract 640 from both sides

$576 = 72h$ Divide both sides by 72

$8 = h$

The height can be no greater than 8 inches.

Chapter 1, Section 1.6

35. a. The volume V of a cone with radius r and height h is $\frac{\pi r^2 h}{3}$. So
$$V = \frac{\pi r^2 (8.4)}{3} = 8.7965 r^2.$$

b. Solve for r in the equation:
$302.4 = 8.7965 r^2$ Divide both sides by 8.7965
$34.3773 = r^2$ Take square roots of both sides
$5.8632 = r$

Section 1.6

1. a. The highest point on the graph corresponds to the highest temperature and the lowest point on the graph corresponds to the lowest temperature. These points are approximately 7°F for the high temperature and –19°F for the low temperature.

b. From the graph we notice that the temperature at noon and at 3 P.M. is about 5°F. Between these times the temperature is greater than 5°F. At 9 A.M. the temperature is –5°F and from midnight to 9 A.M. the temperature is below –5°F. The temperature is also –5°F at approximately 7 P.M. and from 7 P.M to midnight the temperature is below –5°F. Therefore, the temperature is below –5°F from 7 P.M. to 9 A.M.

c. At 7 A.M. the temperature was approximately –10°F. At 2 P.M. the temperature was approximately 6°F. The temperature was approximately 0°F at about 10 A.M. and 5 P.M. The temperature was approximately –12°F at about 6 A.M. and 10 P.M.

d. From 3 A.M. to 6 A.M. the temperature increased from approximately –19°F to –12°F, a change of 7°F. Between 9 A.M. and noon the temperature increased from approximately –5°F to 5°F, a change of 10°F. From 6 P.M. to 9 P.M. the temperature decreased from approximately –1°F to –10°F, a change of 9°F.

e. The greatest increase was from 9 A.M. to noon where the temperature increased about 10°F in 3 hours. The greatest decrease was from 6 P.M. to 9 P.M. when the temperature dropped about 9°F in 3 hours.

3. a. From the graph, the gas mileage at 43 mph is approximately 27 miles per gallon.

b. A gas mileage of 34 miles per gallon is achieved around 51 mph.

c. The best gas mileage is achieved at around 70 mph. The gas mileage would most likely drop at some point since driving at high speeds requires a high engine temperature, and at some point

19

Chapter 1, Section 1.6

    really hot engines run inefficiently.

  d. Other factors to consider are driving conditions (e.g., rain, severe heat or cold) and the maintenance level of the engine (e.g., is it in tune, does it have the proper amount of oil, etc.).

5. a. The only time the car's speed was 0 during the trip was after approximately 12 minutes. This has to be when the car stopped at a traffic signal.

  b. From 15 minutes into the journey until 35 minutes, the car's speed fluctuated between 9 and 25 mph, indicative of city traffic.

  c. The car traveled on the freeway from 40 minutes into the journey until 50 minutes, since this was the only period where the car traveled at speeds in the 50 to 55 mph range.

7. From the graph: $(-3, -2)$, $(1, 6)$, $(-2, 0)$, and $(0, 4)$. Algebraic verification:
$2(-3) + 4 = -6 + 4 = -2$
$2(1) + 4 = 2 + 4 = 6$
$2(-2) + 4 = -4 + 4 = 0$
$2(0) + 4 = 0 + 4 = 4$

9. From the graph: $(-2, 6)$, $(2, 6)$, $(-1, 3)$, and $(0, 2)$. Algebraic verification:
$(-2)^2 + 2 = 4 + 2 = 6$
$(2)^2 + 2 = 4 + 2 = 6$
$(-1)^2 + 2 = 1 + 2 = 3$
$(0)^2 + 2 = 0 + 2 = 2$

11. From the graph: $\left(-1, \frac{-1}{2}\right)$, $\left(\frac{1}{2}, -2\right)$, $\left(4, \frac{1}{3}\right)$, and $(0, -1)$. Algebraic verification:
$\frac{1}{-1-1} = \frac{-1}{2}$
$\frac{1}{\frac{1}{2}-1} = \frac{1}{\frac{-1}{2}} = -2$
$\frac{1}{4-1} = \frac{1}{3}$
$\frac{1}{0-1} = \frac{1}{-1} = -1$

13. From the graph: $(-2, -8)$, $\left(\frac{1}{2}, \frac{1}{8}\right)$, $(0, 0)$, and $(-1, -1)$. Algebraic verification:
$(-2)^3 = 8$
$\left(\frac{1}{2}\right)^3 = \frac{1}{8}$
$(0)^3 = 0$
$(-1)^3 = -1$

15. – 29. Refer to the graphs in the back of the textbook. To graph functions on the TI-82, press the [ZOOM] key then enter 6 to set the range to the standard graphing window. Press the Y= button and then enter the function you wish to graph after the Y₁= prompt. When you are done entering the function press the [GRAPH] key.

20

Chapter 1 Review

31. Refer to the graph in the back of the textbook. Graph the function as described above or in the textbook. Be sure to use the "friendly window" as described. With the graph of the function displayed on your calculator, press the [TRACE] key. Now depress the ◄ and ► keys to move the cursor about. Notice that only the values of $x = -2.0$ and $x = -6.0$ have the y-coordinate of -3.0.

33. Refer to the graph in the back of the textbook. As in Problem 31, with the graph of the function displayed on your calculator, press the [TRACE] key. Now depress the ◄ and ► keys to move the cursor about. Notice that the value $x = 5.0$ has y-coordinate of -1.0 (rounded to one decimal place).

35. Refer to the graph in the back of the textbook. As in Problem 31, with the graph of the function displayed on your calculator, press the [TRACE] key. Now depress the ◄ and ► keys to move the cursor onto the point of intersection of the two graphs. This point is approximately $(5.3, 4.7)$.

37. Refer to the graph in the back of the textbook. As before, use the [TRACE] key and move the cursor to the points where the graph "turns." These points are approximately $(-1.1, -1.1)$ and $(1.1, 5.1)$.

39. – 43. Refer to the graphs in the back of the textbook.

Chapter 1 Review

1. a. Rate times time equals distance, so if the train is traveling at a rate of 48 mph for t hours it will travel $48t$ miles.

b. Romina must travel 2400 miles. By a., we know that after t hours she has traveled $48t$ miles, therefore after t hours she still has $2400 - 48t$ miles left to go.

2. a. If s represents the current salary, then 7% of s is given by the expression $(0.07)s$. So each employee will get a $(0.07)s$ dollar raise.

b. The new salary will be the old salary plus the amount of the raise. Therefore the new salary will be $s + (0.07)s$ or if we combine like terms, $(1.07)s$.

3. a. $P = \dfrac{kT}{V}$

b. Evaluate the expression above at $V = 200$, $T = 400$, and $k = 20$.
$P = \dfrac{20 \cdot 400}{200} = 40$ pounds per square inch.

4. a. $E = kL(T - 65)$

b. Evaluate the expression above at $L = 1000$, $T = 105$, and $k = 0.000012$.
$E = (0.000012) \cdot (1000) \cdot (105 - 65)$
$ = (0.000012) \cdot (1000) \cdot (40)$
$ = 0.48$ feet

Chapter 1 Review

5. $-6^2 = -(6 \cdot 6)$ Evaluate product in parentheses
   $= -(36)$ Evaluate product
   $= -36$

6. $(-6)^2 = (-6) \cdot (-6)$ Evaluate product
   $= 36$

7. $(4-2)[3 - 2(3-4)]$ Evaluate differences in parentheses
   $= (2)[3 - 2(-1)]$ Evaluate product in parentheses
   $= (2)[3 - (-2)]$ Evaluate difference in parentheses
   $= (2)[5]$ Evaluate product
   $= 10$

8. $\dfrac{2[1 + (6-2)]}{-4}$ Evaluate difference in parentheses

   $= \dfrac{2[1 + (4)]}{-4}$ Evaluate sum in parentheses

   $= \dfrac{2[5]}{-4}$ Evaluate products and quotients

   $= \dfrac{-5}{2}$

9. $\dfrac{2 \cdot 3^2}{6} - 3 \cdot 2^2$ Evaluate exponents first

   $= \dfrac{2 \cdot 9}{6} - 3 \cdot 4$ Evaluate products and quotients

   $= 3 - 12$ Evaluate the difference
   $= -9$

10. $-2^3 + 3\left[\dfrac{5^2 + 3}{4 - (-3)}\right] + (-3)^2$ Evaluate exponents first

    $= -8 + 3\left[\dfrac{25 + 3}{4 - (-3)}\right] + 9$ Evaluate sums and differences in parentheses

    $= -8 + 3\left[\dfrac{28}{7}\right] + 9$ Evaluate quotient in parentheses

    $= -8 + 3[4] + 9$ Evaluate product
    $= -8 + 12 + 9$ Evaluate sum
    $= 13$

11. The keying sequence is: [2nd] [√] [(] 18 [x] [−] 4 [x] 3 [x] [(−)] 6 [)] [ENTER]
    which evaluates to 19.9.

Chapter 1 Review

12. The keying sequence is : (26.8 − 32.4) ÷ (− 16 × 25.5) ENTER which evaluates to 0.014.

13. $\frac{1}{2}gt^2 - 6t$, evaluated at $g = 32$ and $t = 2$:

$= \frac{1}{2}(32)(2)^2 - 6(2)$ Evaluate exponent

$= \frac{1}{2}(32)(4) - 6(2)$ Evaluate products

$= 64 - 12$ Evaluate difference

$= 52$

14. $\frac{a - ar^n}{1 - r}$ evaluated at $a = 2.1$, $r = 0.5$, and $n = 3$:

$= \frac{2.1 - (2.1)(0.5)^3}{1 - 0.5}$ Evaluate exponents

$= \frac{2.1 - (2.1)(0.125)}{1 - 0.5}$ Evaluate product in numerator

$= \frac{2.1 - 0.2625}{1 - 0.5}$ Evaluate differences

$= \frac{1.8375}{0.5}$ Evaluate quotient

$= 3.675$

15. $2x - 6 = 4x - 8$ Add 8 to both sides

$2x + 2 = 4x$ Subtract $2x$ from both sides

$2 = 2x$ Divide both sides by 2

$1 = x$

16. $0.40x = 240$ Divide both sides by 0.40

$x = 600$

17. $2(x - 3) + 4(x + 2) = 6$ Distributive property

$2x - 6 + 4x + 8 = 6$ Combine like terms

$6x + 2 = 6$ Subtract 2 from both sides

$6x = 4$ Divide both sides by 6

$x = \frac{2}{3}$

Chapter 1 Review

18. $2x - (x+3) = 2(x+1)$ Distributive property
       $2x - x - 3 = 2x + 2$ Combine like terms
           $x - 3 = 2x + 2$ Subtract x from both sides
              $-3 = x + 2$ Subtract 2 from both sides
                  $-5 = x$

19. $0.30(y + 2) = 2.10$ Distributive property
     $0.30y + 0.60 = 2.10$ Subtract 0.60 from both sides
           $0.30y = 1.50$ Divide both sides by 0.30
                 $y = 5$

20. $0.06y + 0.04(y + 1000) = 60$ Distributive property
     $0.06y + 0.04y + 40 = 60$ Combine like terms
          $0.10y + 40 = 60$ Subtract 40 from both sides
               $0.10y = 20$ Divide both sides by 0.10
                   $y = 200$

21. $5x^2 = 30$ Divide both sides by 5
       $x^2 = 6$ Take square roots of both sides
       $x = \pm\sqrt{6}$

22. $3x^2 - 6 = 0$ Add 6 to both sides
        $3x^2 = 6$ Divide both sides by 3
         $x^2 = 2$ Take square roots of both sides
         $x = \pm\sqrt{2}$

23. $\frac{2x^2}{5} - 7 = 9$ Add 7 to both sides
        $\frac{2x^2}{5} = 16$ Multiply both sides by 5
        $2x^2 = 80$ Divide both sides by 2
         $x^2 = 40$ Take square roots of both sides
         $x = \pm\sqrt{40}$ Simplify radical
         $x = \pm 2\sqrt{10}$

Chapter 1 Review

24. $\dfrac{3x^2-8}{4} = 10$ Multiply both sides by 4
     $3x^2 - 8 = 40$ Add 8 to both sides
     $3x^2 = 48$ Divide both sides by 3
     $x^2 = 16$ Take square roots of both sides
     $x = \pm 4$

25. $\dfrac{y+3}{y+5} = \dfrac{1}{3}$ Fundamental property of proportions
     $3(y+3) = 1(y+5)$ Distributive property
     $3y + 9 = y + 5$ Subtract y from both sides
     $2y + 9 = 5$ Subtract 9 from both sides
     $2y = -4$ Divide both sides by 2
     $y = -2$

26. $\dfrac{y}{6-y} = \dfrac{1}{2}$ Fundamental property of proportions
     $2y = 1(6 - y)$ Distributive property
     $2y = 6 - y$ Add y to both sides
     $3y = 6$ Divide both sides by 3
     $y = 2$

27. $\dfrac{6}{z^2-1} = 2$ Fundamental property of proportions
     $6 = 2(z^2 - 1)$ Distributive property
     $6 = 2z^2 - 2$ Add 2 to both sides
     $8 = 2z^2$ Divide both sides by 2
     $4 = z^2$ Take square roots of both sides
     $\pm 2 = z$

Chapter 1 Review

28. $\dfrac{z^2+2}{z^2-2}=3$ Fundamental property of proportions
$z^2+2=3(z^2-2)$ Distributive property
$z^2+2=3z^2-6$ Add 6 to both sides
$z^2+8=3z^2$ Subtract z^2 from both sides
$8=2z^2$ Divide both sides by 2
$4=z^2$ Take square roots of both sides
$\pm 2=z$

29. $2\sqrt{w}-5=21$ Add 5 to both sides
$2\sqrt{w}=26$ Divide both sides by 2
$\sqrt{w}=13$ Square both sides
$w=169$

30. $16-3\sqrt{w}=-5$ Subtract 16 from both sides
$-3\sqrt{w}=-21$ Divide both sides by -3
$\sqrt{w}=7$ Square both sides
$w=49$

31. $12-\sqrt{5v+1}=3$ Subtract 12 from both sides
$-\sqrt{5v+1}=-9$ Divide both sides by -1
$\sqrt{5v+1}=9$ Square both sides
$5v+1=81$ Subtract 1 from both sides
$5v=80$ Divide both sides by 5
$v=16$

32. $3\sqrt{17-4v}-8=19$ Add 8 to both sides
$3\sqrt{17-4v}=27$ Divide both sides by 3
$\sqrt{17-4v}=9$ Square both sides
$17-4v=81$ Subtract 17 from both sides
$-4v=64$ Divide both sides by -4
$v=-16$

Chapter 1 Review

33. $3N = 5t - 3c$ Add $3c$ to both sides
$3N + 3c = 5t$ Divide both sides by 5
$\dfrac{3N + 3c}{5} = t$

34. $C = 10 + 2p - 2t$ Subtract $(10 + 2p)$ from both sides
$C - 10 - 2p = -2t$ Divide both sides by -2
$\dfrac{C - 10 - 2p}{-2} = t$ Simplify the left side
$\dfrac{10 + 2p - C}{2} = t$

35. $S = 2\pi(R - r)$ Divide both sides by 2π
$\dfrac{S}{2\pi} = R - r$ Add r to both sides
$\dfrac{S}{2\pi} + r = R$ Combine left side terms into one fraction
$\dfrac{S + 2\pi r}{2\pi} = R$

36. $9C = 5(F - 32)$ Divide both sides by 5
$\dfrac{9C}{5} = F - 32$ Add 32 to both sides
$\dfrac{9C}{5} + 32 = F$ Combine left side terms into one fraction
$\dfrac{9C + 160}{5} = F$

37. $V = C\left(1 - \dfrac{t}{n}\right)$ Divide both sides by C
$\dfrac{V}{C} = 1 - \dfrac{t}{n}$ Subtract 1 from both sides
$\dfrac{V}{C} - 1 = \dfrac{-t}{n}$ Multiply both sides by $-n$
$-n\left(\dfrac{V}{C} - 1\right) = t$ Distribute -1 on left side
$n\left(1 - \dfrac{V}{C}\right) = t$

Chapter 1 Review

38. $\dfrac{p}{q} = \dfrac{r}{q+r}$ Fundamental property of proportions

$p(q+r) = rq$ Distributive property

$pq + pr = rq$ Subtract pq from both sides

$pr = rq - pq$ Distributive property

$pr = (r-p)q$ Divide both sides by $(r-p)$

$\dfrac{pr}{r-p} = q$

39. $t = \sqrt{\dfrac{2v}{g}}$ Square both sides

$t^2 = \dfrac{2v}{g}$ Multiply both sides by g

$gt^2 = 2v$ Divide both sides by t^2

$g = \dfrac{2v}{t^2}$

40. $r = \sqrt{\dfrac{A}{\pi}}$ Square both sides

$r^2 = \dfrac{A}{\pi}$ Multiply both sides by π

$\pi r^2 = A$

41. $3x + 1 < 2 - 3(x-1)$ Distributive property

$3x + 1 < 2 - 3x + 3$ Collect like terms on the right

$3x + 1 < 5 - 3x$ Add $3x$ to both sides

$6x + 1 < 5$ Subtract 1 from both sides

$6x < 4$ Divide both sides by 6

$x < \dfrac{2}{3}$

42. $2(x-2) - 3(x+3) \le -24$ Distributive property

$2x - 4 - 3x - 9 \le -24$ Collect like terms on the left

$-x - 13 \le -24$ Add 13 to both sides

$-x \le -11$ Multiply both sides by -1

$x \ge 11$

28

Chapter 1 Review

43. $\dfrac{-x-2.3}{2} > 1.2$ Multiply both sides by 2

 $-x - 2.3 > 2.4$ Add 2.3 to both sides

 $-x > 4.7$ Multiply both sides by -1

 $x < -4.7$

44. $\dfrac{-2x+4}{3} \ge \dfrac{-3x}{2}$ Fundamental property of proportions

 $2(-2x + 4) \ge (-3x)3$ Distributive property

 $-4x + 8 \ge -9x$ Add $4x$ to both sides

 $8 \ge -5x$ Divide both side by -5

 $\dfrac{-8}{5} \le x$

45. a. Let n represent the number of tennis rackets sold. To make n tennis rackets it costs the initial outlay of $36,000 plus $22.00 per racket, or $36{,}000 + 22n$. If the rackets sell for $40.00 apiece, then the revenue from selling n rackets is $40n$. To break even, costs must equal revenues, so we solve for n in:

 $36{,}000 + 22n = 40n$ Subtract $22n$ from both sides

 $\phantom{36{,}000 + 2}36{,}000 = 18n$ Divide both sides by 18

 $\phantom{36{,}000 + 22n}2000 = n$

They must sell 2000 rackets to break even.

b. There are two owners, so in order for each to clear $10,000, the revenue from the rackets must exceed the costs by $20,000. So we solve for n in the equation:

 $36{,}000 + 22n + 20{,}000 = 40n$ Combine like terms

 $\phantom{36{,}000 +}56{,}000 + 22n = 40n$ Subtract $22n$ from both sides

 $\phantom{36{,}000 + 22n +}56{,}000 = 18n$ Divide both sides by 18

 $\phantom{36{,}000 + 22n + 20{,}0}3112 = n$

They need to sell 3112 rackets in order for each owner to make $10,000.

46. a. Let n represent the number of bottles that the manufacturer sells. The cost of manufacturing n bottles is the start-up cost of $16,000 plus $5.00 per bottle, or $16{,}000 + 5n$. The revenue from selling n bottles at $8.00 per bottle is $8n$ dollars. In order to break even we need costs to equal revenues, thus we solve for n in:

 $16{,}000 + 5n = 8n$ Subtract $5n$ from both sides

 $\phantom{16{,}000 + }16{,}000 = 3n$ Divide both sides by 3

 $\phantom{16{,}000 + 5n}5334 = n$

So we must sell 5334 bottles in order to break even.

Chapter 1 Review

b. In order to avoid a loss of no more than $10,000, we need costs to exceed revenues by no more than $10,000. Equivalently, we need:

$16,000 + 5n = 8n + 10,000$ Subtract $5n$ from both sides

$16,000 = 3n + 10,000$ Subtract 10,000 from both sides

$6000 = 3n$ Divide both sides by 3

$2000 = n$

So we need to sell at least 2000 bottles.

47. Let l represent the length of the table and w represent its width. The perimeter of the table is $l + l + w + w$ or $2l + 2w$, which equals 28 feet. So $2l + 2w = 28$. Moreover, the length is 4 feet longer than the width, so $l = w + 4$. Substituting into the equation for the perimeter, we get the equation $2(w + 4) + 2w = 28$, which we solve for w.

$2(w + 4) + 2w = 28$ Distributive property

$2w + 8 + 2w = 28$ Combine like terms

$4w + 8 = 28$ Subtract 8 from both sides

$4w = 20$ Divide both sides by 4

$w = 5$

So the width is 5 feet. The length is 4 feet longer than the width, therefore the length is 9 feet.

48. Let l represent the length of the court and w represent its width. The perimeter of the court is $l + l + w + w$ or $2l + 2w$, which equals 210 feet. So $2l + 2w = 210$. Moreover, the length is 24 feet longer than twice the width, so $l = 2w + 24$. Substituting into the equation for the perimeter, we get the equation $2(2w + 24) + 2w = 210$, which we solve for w.

$2(2w + 24) + 2w = 210$ Distributive property

$4w + 48 + 2w = 210$ Combine like terms

$6w + 48 = 210$ Subtract 48 from both sides

$6w = 162$ Divide both sides by 6

$w = 27$

So the width is 27 feet. The length is 24 feet longer than twice the width, therefore the length is $24 + 2(27) = 78$ feet.

Chapter 1 Review

49. Let A, B, and C represent the angles of the triangle. Since the sum of the angles of any triangle is 180°, $A + B + C = 180$. However, one angle is twice as large as another, say $B = 2A$. And the third angle is 12° more than the sum of the other two, so $C = A + B + 12$.

$C = A + B + 12$ Substitute $2A$ for B
$C = A + 2A + 12$ Combine like terms
$C = 3A + 12$

On the other hand, we have:

$\quad\quad A + B + C = 180$ Substitute $2A$ for B
$\quad\quad A + 2A + C = 180$ Substitute $3A + 12$ for C
$A + 2A + 3A + 12 = 180$ Combine like terms
$\quad\quad\quad\quad 6A + 12 = 180$ Subtract 12 from both sides
$\quad\quad\quad\quad\quad\quad 6A = 168$ Divide both sides by 6
$\quad\quad\quad\quad\quad\quad\quad A = 28$

So $A = 28°$, $B = 2A = 56°$, and $C = A + B + 12 = 96°$.

50. Let A, B, and C represent the angles of the triangle. Since the sum of the angles of any triangle is 180°, $A + B + C = 180$. However, one angle is 25° less than the second, say $A = B - 25$; and 50° less than the third angle, $A = C - 50$. By adding 25 to both sides we see that $A = B - 25$ implies that $A + 25 = B$. Similarly, by adding 50 to both sides we see that $A = C - 50$ implies that $A + 50 = C$. On the other hand:

$\quad\quad\quad A + B + C = 180$ Substitute $A + 25$ for B

$\quad\quad A + A + 25 + C = 180$ Substitute $A + 50$ for C

$A + A + 25 + A + 50 = 180$ Combine like terms

$\quad\quad\quad\quad 3A + 75 = 180$ Subtract 75 from both sides

$\quad\quad\quad\quad\quad\quad 3A = 105$ Divide both sides by 3

$\quad\quad\quad\quad\quad\quad\quad A = 35$

So $A = 35°$, $B = A + 25 = 60°$, and $C = A + 50 = 85°$

51. Let y represent the length of each of the two equal sides and let x represent the length of the third side of the isosceles triangle. We want to find x and y. The perimeter is $y + y + x$ or $2y + x$. On the other hand we are told that $y = (15.6 + x)$ cm and that the perimeter is 65.7 cm. Substituting this information into the formula for the perimeter, we get:

$2(15.6 + x) + x = 65.7$ Distributive property
$\quad 31.2 + 2x + x = 65.7$ Combine like terms
$\quad\quad 31.2 + 3x = 65.7$ Subtract 31.2 from both sides
$\quad\quad\quad\quad 3x = 34.5$ Divide both sides by 3
$\quad\quad\quad\quad\quad x = 11.5$

So the lengths of the sides of the triangle are 11.5 cm, 11.5 + 15.6 = 27.1 cm, and 27.1 cm.

Chapter 1 Review

52. Let x, y, and z represent the lengths of the sides of the triangle, with x the shortest length and z the longest. We want to find x, y, and z. Since the longest side is three times the length of the shortest side, $z = 3x$. Moreover, since the third side is 72.4 cm longer than the shortest, $y = x + 72.4$. The perimeter of the triangle is $x + y + z = 272.9$. Substituting in the equations for y and z in terms of x we get:

$x + (x + 72.4) + 3x = 272.9$ Combine like terms
$\quad\quad\quad 5x + 72.4 = 272.9$ Subtract 72.4 from both sides
$\quad\quad\quad\quad\quad\quad 5x = 200.5$ Divide both sides by 5
$\quad\quad\quad\quad\quad\quad\ x = 40.1$

Therefore, $x = 40.1$ cm, and by the equations relating x with y and z, we have $y = 40.1 + 72.4 = 112.5$ cm, and $z = 3(40.1) = 120.3$ cm.

53. Let x represent the number of applicants last year. We want to find x. The number of students accepted is 75% of x, or $(0.75)x$. On the other hand, this number is 600 students, so we have:

$(0.75)x = 600$ Divide both sides by 0.75

$\quad\quad\ x = 800$

Therefore, 800 students applied and 600 were accepted.

54. Let x represent the secretary's total salary. We want to find x. We know that 80% of the total is $120. That is:

$(0.80)x = 120$ Divide both sides by 0.80

$\quad\quad\ x = 150$

So the secretary's total salary is $150 per week.

55. Let x represent the number of bats manufactured. We want to find x so that the number of acceptable bats is 1710. Since 5% of these bats are unacceptable, the number of bats that are unacceptable is $(0.05)x$. So the number of acceptable bats is $x - (0.05)x$, and we want this number to be 1710. So we have:

$x - (0.05)x = 1710$ Combine like terms

$\quad (0.95)x = 1710$ Divide both sides by 0.95

$\quad\quad\quad x = 1800$

Therefore we should make 1800 bats if we want 1710 of them to be acceptable.

56. Let n represent the number of transistors that were ordered. We wish to find n so that 22,440 were shipped. We are told that the number shipped is 2% over the number ordered, so the number shipped is the number ordered, n, plus 2% of n or $(0.02)n$. That is, the number shipped is $n + (0.02)n$, and we are told that this equals 22,440. So:

$n + (0.02)n = 22,440$ Combine like terms

$\quad (1.02)n = 22,440$ Divide both sides by 1.02

$\quad\quad\quad n = 22,000$

The number of transistors ordered was 22,000.

Chapter 1 Review

57. Let x represent the amount of sales two years ago and let y represent the amount of sales this year. We want to find the percentage increase in sales from two years ago. That is, we want to find a percentage p so that $x + px = y$. We know that last years sales were 9% above two years ago, so last years sales were $x + (0.09)x = (1.09)x$. This years sales were 12% above last years so this years sales $= y = (1.09)x + (0.12)[(1.09)x]$. Combining like terms give the equation $y = (1.2208)x$, or we can write this as $y = x + (0.2208)x$ which means that sales have increased by 22.08% from two years ago.

58. Let x represent the enrollment last year and y represent the enrollment next year. We wish to find the percentage increase in enrollment over the last two years. That is, we want to find a percentage p so that $x + px = y$. Since the increase from last year to this year was 15%, the enrollment this year was $x + (0.15)x = (1.15)x$. And if the enrollment increases by 8% this year, the enrollment next year will be $y = (1.15)x + (0.08)[(1.15)x]$. Combining like terms gives the equation $y = (1.242)x$ or equivalently, $y = x + (0.242)x$. Therefore the increase in enrollment is 24.2%.

59. Let x represent the number of liters of a 20% sugar solution added to the 40 liters of the 32% sugar solution. We want to find x so that the percentage of sugar in the $(x + 40)$ liters of the new solution is 28%. We first compute the amount (in liters) of sugar in the new solution. In x liters of a 20% sugar solution there are $(0.20)x$ liters of sugar, and in 40 liters of a 32% sugar solution there are $(0.32)40 = 12.8$ liters of sugar. So in the new solution there are $(0.20)x + 12.8$ liters of sugar in $(x + 40)$ liters of the solution. We want to find x to make this a 28% solution, so we want to solve for x in the expression:

$(0.28)[x + 40] = (0.20)x + 12.8$ Distributive property

$(0.28)x + 11.2 = (0.20)x + 12.8$ Subtract $(0.20)x$ from both sides

$(0.08)x + 11.2 = 12.8$ Subtract 11.2 from both sides

$(0.08)x = 1.6$ Divide both side by 0.08

$x = 20$

Therefore we should add 20 liters of the 20% solution.

60. Let n represent the number of quarts of 40% antifreeze solution remaining in the radiator after $(8 - n)$ quarts are drained. The number of quarts of antifreeze remaining in the radiator is $(0.40)n$. If we fill the radiator back up to its 8 quart capacity with water then we add no new antifreeze to the mixture, so the number of quarts of antifreeze in the radiator is still $(0.40)n$. We want a 20% antifreeze solution so we need:

$(0.20)8 = (0.40)n$ Simplify left hand side

$1.6 = (0.40)n$ Divide both sides by 0.40

$4 = n$

So we should drain $(8 - 4) = 4$ quarts of antifreeze from the radiator.

Chapter 1 Review

61. The ratio that must remain constant is the ratio of the amount of flour to the amount of honey, $\dfrac{\left(\frac{3}{2}\right)}{\left(\frac{1}{3}\right)}$, or equivalently $\dfrac{9}{2}$. If x represents the amount of honey used when the amount of flour used is $\dfrac{5}{2}$ cups, then we must have:

$$\dfrac{9}{2} = \dfrac{\left(\frac{5}{2}\right)}{x} \quad \text{Simplify the right side}$$
$$\dfrac{9}{2} = \dfrac{5}{2x} \quad \text{Fundamental property of proportions}$$
$$18x = 10 \quad \text{Divide both side by 18}$$
$$x = \dfrac{5}{9}$$

Therefore we should use $\dfrac{5}{9}$ cups of honey.

62. The ratio that must remain constant is the ratio of the amount of peat moss to the amount of nitrohumus, $\dfrac{\left(\frac{3}{4}\right)}{\left(\frac{4}{3}\right)}$ or equivalently, $\dfrac{9}{16}$. If x represents the amount of nitrohumus used when the amount of peat moss used is 7 pounds, then we must have:

$$\dfrac{9}{16} = \dfrac{7}{x} \quad \text{Fundamental property of proportions}$$
$$9x = 112 \quad \text{Divide both sides by 9}$$
$$x = \dfrac{112}{9}$$

Therefore we should use $\dfrac{112}{9}$ = 12 and $\dfrac{4}{9}$ pounds of nitrohumus.

63. The ratio that must remain constant is the ratio of the number of centimeters to the number of kilometers represented, $\dfrac{4.5}{1}$, or equivalently 4.5. If x represents the number of kilometers represented by 10 centimeters, then we must have:

$$4.5 = \dfrac{10}{x} \quad \text{Fudamental property of proportions}$$
$$4.5x = 10 \quad \text{Divide both sides by 4.5}$$
$$x = \dfrac{10}{4.5} \quad \text{Simplify the right side}$$
$$x = \dfrac{20}{9}$$

Therefore the walk is 2 and $\dfrac{2}{9}$ kilometers long.

Chapter 1 Review

64. The ratio that must remain constant is the ratio of the pounds of tin used to the pounds of alloy made, $\frac{2.5}{12}$, or equivalently $\frac{5}{24}$. If x represents the pounds of tin used to make 90 pounds of alloy then we must have:

$\frac{5}{24} = \frac{x}{90}$ Fundamental property of proportions
$450 = 24x$ Divide both sides by 24
$18.75 = x$

Therefore 18.75 pounds of tin are needed to make 90 pounds of alloy.

65. Let x represent the length shown below.

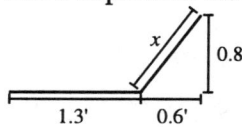

By the Pythagorean theorem $x = \sqrt{(0.6)^2 + (0.8)^2} = 1.0$ feet long, so the length of the strip before it was bent was $1.3 + 1.0 = 2.3$ feet long.

66. Let x represent the length of the rafter. From the diagram we see that we have the right triangle shown below with all lengths converted to inches.

By the Pythagorean theorem $x - 2 = \sqrt{(168)^2 + (66)^2}$ or
$x - 2 = 180.5$ Add 2 to both sides
$x = 182.5$

The rafter is 182.5 inches or 15.21 feet long.

67. The ratio that remains constant here is the height of an object to the length of its shadow. In the case of the friend, this ratio is $\frac{6.5}{14.2}$ with all lengths in feet. If x represents the height of the tree, then we expect the ratio $\frac{x}{39.4}$ to be equal to the ratio $\frac{6.5}{14.2}$. So to find the height of the tree we solve for x in:

$\frac{6.5}{14.2} = \frac{x}{39.4}$ Fundamental property of proportions
$(6.5)(39.4) = (14.2)x$ Simplify the left side
$256.1 = (14.2)x$ Divide both sides by 14.2
$18.04 = x$

The tree is 18.04 feet tall.

Chapter 1 Review

68. Let d be the distance across the lake. Looking at the figure in the text we see that by similar triangles, $\frac{2.3}{6.7} = \frac{d}{2.1+6.7}$. We now solve for d.

$\frac{2.3}{6.7} = \frac{d}{2.1+6.7}$ Simplify denominator on the right

$\frac{2.3}{6.7} = \frac{d}{8.8}$ Fundamental property of proportions

$(2.3)(8.8) = (6.7)d$ Simplify the left side

$20.24 = (6.7)d$ Divide both sides by 6.7

$3.02 = d$

So the lake is 3.02 miles across.

69. Let s be the height in feet of the crate. Since the base of the crate is 6 feet by 6 feet, its surface area is $2(6^2) + 4(6s) = 72 + 24s$ ft^2. Wood costs $0.10 per ft^2, so $72 + 24s$ ft^2 costs $(72 + 24s)(0.10)$ dollars. However, the cost of the crate is $16.80, so we can find the dimensions of the crate by solving for s in the equation:

$(72 + 24s)(0.10) = 16.80$ Divide both sides by 0.10

$72 + 24s = 168$ Subtract 72 from both sides

$24s = 96$ Divide both sides by 24

$s = 4$

The height of the crate is 4 feet. The dimensions of the crate are 6 ft × 6 ft × 4 ft.

70. Let s be the length in inches of the planter. Since the square cross section is 8 in. × 8 in., we see from the figure below that the surface area is $2(8^2) + 3(8s) = 128 + 24s$ in.^{2}.

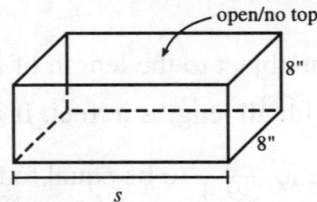

Redwood costs $0.005 per in.^{2}, so $128 + 24s$ in.^{2} costs $(128 + 24s)(0.005)$ dollars. However the cost of the planter is $5.44, so we can find the dimensions of the planter by solving for s in the equation:

$(128 + 24s)(0.005) = 5.44$ Divide both sides by 0.005

$128 + 24s = 1088$ Subtract 128 from both sides

$24s = 960$ Divide both sides by 24

$s = 40$

The length of the planter is 40 inches.

Chapter 1 Review

71. The formula for the volume of a pyramid is $V = \frac{1}{3}s^2h$. For the Great Pyramid, we evaluate the expression above at $s = 160$ yards and $h = 250$ yards, so
$V = \frac{1}{3}(160)^2(250) = 2,133,333.\overline{3}$ yards^3.

72. a. Let a be the length of the altitude of one of the faces of the Great Pyramid. Imagine cutting the pyramid in half as shown and looking at the cross section below.

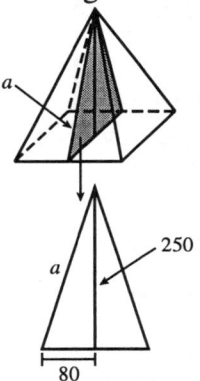

By the Pythagorean theorem,
$a = \sqrt{250^2 + 80^2} = 262.49$ yards.

b. The area of a triangle is $A = \frac{1}{2}sa$ where s is the length of the base and a is the length of the altitude. Hence the surface area of one face of the Great Pyramid is
$A = \frac{1}{2}(160)(262.49) = 20,999.2$ yards^2. There are, however, four triangular faces on the Great Pyramid, so the total surface area is $4(20,999.2) = 83,996.8$ yards^2.

73. a. The greatest height the graph achieves corresponds to the highest snow level. By inspecting the graph we notice that this value is approximately 9 feet.

b. Notice that the graph of the snow level climbs above 4 feet on roughly the fifth day of the month and remains above this level through the fourteenth. Therefore the snow level was above 4 feet for 10 days.

c. On January 7 the snow level was approximately 5 feet and on January 9 it was approximately 8 feet so in this time period the snow level increased 3 feet.

d. By inspecting the graph we notice that the greatest increase occurs from January 8 to January 9 when the snow level increased approximately 2.5 feet in one day. The greatest decrease occurs from January 13 to January 14 when the snow level dropped approximately 1 foot in one day.

74. a. The highest maximum temperature is the point with the highest y-coordinate. This occurs on day 3 and corresponds to a maximum temperature that day of approximately 32°F. The lowest maximum is the point with the smallest y-coordinate. This occurs on day 11 and represents a maximum temperature that day of approximately 11°F.

b. The number of days where the maximum temperature was below 20°F is the number of points where the y-coordinate is below 20. This happens for 4 days, January 9 through January 12.

Chapter 1 Review

c. Since the maximum temperatures decline so noticeably starting on January 9, this day is most likely the day that a cold front passed over the resort.

d. On January 8 the maximum temperature was approximately 29°F and on January 9 it was about 14°F, a drop of 15°F.

75. a. From the graph: (2, –6), (–1, 3), (0, 2) and (1, 1).

b. Algebraic verification:
$2 - (2)^3 = 2 - 8 = -6$
$2 - (-1)^3 = 2 - (-1) = 3$
$2 - (0)^3 = 2 - 0 = 2$
$2 - (1)^3 = 2 - 1 = 1$

76. a. From the graph: (1, 0), (–3, 2), (0, 1), and (–8, 3).

b. Algebraic verification:
$\sqrt{1 - (1)} = \sqrt{0} = 0$
$\sqrt{1 - (-3)} = \sqrt{4} = 2$
$\sqrt{1 - (0)} = \sqrt{1} = 1$
$\sqrt{1 - (-8)} = \sqrt{9} = 3$

77. By experimenting with different graphing windows, notice that the graph of $y = 0.5x^4 - 5x^2 + 4.5$ looks very nice with the range set to Xmin = –5, Xmax = 5, Xscl = 1, Ymin = –10, Ymax = 10, and Yscl = 1.

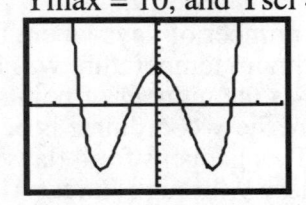

a. Using the TRACE key as in exercises 31. through 38. in Section 1.6, we see that $y = 0$ when $x = \pm 1$ and $x = \pm 3$.

b. Again using the TRACE key, we see that $y = 8$ when $x = \pm 3.3$.

c. Since the horizontal line $y = -2$ intersects the graph in four points (as shown below), there are four points where the graph has y-coordinate –2.

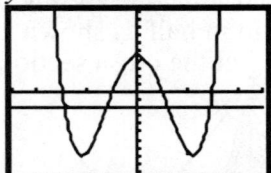

d. Again using TRACE, we identify the turning points as (–2.3, –8), (0, 4.5), and (2.3, –8).

78. Graph the function in the standard window.

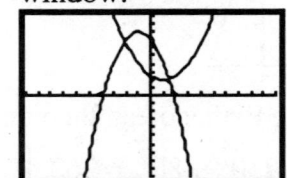

a. Using the TRACE key, we see that the minimum point of the first graph occurs at approximately (1.0, 2.0).

b. Using the TRACE key, we see that the maximum point of the second graph occurs at approximately (–1.0, 8.0).

c. Using the TRACE key, we see that the intersection points of the two graphs occur at approximately (1.4, 2.1) and (–2.1, 6.8).

Chapter 2

Section 2.1

1. a. $h = 2t + 6$

  b. Choose several values of t and calculate corresponding values for h to obtain the following ordered pairs:

t	h	(t, h)
0	6	(0, 6)
1	8	(1, 8)
2	10	(2, 10)
3	12	(3, 12)

  Refer to the graph in the back of the textbook.

  c. 3 weeks = 21 days. Substitute $t = 21$ into the equation and solve for h.
  $h = 2(21) + 6 = 48$
  The corn is 48 in. or 4 ft tall.

  d. 6 ft = 72 in. Substitute $h = 72$ into the equation and solve for t.
  $72 = 2t + 6$
  $66 = 2t$
  $33 = t$
  It will be 33 days.

3. a. $A = 250 - 15w$

  b. Choose several values of w and calculate corresponding values for A to obtain the following:

w	A	(w, A)
0	250	(0, 250)
5	175	(5, 175)
10	100	(10, 100)
15	25	(15, 25)

  Refer to the graph in the back of the textbook.

  c. Using the graph: when $w = 3$, $A = 205$, and when $w = 8$, $A = 130$.
  The decrease in oil is the difference of the two A-values.
  decrease in fuel oil = 205 − 130
  $\qquad$ = 75 gallons
  The decrease is illustrated by the vertical arrow in the graph.

  d. Using the graph: when $A = 175$, $w = 5$. Therefore, the tank will contain more than 175 gallons up to the fifth week. This is illustrated by the horizontal arrow in the graph

5. a. $P = 40t - 800$

  b. Set $t = 0$
  $P = 40(0) - 800 = -800$
  The P-intercept is −800.
  Set $P = 0$
  $0 = 40t - 800$
  $800 = 40t$
  $20 = t$
  The t-intercept is 20. Refer to the graph in the back of the textbook.

Chapter 2, Section 2.1

    c. The P-intercept, -800, is the initial ($t = 0$) value of the profit. Phil and Ernie start out $800 in debt. The t-intercept, 20, is the number of hours required for Phil and Ernie to break even.

7. a. The owner spends $0.6x$ on regular unleaded and $0.8y$ on premium unleaded.
$0.6x + 0.8y = 4800$

    b. Set $x = 0$
$0.6(0) + 0.8y = 4800$
$0.8y = 4800$
$y = 6000$
The y-intercept is 6000.
Set $y = 0$
$0.6x + 0.8(0) = 4800$
$0.6x = 4800$
$x = 8000$
The x-intercept is 8000.
Refer to the graph in the back of the textbook.

    c. The y-intercept, 6000, is the amount of premium the gas station owner can buy if he buys no regular. The x-interecept, 8000, is the amount of regular he can buy if he buys no premium.

9. a. $I = 0.03s + 10,000$

    b. Choose several values of s and calculate corresponding values of I to obtain the following ordered pairs.

s	I
0	10,000
100,000	13,000
200,000	16,000
300,000	19,000

(s, I)
(0, 10,000)
(100,000, 13,000)
(200,000, 16,000)
(300,000, 19,000)

Refer to the graph in the back of the textbook.

    c. Substitute $I = 16,000$ into the equation and solve for s.
$16,000 = 0.03s + 10,000$
$6,000 = 0.03s$
$200,000 = s$
Substitute $I = 22,000$ into the equation and solve for s.
$22,000 = 0.03s + 10,000$
$12,000 = 0.03s$
$400,000 = s$
Therefore, annual sales are between $200,000 and $400,000.

    d. Using the graph: when $s = 500,000$, $I = 25,000$, and when $s = 700,000$, $I = 31,000$. The increase in salary is the difference of the two I values.
increase in salary
$= 31,000 - 25,000 = \$6,000$

11. Set $x = 0$.
$0 + 2y = 8$
$y = 4$
The y-intercept is the point $(0, 4)$.
Set $y = 0$.
$x + 2(0) = 8$
$x = 8$
The x-intercept is the point $(8, 0)$.
Refer to the graph in the back of the textbook.

Chapter 2, Section 2.1

13. Set $x = 0$.
$3(0) - 4y = 12$
$-4y = 12$
$y = -3$
The y-intercept is the point $(0, -3)$.
Set $y = 0$.
$3x - 4(0) = 12$
$3x = 12$
$x = 4$
The x-intercept is the point $(4, 0)$.
Refer to the graph in the back of the textbook.

15. Set $x = 0$.
$\frac{0}{9} - \frac{y}{4} = 1$
$\frac{-y}{4} = 1$
$y = -4$
The y-intercept is the point $(0, -4)$.
Set $y = 0$.
$\frac{x}{9} - \frac{0}{4} = 1$
$\frac{x}{9} = 1$
$x = 9$
The x-intercept is the point $(9, 0)$.
Refer to the graph in the back of the textbook.

17. Set $x = 0$.
$\frac{2(0)}{8} + \frac{3y}{11} = 1$
$\frac{3y}{11} = 1$
$y = \frac{11}{3}$
The y-intercept is the point $\left(0, \frac{11}{3}\right)$.
Set $y = 0$.

$\frac{2x}{3} + \frac{3(0)}{11} = 1$
$\frac{2x}{3} = 1$
$x = \frac{3}{2}$
The x-intercept is the point $\left(\frac{3}{2}, 0\right)$.

19. Set $x = 0$.
$20(0) = 30y - 45000$
$0 = 30y - 45000$
$45000 = 30y$
$1500 = y$
The y-intercept is the point $(0, 1500)$.
Set $y = 0$.
$20x = 30(0) - 45{,}000$
$20x = -45{,}000$
$x = -2250$
The x-intercept is the point $(-2250, 0)$.
Refer to the graph in the back of the textbook.

21. Set $x = 0$.
$0.4(0) + 1.2y = 4.8$
$1.2y = 4.8$
$y = 4$
The y-intercept is the point $(0, 4)$.
Set $y = 0$.
$0.4x + 1.2(0) = 4.8$
$0.4x = 4.8$
$x = 12$
The x-intercept is the point $(12, 0)$.
Refer to the graph in the back of the textbook.

Chapter 2, Section 2.1

23. Choose several values of x and calculate corresponding values for y to obtain the following ordered pairs.

x	y	(x, y)
−2	2	(−2, 2)
−1	1	(−1, 1)
0	0	(0, 0)
1	−1	(1, −1)
2	−2	(2, −2)

Refer to the graph in the back of the textbook.

25. Choose several values of x and calculate corresponding values for y to obtain the following ordered pairs.

x	y	(x, y)
−2	−4	(−2, −4)
−1	−2	(−1, −2)
0	0	(0, 0)
1	2	(1, 2)
2	4	(2, 4)

Refer to the graph in the back of the textbook.

27. Choose several values of x and calculate corresponding values for y to obtain the following ordered pairs.

x	y	(x, y)
−6	−2	(−6, −2)
−3	−1	(−3, −1)
0	0	(0, 0)
3	1	(3, 1)
6	2	(6, 2)

Refer to the graph in the back of the textbook.

29. a. Let d represent distance in miles and let t represent the number of hours elapsed.
$d = 50t$

b. Choose several values of t and calculate corresponding values for d to obtain the following ordered pairs.

t	d	(t, d)
0	0	(0, 0)
1	50	(1, 50)
2	100	(2, 100)
3	150	(3, 150)

Refer to the graph in the back of the textbook.

31. a. Let T represent the number of hours and let h represent the number of houses.
$T = \dfrac{h}{24}$

Chapter 2, Section 2.2

b. Choose several values of h and calculate the corresponding values for T to obtain the following ordered pairs.

h	T	(h, T)
0	0	(0, 0)
24	1	(24, 1)
48	2	(48, 2)
72	3	(72, 3)

Refer to the graph in the back of the textbook.

Section 2.2

1. Carl's average speed is
$$\frac{100 \text{ meters}}{10 \text{ seconds}} = 10 \text{ m/s}$$
and Anthony's average speed is
$$\frac{200 \text{ meters}}{19.6 \text{ seconds}} \approx 10.2 \text{ m/s}$$
Anthony has the faster average speed.

3. Øksendahl's brand costs
$$\frac{\$6.98}{30 \text{ ounces}} = \$0.23/\text{oz}$$
and Fran and Jenny's brand costs
$$\frac{\$7.19}{32 \text{ ounces}} = \$0.22/\text{oz}$$
Øksendahl's brand costs more per ounce.

5. Corey's fuel economy to Las Vegas and back is
$$\frac{512 \text{ miles}}{16 \text{ gallons}} = 32 \text{ mi/gal}$$
and his fuel economy to Los Angeles is
$$\frac{429 \text{ miles}}{13 \text{ gallons}} = 33 \text{ mi/gal}$$
He got better fuel economy on his trip to Los Angeles.

7. For Stone Canyon Drive, the ratio of vertical gain to horizontal distance is
$$\frac{840 \text{ feet}}{1500 \text{ feet}} = 0.56,$$
and for Highway 33, the ratio is
$$\frac{1150 \text{ feet}}{2000 \text{ feet}} = 0.575.$$
Highway 33 is steeper.

9. For the truck ramp for Acme Movers, the ratio of vertical gain to horizontal distance is
$$\frac{4 \text{ feet}}{9 \text{ feet}} \approx 0.44,$$
and for a truck, the ratio is
$$\frac{3 \text{ cm}}{7 \text{ cm}} \approx 0.43.$$
The truck ramp for Acme Movers is steeper.

11. $m = \frac{\Delta y}{\Delta x} = \frac{2}{8} = \frac{1}{4}$

13. $m = \frac{\Delta y}{\Delta x} = \frac{-4}{6} = -\frac{2}{3}$

15. $m = \frac{\Delta y}{\Delta x} = \frac{-4}{4} = -1$

17. $\frac{\text{vertical change}}{\text{horizontal change}} = \text{slope}$

    vertical change = horizontal change × slope

    a. $6 \times \frac{7}{3} = 14$

    b. $10 \times \frac{7}{3} = \frac{70}{3}$ or $23\frac{1}{3}$

    c. $-24 \times \frac{7}{3} = -56$

Chapter 2, Section 2.2

19. $\dfrac{\text{vertical distance}}{\text{horizontal space}} = \text{slope}$

$\dfrac{10 \text{ ft}}{\text{horizontal space}} = \dfrac{7}{10}$

$\text{horizontal space} = \dfrac{100 \text{ ft}}{7}$

$\approx 14.29 \text{ ft}$

21. Set $x = 0$.
$3(0) - 4y = 12$
$-4y = 12$
$y = -3$
The y-intercept is the point $(0, -3)$.
Set $y = 0$.
$3x - 4(0) = 12$
$3x = 12$
$x = 4$
The x-intercept is the point $(4, 0)$.
Refer to the graph in the back of the textbook.
$m = \dfrac{\Delta y}{\Delta x} = \dfrac{3}{4}$

23. Set $x = 0$.
$2y + 6(0) = -18$
$2y = -18$
$y = -9$
The y-intercept is the point $(0, -9)$.
Set $y = 0$.
$2(0) + 6x = -18$
$6x = -18$
$x = -3$
The x-intercept is the point $(-3, 0)$.
Refer to the graph in the back of the textbook.
$m = \dfrac{\Delta y}{\Delta x} = \dfrac{9}{-3} = -3$

25. Set $x = 0$.
$\dfrac{0}{5} - \dfrac{y}{8} = 1$
$\dfrac{-y}{8} = 1$
$y = -8$
The y-intercept is the point $(0, -8)$.
Set $y = 0$.
$\dfrac{x}{5} - \dfrac{0}{8} = 1$
$\dfrac{x}{5} = 1$
$x = 5$
The x-intercept is the point $(5, 0)$.
Refer to the graph in the back of the textbook.
$m = \dfrac{\Delta y}{\Delta x} = \dfrac{8}{5}$

27. If the tables represent variables that are related by a linear equation, the slopes from one point to the next must be constant.

a. $m = \dfrac{17-12}{3-2} = 5$

$m = \dfrac{22-17}{4-3} = 5$

$m = \dfrac{27-22}{5-4} = 5$

$m = \dfrac{32-27}{6-5} = 5$

The table represents variables that are related by a linear equation.

Chapter 2, Section 2.2

b. $m = \dfrac{9-4}{3-2} = 5$

$m = \dfrac{16-9}{4-3} = 7$

$m = \dfrac{25-16}{5-4} = 9$

$m = \dfrac{36-25}{6-5} = 11$

The table does not represent variables that are related by a linear equation.

c. $m = \dfrac{18-20}{-3-(-6)} = \dfrac{-2}{3}$

$m = \dfrac{16-18}{0-(-3)} = \dfrac{-2}{3}$

$m = \dfrac{14-16}{3-0} = \dfrac{-2}{3}$

$m = \dfrac{12-14}{6-3} = \dfrac{-2}{3}$

The table represents variables that are related by a linear equation.

d. $m = \dfrac{3-0}{10-5} = \dfrac{3}{5}$

$m = \dfrac{6-3}{15-10} = \dfrac{3}{5}$

$m = \dfrac{12-6}{20-15} = \dfrac{6}{5}$

$m = \dfrac{24-12}{25-20} = \dfrac{12}{5}$

The table does not represent variables that are related by a linear equation.

29. a. Refer to the graph in the back of the textbook.

b. For example, use (0, 0) and (1, 8).

$m = \dfrac{\Delta s}{\Delta t} = \dfrac{8}{1} = 8$ dollars per hour.

c. The slope gives the typist's rate of pay, in dollars per hour.

31. a. For example, use (10, 20) and (20, 35).

$m = \dfrac{35-20}{20-10} = \dfrac{15}{10}$

$= 1.5$ m per min.

b. Speed of train

33. a. For example, use (0, 2000) and (4, 7000).

$m = \dfrac{7000-2000}{4-0} = \dfrac{5000}{4}$

$= 1250$ barrels per day.

b. Rate of pumping

35. a. For example, use (0, 48) and (8, 0).

$m = \dfrac{0-48}{8-0} = \dfrac{-48}{8}$

$= -6$ L per day.

b. Rate of water consumption

37. a. For example, use (0,0) and (5, 60).

$m = \dfrac{60-0}{5-0} = \dfrac{60}{5}$

$= 12$ in. per ft.

b. Conversion from feet to inches

39. a. For example, use (0, 0) and (5, 20).

$m = \dfrac{20-0}{5-0} = \dfrac{20}{5}$

$= 4$ dollars per kg.

b. Unit cost of beans per kilogram

Chapter 2, Section 2.3

Section 2.3

1. $2x + y = 6$
   $y = 6 - 2x$

3. $8 - y + 3x = 0$
   $-y = -3x - 8$
   $y = 3x + 8$

5. $x + 2y = 5$
   $2y = -x + 5$
   $y = -\frac{1}{2}x + \frac{5}{2}$

7. $3x - 4y = 6$
   $-4y = -3x + 6$
   $y = \frac{3}{4}x - \frac{3}{2}$

9. $0.2x + 0.5y = 1$
   $0.5y = -0.2x + 1$
   $y = \frac{-2}{5}x + 2$
   or $y = -0.4x + 2$

11. $7x + 3y = y - 32$
    $7x + 2y = -32$
    $2y = -7x - 32$
    $y = -\frac{7}{2}x - 16$

13. a. Set $x = 0$.
    $0 + y = 100$
    $y = 100$
    The y-intercept is the point $(0, 100)$.
    Set $y = 0$.
    $x + 0 = 100$
    $x = 100$
    The x-intercept is the point $(100, 0)$.

    b. Xmin = –20, Xmax = 120, Ymin = –20, and Ymax = 120.

    c. $x + y = 100$
    $y = -x + 100$

    d. Refer to the graph in the back of the textbook.

15. a. Set $x = 0$.
    $25(0) - 36y = 1$
    $-36y = 1$
    $y = -\frac{1}{36}$
    The y-intercept is the point $\left(0, -\frac{1}{36}\right)$.
    Set $y = 0$.
    $25x - 36(0) = 1$
    $25x = 1$
    $x = \frac{1}{25}$
    The x-intercept is the point $\left(\frac{1}{25}, 0\right)$.

    b. Xmin = –0.1, Xmax = 0.1, Ymin = –0.1, and Ymax = 0.1.

    c. $25x - 36y = 1$
    $-36y = -25x + 1$
    $y = \frac{25}{36}x - \frac{1}{36}$

    d. Refer to the graph in the back of the textbook.

17. a. Set $x = 0$.
    $\frac{y}{12} - \frac{0}{47} = 1$
    $\frac{y}{12} = 1$
    $y = 12$
    The y-intercept is the point $(0, 12)$.

Chapter 2, Section 2.3

Set $y = 0$.
$$\frac{0}{12} - \frac{x}{47} = 1$$
$$\frac{-x}{47} = 1$$
$$x = -47$$
The x-intercept is the point $(-47, 0)$.

b. Xmin = -50, Xmax = 10, Ymin = -5, and Ymax = 15.

c. $\frac{y}{12} - \frac{x}{47} = 1$
$$\frac{y}{12} = \frac{x}{47} + 1$$
$$y = \frac{12}{47}x + 12$$

d. Refer to the graph in the back of the textbook.

19. a. Set $x = 0$.
$$-2(0) = 3y + 84$$
$$0 = 3y + 84$$
$$-3y = 84$$
$$y = -28$$
The y-intercept is the point $(0, -28)$.
Set $y = 0$.
$$-2x = 3(0) + 84$$
$$-2x = 84$$
$$x = -42$$
The x-intercept is the point $(-42, 0)$.

b. Xmin = -50, Xmax = 10, Ymin = -30, and Ymax = 5.

c. $-2x = 3y + 84$
$$-3y = 2x + 84$$
$$y = -\frac{2}{3}x - 28$$

d. Refer to the graph in the back of the textbook.

21. a. Set $x = 0$.
$$y - 42 = \frac{1}{3}(0 - 12)$$
$$y - 42 = -4$$
$$y = 38$$
The y-intercept is the point $(0, 38)$.

Set $y = 0$.
$$0 - 42 = \frac{1}{3}(x - 12)$$
$$-42 = \frac{x}{3} - 4$$
$$-\frac{x}{3} = 42 - 4$$
$$-\frac{x}{3} = 38$$
$$x = -114$$
The x-intercept is the point $(-114, 0)$.

b. Xmin = -120, Xmax = 10, Ymin = -10, and Ymax = 40.

c. $y - 42 = \frac{1}{3}(x - 12)$
$$y - 42 = \frac{1}{3}x - 4$$
$$y = \frac{1}{3}x + 38$$

d. Refer to the graph in the back of the textbook.

23. a. Set $x = 0$.
$$y - 3 = \frac{-13}{22}(0 + 5)$$
$$y - 3 = \frac{-65}{22}$$
$$y = \frac{1}{22}$$
The y-intercept is the point $\left(0, \frac{1}{22}\right)$.

Chapter 2, Section 2.3

Set $y = 0$.
$0 - 3 = \frac{-13}{22}(x + 5)$

$-3 = \frac{-13}{22}x - \frac{65}{22}$

$\frac{13}{22}x = 3 - \frac{65}{22}$

$\frac{13}{22}x = \frac{1}{22}$

$x = \frac{1}{13}$

The x-intercept is the point $\left(\frac{1}{13}, 0\right)$.

b. Xmin = –0.1, Xmax = 0.1, Ymin = –0.1, and Ymax = 0.1.

c. $y - 3 = \frac{-13}{22}(x + 5)$

$y - 3 = \frac{-13}{22}x - \frac{65}{22}$

$y = -\frac{13}{22}x + \frac{1}{22}$

d. Refer to the graph in the back of the textbook.

25. $m = \frac{14 - 2}{2 - (-3)} = \frac{12}{5}$

27. $m = \frac{-1 - (-3)}{-2 - 2} = \frac{2}{-4} = -\frac{1}{2}$

29. $m = \frac{3 - (-3)}{-6 - (-6)} = \frac{6}{0} =$ undefined

31. $m = \frac{6.8 - (-4.2)}{-3.1 - 7.6} = \frac{11}{-10.7} = \frac{-110}{107}$
≈ -1.028

33. $m = \frac{-\frac{1}{2} - \left(-\frac{1}{8}\right)}{\frac{5}{6} - \frac{3}{4}} = \frac{-\frac{4}{8} + \frac{1}{8}}{\frac{10}{12} - \frac{9}{12}}$

$= \frac{-\frac{3}{8}}{\frac{1}{12}} = -\frac{9}{2}$

35. $m = \frac{-\frac{2}{3} - 2\frac{1}{3}}{4\frac{1}{3} - (-2)} = \frac{-\frac{2}{3} - \frac{7}{3}}{\frac{13}{3} + \frac{6}{3}} = \frac{-\frac{9}{3}}{\frac{19}{3}}$

$= -\frac{9}{19}$

37. $m = \frac{-1.2 - (-3.6)}{1.4 - (-4.8)} = \frac{2.4}{6.2} = \frac{12}{31}$
≈ 0.387

39. $m = \frac{7000 - (-2000)}{5000 - 5000} = \frac{9000}{0}$
$=$ undefined

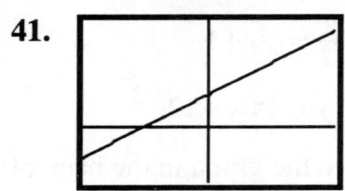

41.

a. For example, use (–2.5, 0) and (0, 6.25).
$m = \frac{6.25 - 0}{0 - (-2.5)} = \frac{6.25}{2.5} = 2.5$

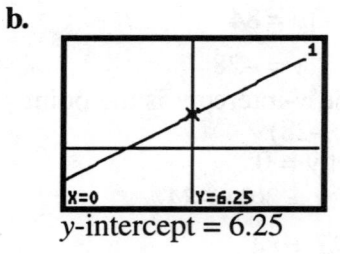

b.

X=0   Y=6.25
y-intercept = 6.25

43.

a. For example, use (0, 63) and (7.5, 0).
$m = \frac{0 - 63}{7.5 - 0} = \frac{-63}{7.5} = -8.4$

48

Chapter 2, Section 2.3

b.

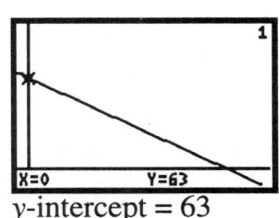

[graph showing y-intercept at (0, 63), line descending]
X=0 Y=63
y-intercept = 63

45. a. Yes; recall that $C = 2\pi r$ which is an equation for a straight line

b. Using the points (0, 0) and (1, 2π), the slope is $\frac{2\pi - 0}{1 - 0} = 2\pi$.

47. a. Refer to the graph in the back of the textbook, where Xmin = –40, Xmax = 40, Ymin = –50, and Ymax =100.

b. Substitute $C = 10$ into the equation and solve for F.
$F = \frac{9}{5}(10) + 32 = 18 + 32 = 50$
The temperature is 50°F.

c. Substitute $F = -4$ into the equation and solve for C.
$-4 = \frac{9}{5}C + 32$
$-36 = \frac{9}{5}C$
$-20 = C$
The temperature is –20°C.

d. For example, use (10, 50) and (–20, –4) as two points on the graph.
$m = \frac{-4 - 50}{-20 - 10} = \frac{-54}{-30} = \frac{9}{5}$

e. The Fahrenheit temperature increases $\frac{9}{5}$ of a degree for every degree increase in the Celsius temperature.

f. The C-intercept gives the Celsius temperature at 0°F, and the F-intercept gives the Fahrenheit temperature at 0°C.

49. a. Refer to the graph in the back of the textbook.

b. $m = \frac{1}{2}, 2, -2$
y-intercepts = 0, 0, 0

c. The coefficient of x in each equation gives the slope of the line. The y-intercepts are all 0. The graph of $y = 2x$ is a reflection of $y = -2x$ across the y-axis. The graph of $y = -2x$ is perpendicular to $y = \frac{1}{2}x$.

51. a. Refer to the graph in the back of the textbook.

b. $m = 2, 3, 4$
y-intercepts = 0, 0, 0

c. The coefficient of x in each equation gives the slope of the line. The y-intercepts are all 0.

53. a. Refer to the graph in the back of the textbook.

b. $m = -1, -\frac{1}{2}, -\frac{1}{4}$
y-intercepts = 0, 0, 0

c. The coefficient of x in each equation gives the slope of the line. The y-intercepts are all 0.

55. a. Refer to the graph in the back of the textbook.

b. $m = -2, -2, -2$
y-intercept = 0, 10, –25

c. The slopes are all –2 which is the coefficient of x in each equation. The constant term is the y-intercept.

Chapter 2, Section 2.4

57. a. Refer to the graph in the back of the textbook.

b. $m = \frac{2}{3}, \frac{2}{3}, \frac{2}{3}$
y-intercepts $= -12, -24, -36$

c. The slopes are all $\frac{2}{3}$ which is the coefficient of x in each equation. The constant term is the y-intercept.

Section 2.4

1. a. $3x + 2y = 1$
$2y = -3x + 1$
$y = -\frac{3}{2}x + \frac{1}{2}$

b. $m = -\frac{3}{2}, b = \frac{1}{2}$

3. a. $x - 3y = 2$
$-3y = -x + 2$
$y = \frac{1}{3}x - \frac{2}{3}$

b. $m = \frac{1}{3}, b = -\frac{2}{3}$

5. a. $\frac{1}{4}x + \frac{3}{2}y = \frac{1}{6}$
$\frac{3}{2}y = -\frac{1}{4}x + \frac{1}{6}$
$y = -\frac{1}{6}x + \frac{1}{9}$

b. $m = -\frac{1}{6}, b = \frac{1}{9}$

7. a. $4.2x - 0.3y = 6.6$
$-0.3y = -4.2x + 6.6$
$y = 14x - 22$

b. $m = 14, b = -22$

9. a. $y + 29 = 0$
$y = -29$

b. $m = 0, b = -29$

11. a. $250x + 150y = 2450$
$150y = -250x + 2450$
$y = -\frac{5}{3}x + \frac{49}{3}$

b. $m = \frac{-5}{3}, b = \frac{49}{3}$

13. a. Plot the y-intercept $(0, -2)$. Move 3 units in the y-direction and 1 unit in the x-direction to arrive at $(1, 1)$. Draw the line through the two points. Refer to the graph in the back of the textbook.

b. $y = 3x - 2$

15. a. Plot the y-intercept $(0, 4)$. Move -2 units in the y-direction and 1 unit in the x-direction to arrive at $(1, 2)$. Draw the line through the two points. Refer to the graph at the back of the textbook.

b. $y = -2x + 4$

17. a. Plot the y-intercept $(0, -6)$. Move 5 units in the y-direction and 3 units in the x-direction to arrive at $(3, -1)$. Draw the line through the two points. Refer to the graph in the back of the textbook.

b. $y = \frac{5}{3}x - 6$

Chapter 2, Section 2.4

19. a. Plot the y-intercept (0, 3). Move −1 unit in the y-direction and 2 units in the x-direction to arrive at (2, 2). Draw the line through the two points. Refer to the graph in the back of the textbook.

b. $y = -\frac{1}{2}x + 3$

21. $m = 1.5, b = 5$
$d = 1.5t + 5$

23. $m = 1250, b = 2000$
$B = 1250t + 2000$

25. $m = -6, b = 48$
$W = -6t + 48$

27. $m = 12, b = 0$
$i = 12f$

29. $m = 4, b = 0$
$C = 4b$

31. a. $m = 4, b = 40$
b. $y = 4x + 40$

33. a. $m = -80, b = 2000$
b. $P = -80t + 2000$

35. a. $m = \frac{1}{4}, b = 0$
b. $V = \frac{1}{4}d$

37. $y - (-5) = -3(x - 2)$
$y + 5 = -3x + 6$
$y = -3x + 1$

39. $y - (-1) = \frac{5}{3}(x - 2)$
$y + 1 = \frac{5}{3}x - \frac{10}{3}$
$y = \frac{5}{3}x - \frac{13}{3}$

41. $y - (-3.5) = -0.27(x - (-6.4))$
$y + 3.5 = -0.27(x + 6.4)$
$y + 3.5 = -0.27x - 1.728$
$y = -0.27x - 5.228$

43. $m = \frac{3-2}{3-(-4)} = \frac{1}{7}$
$y - 3 = \frac{1}{7}(x - 3)$
$y - 3 = \frac{1}{7}x - \frac{3}{7}$
$y = \frac{1}{7}x + \frac{18}{7}$

45. $m = \frac{5-(-6)}{2-(-2)} = \frac{11}{4}$
$y - 5 = \frac{11}{4}(x - 2)$
$y - 5 = \frac{11}{4}x - \frac{11}{2}$
$y = \frac{11}{4}x - \frac{1}{2}$

47. $m = \frac{-10.8 - 9.6}{-2.4 - 15.3} = \frac{-20.4}{-17.7} = \frac{68}{59}$
$y - 9.6 = \frac{68}{59}(x - 15.3)$
$y - 9.6 = \frac{68}{59}x - \frac{1040.4}{59}$
$y = 1.15x - 8.03$

49. a. $m = \frac{15,000 - 9000}{125 - 50} = \frac{6000}{75} = 80$
$C - 9000 = 80(b - 50)$
$C - 9000 = 80b - 4000$
$C = 80b + 5000$

b. Plot the points (50, 9000) and (125, 15,000). Draw the line through the two points. Refer to the graph in the back of the textbook.

Chapter 2, Section 2.5

    c. $m = 80$ dollars/bike is the cost of making each bike.

51. a. $m = \dfrac{154-110}{70-50} = \dfrac{44}{20} = 2.2$

$p - 110 = 2.2(k - 50)$
$p - 110 = 2.2k - 110$
$p = 2.2k$

    b. Plot the points (70, 154) and (50, 110). Draw the line through the two points. Refer to the graph in the back of the textbook.

    c. $m = 2.2$ lb/kg is the factor for conversion from kilograms to pounds.

53. a. $m = \dfrac{590-265}{5} = \dfrac{325}{5} = 65$

$d - 265 = 65t$
$d = 65t + 265$

    b. Plot the points (0, 265) and (5, 590). Draw the line through the two points. Refer to the graph in the back of the textbook.

    c. $m = 65$ mph is their average speed.

Section 2.5

1. a. $y = -3$
Refer to the graph in the back of the textbook.

  b. $m = 0$

3. a. $2x = 8$
$x = 4$
Refer to the graph in the back of the textbook.

  b. Undefined

5. a. $x = 0$
Refer to the graph in the back of the book.

  b. Undefined

7. a. l_1 negative, l_2 negative, l_3 positive, l_4 zero

  b. l_1, l_2, l_4, l_3

9. a. $y = \dfrac{3}{5}x - 7$, so $m_1 = \dfrac{3}{5}$.

$3x - 5y = 2$
$-5y = -3x + 2$
$y = \dfrac{3}{5}x - \dfrac{2}{5}$,
so $m_2 = \dfrac{3}{5}$.
The given lines are parallel.

  b. $y = 4x + 3$, so $m_1 = 4$.

$y = \dfrac{1}{4}x - 3$, so $m_2 = \dfrac{1}{4}$.
The given lines are neither parallel nor perpendicular.

  c. $6x + 2y = 1$
$2y = -6x + 1$
$y = -3x + \dfrac{1}{2}$,
so $m_1 = -3$.

$x = 1 - 3y$
$3y = -x + 1$
$y = -\dfrac{1}{3}x + \dfrac{1}{3}$,
so $m_2 = \dfrac{-1}{3}$.
The given lines are neither parallel nor perpendicular.

d. $2y = 5$

$y = \frac{5}{2}$,

so $m_1 = 0$.

$5y = -2$

$y = \frac{-2}{5}$,

so $m_2 = 0$.
The given lines are parallel.

11. a.

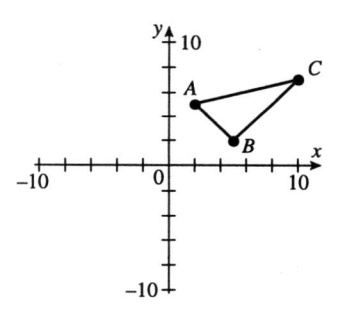

b. Slope $\overline{AB}$: $\frac{2-5}{5-2} = \frac{-3}{3} = -1$,

Slope $\overline{BC}$: $\frac{7-2}{10-5} = \frac{5}{5} = 1$,

Slope $\overline{AC}$: $\frac{7-5}{10-2} = \frac{2}{8} = \frac{1}{4}$.

Hence $\overline{AB} \perp \overline{BC}$, so the triangle is a right triangle.

13. a.

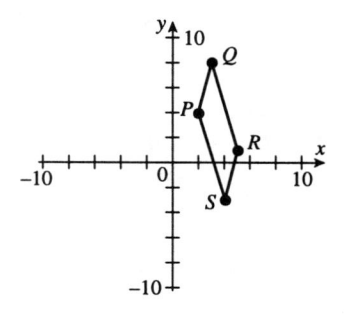

b. Slope $\overline{PQ}$: $\frac{8-4}{3-2} = \frac{4}{1} = 4$,

Slope $\overline{QR}$: $\frac{1-8}{5-3} = \frac{-7}{2} = -\frac{7}{2}$,

Slope $\overline{RS}$: $\frac{-3-1}{4-5} = \frac{-4}{-1} = 4$,

Slope $\overline{SP}$: $\frac{4-(-3)}{2-4} = \frac{7}{-2} = -\frac{7}{2}$.

Hence $\overline{PQ} \parallel \overline{RS}$ and
$\overline{QR} \parallel \overline{SP}$, so the quadrilateral is a parallelogram.

15. a. $x - 2y = 5$

$-2y = -x + 5$

$y = \frac{1}{2}x - \frac{5}{2}$

$m_1 = \frac{1}{2}$ so $m_2 = \frac{1}{2}$.

$y - (-1) = \frac{1}{2}(x - 2)$

$y + 1 = \frac{1}{2}x - 1$

$y = \frac{1}{2}x - 2$ or $x - 2y = 4$

b. Refer to the graph in the back of the textbook.

Chapter 2 Section 2.6

17. a. $2y - 3x = 5$
$2y = 3x + 5$
$y = \frac{3}{2}x + \frac{5}{2}$
$m_1 = \frac{3}{2}$,
so $m_2 = -\frac{1}{m_1} = -\frac{2}{3}$.
$y - 4 = -\frac{2}{3}(x - 1)$
$y - 4 = -\frac{2}{3}x + \frac{2}{3}$
$y = -\frac{2}{3}x + \frac{14}{3}$ or $2x + 3y = 14$

b. Refer to the graph in the back of the textbook.

19. a. Slope $\overline{AB}$: $\frac{-4-2}{-2-(-5)} = \frac{-6}{3} = -2$
$y - 2 = -2(x - (-5))$
$y - 2 = -2(x + 5)$
$y - 2 = -2x - 10$
$y = -2x - 8$

b. Slope $\overline{BC}$: $-\frac{1}{\text{slope } \overline{AB}} = -\frac{1}{-2} = \frac{1}{2}$
$y - (-4) = \frac{1}{2}(x - (-2))$
$y + 4 = \frac{1}{2}(x + 2)$
$y + 4 = \frac{1}{2}x + 1$
$y = \frac{1}{2}x - 3$

21. Slope $\overline{CP}$: $\frac{4-6}{2-(-1)} = \frac{-2}{3} = -\frac{2}{3}$
Slope of the tangent line:
$-\frac{1}{\text{slope } \overline{CP}} = -\frac{1}{-\frac{2}{3}} = \frac{3}{2}$
$y - 6 = \frac{3}{2}(x - (-1))$
$y - 6 = \frac{3}{2}(x + 1)$
$y - 6 = \frac{3}{2}x + \frac{3}{2}$
$y = \frac{3}{2}x + \frac{15}{2}$

23. a. Right angles are equal.
b. Alternate interior angles are equal.
c. Two angles of one triangle equal two angles of the other.
d. Definition of slope
e. Corresponding sides of similar triangles are proportional.

Section 2.6

1. a. $x = 0.6$
Solve algebraically:
$1.4x - 0.64 = 0.2$
$1.4x = 0.84$
$x = 0.6$

b. $x = -0.4$
Solve algebraically:
$1.4x - 0.64 = -1.2$
$1.4x = -0.56$
$x = -0.4$

Chapter 2, Section 2.6

c. $x > 0.6$
Solve algebraically:
$1.4x - 0.64 > 0.2$
$1.4x > 0.84$
$x > 0.6$

d. $x < -0.4$
Solve algebraically:
$-1.2 > 1.4x - 0.64$
$-0.56 > 1.4x$
$-0.4 > x$

3. a. $x = 6$
Solve algebraically:
$-36x + 226 = 10$
$-36x = -216$
$x = 6$

b. $x = 1$
Solve algebraically:
$190 = -36x + 226$
$36x = 36$
$x = 1$

c. $x \geq 6$
Solve algebraically:
$-36x + 226 \leq 10$
$-36x \leq -216$
$x \geq 6$

d. $x \leq 1$
Solve algebraically:
$190 \leq -36x + 226$
$36x \leq 36$
$x \leq 1$

5. a. $x \approx 12$
Solve algebraically:
$\sqrt{x} - 2 = 1.5$
$\sqrt{x} = 3.5$
$x = 12.25 \approx 12$

b. $x \approx 18$
Solve algebraically:
$\sqrt{x} - 2 = 2.25$
$\sqrt{x} = 4.25$
$x = 18.0625 \approx 18$

c. $x < 9$
Solve algebraically:
$\sqrt{x} - 2 < 1$
$\sqrt{x} < 3$
$x < 9$

d. Approximately $x > 3$
Solve algebraically:
$\sqrt{x} - 2 > -0.25$
$\sqrt{x} > 1.75$
$x > 3.0625$

7. a. $t \approx -3.1$
Check algebraically:
$-10(-3.1 + 1)^3 + 10$
$= -10(-2.1)^3 + 10$
$= -10(-9.261) + 10$
$= 92.61 + 10$
$= 102.61 \approx 100$

Chapter 2 Section 2.6

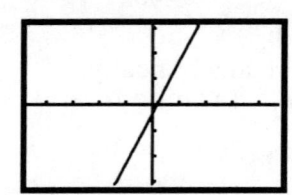

    b. $t \approx 1.5$
    Check algebraically:
    $-10(1.5+1)^3 + 10$
    $= -10(2.5)^3 + 10$
    $= -10(15.625) + 10$
    $= -156.25 + 10$
    $= -146.25 \approx -140$

    c. Approximately $t < 0.8$
    Check algebraically:
    $-10(0.8+1)^3 + 10$
    $= -10(1.8)^3 + 10$
    $= -10(5.832) + 10$
    $= -58.32 + 10$
    $= -48.32 \approx -50$

    d. Approximately $-2.4 < t < 0.4$
    Check algebraically:
    $-10(-2.4+1)^3 + 10$
    $= -10(-1.4)^3 + 10$
    $= -10(-2.744) + 10$
    $= 27.44 + 10$
    $= 37.44 \approx 40$

    $-10(0.4+1)^3 + 10$
    $= -10(1.4)^3 + 10$
    $= -10(2.744) + 10$
    $= -27.44 + 10$
    $= -17.44 \approx -20$

    e. $t < -1$ and $t > -1$

9. a. $q = -2$, $q = 2$
    b. $q \approx -2.8$, $q = 0$, $q \approx 2.8$
    c. $-2.5 < q < -1.25$, $1.25 < q < 2.5$
    d. $-2 < q < 0, 2 < q$
    e. $0 \leq M \leq 26$

11. a. $u = -1$
    b. $u = 0$
    c. $u < 1$
    d. $-0.5 < u < 0$

13. a. 1991
    b. 1 year (1992)
    c. 1 year (1993)
    d. $12{,}000 - 4000 = 8000$

15. a. Approximately $365
    b. $2 or $8
    c. $3.25 < d < 6.75$

17.

    a. $x = 4$
    Verify:
    $2x - 3 = 5$
    $2x = 8$
    $x = 4$

    b. $x = -5$
    Verify:
    $2x - 3 = -13$
    $2x = -10$
    $x = -5$

    c. $x > 1$
    Verify:
    $2x - 3 > -1$
    $2x > 2$
    $x > 1$

Chapter 2, Section 2.6

    d. $x < 14$
       Verify:
$$2x - 3 < 25$$
$$2x < 28$$
$$x < 14$$

19.

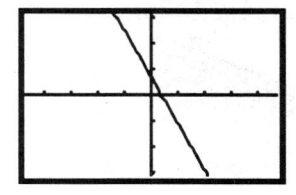

    a. $x = 11$
       Verify:
$$6.5 - 1.8x = -13.3$$
$$-1.8x = -19.8$$
$$x = 11$$

    b. $x = -10$
       Verify:
$$6.5 - 1.8x = 24.5$$
$$-1.8x = 18$$
$$x = -10$$

    c. $x \geq -5$
       Verify:
$$6.5 - 1.8x \leq 15.5$$
$$-1.8x \leq 9$$
$$x \geq -5$$

    d. $x \leq 8$
       Verify:
$$6.5 - 1.8x \geq -7.9$$
$$-1.8x \geq -14.4$$
$$x \leq 8$$

21.

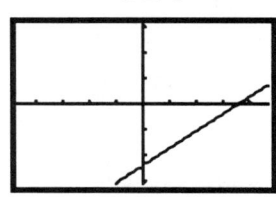

    a. $x = 20$
       Check algebraically:
$$\tfrac{2}{3}(20) - 24 = \tfrac{40}{3} - \tfrac{72}{3} = \tfrac{-32}{3}$$
$$= -10\tfrac{2}{3}$$

    b. $x \leq 7$
       Check algebraically:
$$\tfrac{2}{3}(7) - 24 = \tfrac{14}{3} - \tfrac{72}{3} = \tfrac{-58}{3}$$
$$= -19\tfrac{1}{3}$$

23.

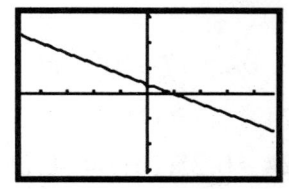

    a. $x = 4$
       Check algebraically:
$$-0.4(4) + 3.7 = -1.6 + 3.7 = 2.1$$

    b. $x < 22$
       Check algebraically:
$$-0.4(22) + 3.7 = -8.8 + 3.7$$
$$= -5.1$$

25.

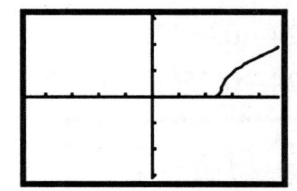

    a. $x = 41$
       Check algebraically:
$$4\sqrt{41 - 25} = 4\sqrt{16} = 4 \cdot 4 = 16$$

    b. $29 < x \leq 61$
$$4\sqrt{29 - 25} = 4\sqrt{4} = 4 \cdot 2 = 8$$
$$4\sqrt{61 - 25} = 4\sqrt{36} = 4 \cdot 6 = 24$$

Chapter 2 Review

27.

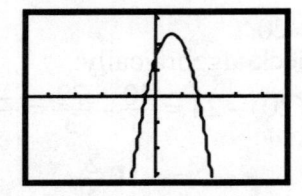

a. $x = -5$, $x = 17$
Check algebraically:
$24 - 0.25(-5-6)^2$
$= 24 - 0.25(-11)^2$
$= 24 - 0.25(121) = -6.25$

$24 - 0.25(17-6)^2$
$= 24 - 0.25(11)^2$
$= 24 - 0.25(121) = -6.25$

b. $-1 < x < 13$
Check algebraically:
$24 - 0.25(-1-6)^2$
$= 24 - 0.25(-7)^2$
$= 24 - 0.25(49) = 11.75$
$24 - 0.25(13-6)^2$
$= 24 - 0.25(7)^2$
$= 24 - 0.25(49) = 11.75$

29.

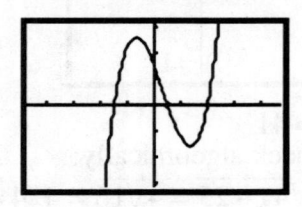

a. $x = -15$, $x = 5$, $x = 20$

b.

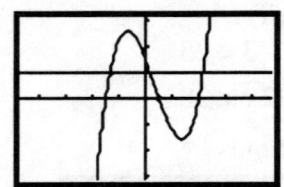

Three solutions:
approximately $-13, 2, 22$

Chapter 2 Review

1. a. $C = 20n + 2000$

b. For example, plot the points (0, 2000) and (100, 4000), then draw a line through the points. Refer to the graph in the back of the textbook.

c. Substitute $n = 1000$ into the equation.
$C = 20(1000) + 2000$
$= 20,000 + 2000$
$= 22,000$
The cost is $22,000.
This is the point (1000, 22,000) on the graph.

d. Substitute $C = 10,000$ into the equation.
$10,000 = 20n + 2000$
$8000 = 20n$
$400 = n$
The number of calculators produced is 400.
This is the point (400, 10,000) on the graph.

2. a. 5 lb = 80 oz
$W = 18m + 80$

Chapter 2 Review

b. For example, plot the points (0, 80) and (5, 170), then draw a line through the points. Refer to the graph in the back of the textbook.

c. Substitute $m = 9$ into the equation
$w = 18(9) + 80 = 162 + 80 = 242$
Megan weighed 242 ounces or 15 pounds, 2 ounces.
This is the point (9, 242) on the graph.

d. 9 lb = 144 oz
Substitute $w = 144$ into the equation.
$144 = 18m + 80$
$64 = 18m$
$\frac{32}{9} = m$
Megan weighed 9 pounds at $\frac{32}{9} = 3\frac{5}{9}$ months or about 3 months, 17 days. This is the point $\left(3\frac{5}{9},\ 144\right)$ on the graph.

3. a. $R = 1660 - 20t$

b. Substitute $t = 0$ into the equation.
$R = 1660 - 20(0) = 1660$
The R-intercept is 1660.
Substitute $R = 0$ into the equation.
$0 = 1660 - 20t$
$20t = 1660$
$t = 83$
The t-intercept is 83.
Plot the points (0, 1660) and (83, 0), then draw a line through the points. Refer to the graph in the back of the book.

c. When $t = 0$ (1976) R was 1660. R will be 0 when $t = 83$ (in 2059).

4. a. $R = 500 - 8t$

b. Substitute $t = 0$ into the equation.
$R = 500 - 8(0) = 500$
The R-intercept is 500.
Substitute $R = 0$ into the equation.
$0 = 500 - 8t$
$8t = 500$
$t = 62.5$
The t-intercept is 62.5.
Plot the points (0, 500) and (62.5, 0) and draw a line through the points. Refer to the graph in the back of the textbook.

c. When $t = 0$ (1976), R was 500. R will be 0 when $t = 62.5$ (in 2038).

5. a. $5A + 2C = 1000$

b. Substitute $A = 0$ into the equation.
$5(0) + 2C = 1000$
$2C = 1000$
$C = 500$
The C-intercept is 500.
Substitute $C = 0$ into the equation.
$5A + 2(0) = 1000$
$5A = 1000$
$A = 200$
The A-intercept is 200.
For example, let C be the first coordinate. Plot the points (0, 200) and (500, 0), then draw a line through the points. Refer to the graph in the back of the textbook.

59

Chapter 2 Review

   c. Substitute $A = 120$ into the equation.
$$5(120) + 2C = 1000$$
$$600 + 2C = 1000$$
$$2C = 400$$
$$C = 200$$
He must sell 200 children's tickets.

   d. The A-intercept, 200, is the number of adult tickets that must be sold if no children's tickets are sold. The C-intercept, 500, is the number of children's tickets that must be sold if no adult tickets are sold.

6. a. $60C + 100T = 1200$

   b. Substitute $C = 0$ into the equation.
$$60(0) + 100T = 1200$$
$$100T = 1200$$
$$T = 12$$
The T-intercept is 12.
Substitute $T = 0$ into the equation.
$$60C + 100(0) = 1200$$
$$60C = 1200$$
$$C = 20$$
The C-intercept is 20.
For example, let T be the first coordinate. Plot the points $(12, 0)$ and $(0, 20)$, then draw a line through the points. Refer to the graph in the back of the textbook.

   c. Substitute $C = 10$ into the equation.
$$60(10) + 100T = 1200$$
$$600 + 100T = 1200$$
$$100T = 600$$
$$T = 6$$
She can spend 6 days.

   d. The T-intercept, 12, is the number of days in Saint-Tropez if she spends no days in Atlantic City. The C-intercept, 20, is the number of days in Atlantic City if she spends no days in Saint-Tropez.

7. For example, plot the points $(0, -4)$ and $(3, 0)$, then draw a line through the points. Refer to the graph in the back of the textbook.

8. For example, plot the points $(0, -12)$ and $(6, 0)$, then draw a line through the points. Refer to the graph in the back of the textbook.

9. For example, plot the points $(0, 500)$ and $(-400, 0)$, then draw a line through the points. Refer to the graph in the back of the textbook.

10. For example, plot the points $(0, 4)$ and $(6, 0)$, then draw a line through the points. Refer to the graph in the back of the textbook.

11. For example, plot the points $(0, 0)$ and $(4, 3)$, then draw a line through the points. Refer to the graph in the back of the textbook.

12. For example, plot the points $(0, 0)$, and $(-4, 1)$, then draw a line through the points. Refer to the graph in the back of the textbook.

13. $4x = -12$

   $x = -3$
   Graph the vertical line through $(-3, 0)$. Refer to the graph in the back of the textbook.

14. $2y - 6 = 0$

   $2y = 6$

   $y = 3$
   Graph the horizontal line through $(0, 3)$. Refer to the graph in the back of the book.

15. a. $m = \dfrac{750 - 800}{10 - 0} = \dfrac{-50}{10} = -5$
   B-intercept $= 800$
   $B = 800 - 5t$

   b. For example, plot the points $(0, 800)$ and $(10, 750)$, then draw a line through the points. Refer to the graph in the back of the textbook.

   c. $m = -5$ thousands of barrels per minute gives the rate at which the oil is leaking.

16. a. $m = \dfrac{11.5 - 12}{10 - 0} = \dfrac{-0.5}{10} = -0.05$
   h-intercept $= 12$
   $h = 12 - 0.05t$

   b. For example, plot the points $(0, 12)$ and $(10, 11.5)$, then draw a line through the points. Refer to the graph in the back of the book.

   c. $m = -0.05$ in. per minute gives the rate at which the height of the candle is changing.

17. a. $m = \dfrac{1500 - 1000}{10,000 - 5000} = \dfrac{500}{5000} = 0.1$
   $F - 1000 = 0.1(C - 5000)$
   $F - 1000 = 0.1C - 500$
   $F = 0.1C + 500$

   b. For example, plot the points $(5000, 1000)$ and $(10,000, 1500)$, then draw a line through the points. Refer to the graph in the back of the book.

   c. $m = 0.1$ gives the percent of her fee with respect to the cost. The decorator charges 10% of the cost of the job (plus a flat $500 fee).

18. a. $m = \dfrac{235 - 135}{10,000 - 5000} = \dfrac{100}{5000}$
   $= 0.02$
   $R - 135 = 0.02(V - 5000)$
   $R - 135 = 0.02V - 100$
   $R = 0.02V + 35$

   b. For example, plot the points $(5000, 135)$ and $(10,000, 235)$, then draw a line through the points. Refer to the graph in the back of the textbook.

   c. $m = 0.02$ gives the percent of the fee with respect to the value of the car. The registration fee is 2% of the value of the car (plus a flat $35 fee).

19. $m = \dfrac{-2 - 4}{3 - (-1)} = \dfrac{-6}{4} = -\dfrac{3}{2}$

20. $m = \dfrac{-6 - 0}{2 - 5} = \dfrac{-6}{-3} = 2$

21. $m = \dfrac{4.8 - 1.4}{-2.1 - 6.2} = \dfrac{3.4}{-8.3} \approx -0.40$

22. $m = \dfrac{3.2 - (-6.4)}{-5.6 - 0} = \dfrac{9.6}{-5.6} \approx -1.7$

Chapter 2 Review

23. First find the linear equation.
$$m = \frac{-4.8-(-3)}{-5-(-2)} = \frac{-1.8}{-3} = 0.6$$
$$V-(-4.8) = 0.6(d-(-5))$$
$$V+4.8 = 0.6d+3$$
$$V = 0.6d-1.8$$
Substitute $V = -1.2$ into the equation.
$$-1.2 = 0.6d-1.8$$
$$0.6 = 0.6d$$
$$1 = d$$
Substitute $d = 10$ into the equation.
$V = 0.6(10) - 1.8 = 6 - 1.8 = 4.2$
The missing values are $d = 1$ and $V = 4.2$.

24. First, find the linear equation.
$$m = \frac{56-(-8)}{-4-(-8)} = \frac{64}{4} = 16$$
$$S-(-8) = 16(q-(-8))$$
$$S+8 = 16q+128$$
$$S = 16q+120$$
Substitute $q = 3$ into the equation.
$S = 16(3) + 120 = 48 + 120 = 168$
Substitute $S = 200$ into the equation.
$$200 = 16q+120$$
$$80 = 16q$$
$$5 = q$$
The missing values are $S = 168$ and $q = 5$.

25. Let x be the distance from the end to the base.
$$\frac{20}{x} = 0.25$$
$$20 = 0.25x$$
$$80 = x$$
The end should be 80 ft from the base.

26. Let y be the height to the center of the roof.
$$\frac{y}{\frac{1}{2}(40)} = 1.2$$
$$\frac{y}{20} = 1.2$$
$$y = 24$$
The center of the roof is 24 ft tall.

27. $2x - 4y = 5$
$$-4y = -2x+5$$
$$y = \frac{1}{2}x - \frac{5}{4}$$
$$m = \frac{1}{2};\ b = -\frac{5}{4}$$

28. $\frac{1}{2}x + \frac{2}{3}y = \frac{5}{6}$
$$\frac{2}{3}y = -\frac{1}{2}x + \frac{5}{6}$$
$$y = -\frac{3}{4}x + \frac{5}{4}$$
$$m = -\frac{3}{4};\ b = \frac{5}{4}$$

29. $8.4x + 2.1y = 6.3$
$$2.1y = -8.4x + 6.3$$
$$y = -4x + 3$$
$$m = -4;\ b = 3$$

30. $y - 3 = 0$
$$y = 3$$
$$m = 0;\ b = 3$$

31. a. Plot the point $(-4, 6)$. Move 2 units in the y-direction and -3 units in the x-direction to arrive at $(-7, 8)$. Draw a line through the points. Refer to the graph in the back of the textbook.

Chapter 2 Review

b. $y - 6 = -\frac{2}{3}(x - (-4))$

$y - 6 = -\frac{2}{3}(x + 4)$

$y - 6 = -\frac{2}{3}x - \frac{8}{3}$

$y = -\frac{2}{3}x + \frac{10}{3}$

32. a. Plot the point (2, –5). Move 3 units in the y-direction and 2 units in the x-direction to arrive at (4, –2). Draw a line through the points. Refer to the graph at the back of the textbook.

b. $y - (-5) = \frac{3}{2}(x - 2)$

$y + 5 = \frac{3}{2}x - 3$

$y = \frac{3}{2}x - 8$

33. $m = \frac{4-(-5)}{-2-3} = \frac{9}{-5} = -\frac{9}{5}$

$y - (-5) = -\frac{9}{5}(x - 3)$

$y + 5 = -\frac{9}{5}x + \frac{27}{5}$

$y = -\frac{9}{5}x + \frac{2}{5}$

34. $m = \frac{-2-8}{4-0} = \frac{-10}{4} = -\frac{5}{2}$

$y - 8 = -\frac{5}{2}(x - 0)$

$y = -\frac{5}{2}x + 8$

35. a. $m = \frac{6780 - 4800}{15 - 0} = \frac{1980}{15} = 132$

$P - 4800 = 132(t - 0)$

$P = 4800 + 132t$

b. $m = 132$ people per year gives the rate of population growth.

36. a. $m = \frac{308 - 112}{7 - 0} = \frac{196}{7} = 28$

$M - 112 = 28(g - 0)$

$M = 28g + 112$

b. $m = 28$ mpg gives the gas mileage (rate of gas consumption).

37. a. $m = \frac{1-3}{1-0} = \frac{-2}{1} = -2$

$b = 3$

b. $y = -2x + 3$

38. a. $m = \frac{1-(-5)}{4-0} = \frac{6}{4} = \frac{3}{2}$

$b = -5$

b. $y = \frac{3}{2}x - 5$

39. $y = \frac{1}{2}x + 3$

$m_1 = \frac{1}{2}$

$x - 2y = 8$

$-2y = -x + 8$

$y = \frac{1}{2}x - 4$

$m_2 = \frac{1}{2}$

$m_1 = m_2$
The lines are parallel.

63

Chapter 2 Review

40. $4x - y = 6$
$-y = -4x + 6$
$y = 4x - 6$
$m_1 = 4$
$x + 4y = -2$
$4y = -x - 2$
$y = -\frac{1}{4}x - 2$
$m_2 = -\frac{1}{4}$
$m_1 = -\frac{1}{m_2}$
The lines are perpendicular.

41. $2x + 3y = 6$
$3y = -2x + 6$
$y = -\frac{2}{3}x + 2$
$m = -\frac{2}{3}$
$y - 4 = -\frac{2}{3}(x - 1)$
$y - 4 = -\frac{2}{3}x + \frac{2}{3}$
$y = -\frac{2}{3}x + \frac{14}{3}$

42. From Exercise 41, the slope of $2x + 3y = 6$ is $-\frac{2}{3}$. A perpendicular line has a slope of $\frac{-1}{-\frac{2}{3}} = \frac{3}{2}$.
$y - 4 = \frac{3}{2}(x - 1)$
$y - 4 = \frac{3}{2}x - \frac{3}{2}$
$y = \frac{3}{2}x + \frac{5}{2}$

43. slope of $\overline{AB}$: $\frac{-4-2}{7-3} = \frac{-6}{4} = \frac{-3}{2}$
slope of $\overline{BC}$: $-\frac{1}{\frac{-3}{2}} = \frac{2}{3}$
$y - (-4) = \frac{2}{3}(x - 7)$
$y + 4 = \frac{2}{3}x - \frac{14}{3}$
$y = \frac{2}{3}x - \frac{26}{3}$

44. slope of $\overline{PQ}$: $\frac{-5-(-1)}{-2-(-8)} = \frac{-4}{6} = \frac{-2}{3}$
slope of $\overline{QR}$: $-\frac{1}{-\frac{2}{3}} = \frac{3}{2}$
$y - (-5) = \frac{3}{2}(x - (-2))$
$y + 5 = \frac{3}{2}(x + 2)$
$y + 5 = \frac{3}{2}x + 3$
$y = \frac{3}{2}x - 2$

45.
a. $x = -5, x = 7$
b. $-10 \le x \le 12$

46.
a. $x = 10, x = 30$
b. $x < 15, x > 25$

Chapter 2 Review

47. a. $x = -1, x = 1$
   b. $-3 < x < 2$ or $2 < x < 3$
48. a. $x = 0.5, x = 2$
   b. $x < 0.25$ or $x > 4$

Chapter 3

Section 3.1

1. It appears from the graph that the solution is $x = 3, y = 0$.
   Algebraic verification:
   $2.3(3) - 3.7(0) = 6.9 - 0 = 6.9$
   $1.1(3) + 3.7(0) = 3.3 + 0 = 3.3$

3. It appears from the graph that the solution is $s = 50, t = 70$.
   Algebraic verification:
   $35(50) - 17(70) = 1750 - 1190$
   $= 560$
   $24(50) + 15(70) = 1200 + 1050$
   $= 2250$

5. It appears from the graph that the solution is $x = 7, y = 2$.

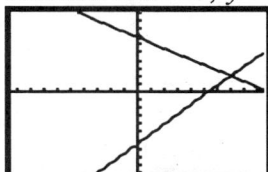

   Algebraic verification:
   $-0.7(7) + 6.9 = -4.9 + 6.9 = 2$
   $1.2(7) - 6.4 = 8.4 - 6.4 = 2$

7. It appears from the graph that the solution is $(-2, 3)$.

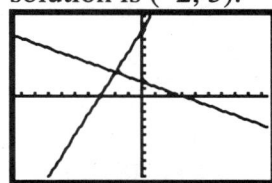

   Algebraic verification:
   $2.6(-2) + 8.2 = 5.2 + 8.2 = 3$
   $1.8 - 0.6(-2) = 1.8 - (-1.2) = 3$

9. It appears from the graph that the solution is $(2, 3)$.

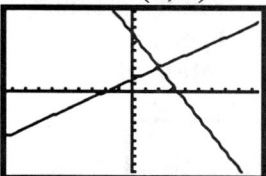

   Algebraic verification:
   $7.2 - 2.1(2) = 7.2 - 4.2 = 3$
   $-2.8(2) + 3.7(3) = -5.6 + 11.1 = 5.5$

11. Graph $Y_1 = 2630 + 32x$ and $Y_2 = -21x - 1610$ using Xmin $= -100$, Xmax $= 10$, Ymin $= -100$, and Ymax $= 500$. Then use ZOOM and TRACE to find the solution of $(-80, 70)$. So $n = 70$ and $m = -80$.

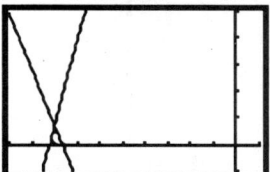

13. Graph $Y_1 = \dfrac{-55.2 - 38x}{2.3}$ and $Y_2 = 15x + 121$ using Xmin $= -10$, Xmax $= 1$, Ymin $= -10$, and Ymax $= 100$. Then use ZOOM and TRACE to find the solution of $(-4.6, 52)$. So $a = -4.6$ and $b = 52$.

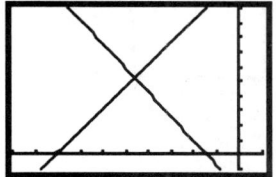

Chapter 3, Section 3.1

15. Graph $Y_1 = \dfrac{707 - 64x}{58}$ and $Y_2 = \dfrac{496 - 82x}{-21}$ using Xmin = −1, Xmax = 10, Ymin = −1, and Ymax = 10. Then use ZOOM and TRACE to find the solution of (7.15, 4.3).

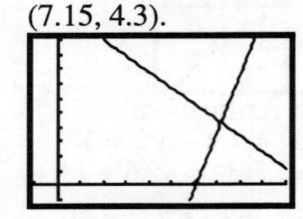

17. Graph $Y_1 = 3x + 6$ and $Y_2 = \dfrac{14 - 6x}{-2}$ in the standard window. The lines are parallel, so the system is inconsistent.

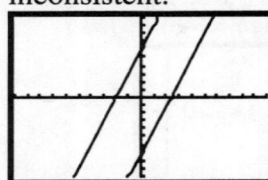

19. Graph $Y_1 = \dfrac{x + 1}{2}$ and $Y_2 = \dfrac{4 + 4x}{8}$ in the standard window. The lines are identical so the system is dependent.

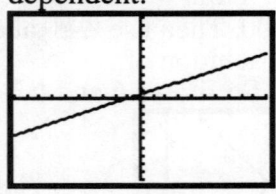

21. Graph $Y_1 = 6 + 3x$ and $Y_2 = \dfrac{8 - x}{2}$ in the standard window. The lines intersect, so the system is consistent.

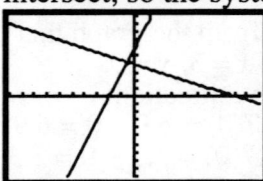

23. Graph $Y_1 = \dfrac{8 + 5x}{2}$ and $Y_2 = \dfrac{\frac{15}{2}x + 9}{3}$ in the standard window. The lines are parallel, so the system is inconsistent.

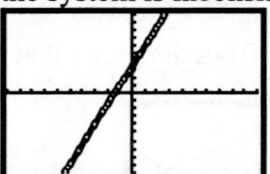

25. Let x = the number of paperback copies and y = the number of hardback copies. Then $4.95x + 9.95y = 300$ and $x = 3y$. Substitute the value for x in terms of y:

$$4.95(3y) + 9.95y = 300$$
$$14.85y + 9.95y = 300$$
$$24.8y = 300$$
$$y = 12.10$$

The manager should order 12 hardbacks and 36 paperbacks of the cookbook.

Chapter 3, Section 3.1

27. The equilibrium price occurs when
$50x = 2100 - 20x$

$70x = 2100$

$x = 30$

Yasuo should sell the wheat at 30 cents per bushel and produce 1500 bushels.

29. a. The cost is the initial cost ($200) plus the cost per pendant ($4), so the cost of production is
$C(x) = 200 + 4x$.

b. The revenue is $12 per pendant sold, so $R(x) = 12x$.

c.

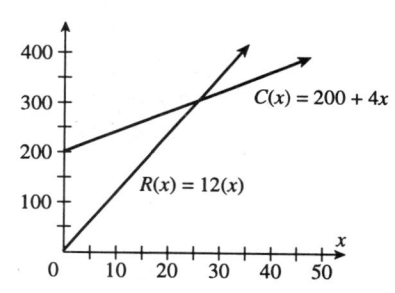

d. From the graph, it appears that the solution is $x = 25$. To verify this, check:
$C(250 = 200 + 4(25)$

$= 200 + 100$ and

$= 300$
$R(25) = 12(25) = 300$.
$C(25) = R(25)$. The company needs to sell 25 pendants to break even.

31. a. Let m = the number of minutes of long-distance calls, $D(m)$ = the amount that Dash Phone Company charges for m minutes, and $F(m)$ = the amount that Friendly Phone company charges for m minutes.

Then $D(m) = 10 + 0.09m$ and $F(m) = 15 + 0.05m$

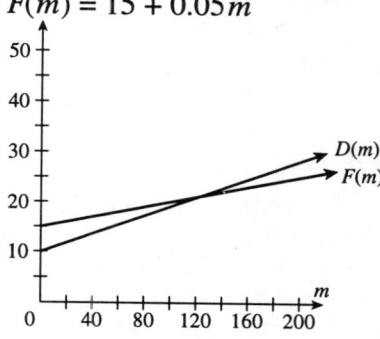

b. Equal bills occur when $D(m) = F(m)$. From the graph, it appears that the solution occurs when $m = 125$. Verification:
$D(125) = 10 + 0.09(125)$

$= 10 + 11.25$

$= 21.25$
$F(125) = 15 + 0.05(125)$

$= 15 + 6.25$

$= 21.25$

33. Let x = the number of adults and y = the number of students. Then
$x + y = 82$ and
$1.50x + 0.85y = 93.10$

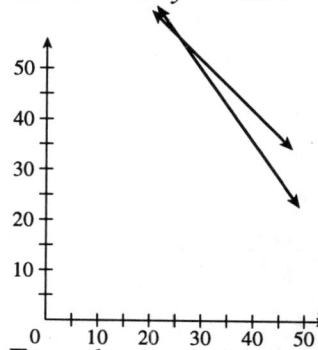

From the graph, it appears that the solution is $x = 36$ and $y = 46$.

69

Chapter 3, Section 3.2

Verification:
$36 + 46 = 82$ and
$1.50(36) + 0.85(46)$
$= 54 + 39.1 = 93.1$

Section 3.2

1. Given $2x - 3y = 6$ and $x + 3y = 3$, solve the second equation for x to get $x = 3 - 3y$. Substitute this into the first equation:
$2(3 - 3y) - 3y = 6$
$6 - 6y - 3y = 6$
$6 - 9y = 6$
$-9y = 0$
$y = 0$
Substitute this value for y into the first equation:
$2x - 3(0) = 6$
$2x = 6$
$x = 3$
The solution is $x = 3$, $y = 0$.

3. Given $3m + n = 7$ and $2m = 5n - 1$, solve the first equation for n to get $n = 7 - 3m$. Substitute this into the second equation:
$2m = 5(7 - 3m) - 1$
$2m = 35 - 15m - 1$
$17m = 35 - 1$
$17m = 34$
$m = 2$
Substitute this value for m into the first equation:
$3(2) + n = 7$
$6 + n = 7$
$n = 1$
The solution is $m = 2$, $n = 1$.

5. Given $2u - 3v = -4$ and $5u + 2v = 9$, multiply the first equation by 2 and the second equation by 3:
$4u - 6v = -8$
$15u + 6v = 27$
Add these equations:
$19u = 19$
$u = 1$
Put this value into the first equation:
$2(1) - 3v = -4$
$2 - 3v = -4$
$-3v = -6$
$v = 2$
The solution is $u = 1$, $v = 2$.

7. Given $3y = 2x - 8$ and $4y + 11 = 3x$, multiply the first equation by 4 and the second equation by -3:
$12y = 8x - 32$
$-12y - 33 = -9x$
Add these equations:
$-33 = 8x - 32 - 9x$
$-33 = -x - 32$
$-1 = -x$
$1 = x$
Put this value into the first equation:
$3y = 2(1) - 8$
$3y = 2 - 8$
$3y = -6$
$y = -2$
The solution is $x = 1$, $y = -2$.

Chapter 3, Section 3.2

9. Given $\frac{2}{3}A - B = 4$ and $A - \frac{3}{4}B = 6$,
multiply the first equation by −3 and the second equation by 4:
$-2A + 3B = -12$
$4A - 3B = 24$
Add these equations:
$2A = 12$
$A = 6$
Put this value into the first equation:
$\frac{2}{3}(6) - B = 4$
$4 - B = 4$
$-B = 0$
$B = 0$
The solution is $A = 6$, $B = 0$.

11. Given
$\frac{M}{4} = \frac{N}{3} - \frac{5}{12}$ and
$\frac{N}{5} = \frac{1}{2} - \frac{M}{10}$
Multiply the first equation by 12 and the second equation by 10:
$3M = 4N - 5$
$2N = 5 - M$
We can solve the second of these equations for M to get $M = 5 - 2N$. Put this into the first of the new equations:
$3(5 - 2N) = 4N - 5$
$15 - 6N = 4N - 5$
$20 - 6N = 4N$
$20 = 10N$
$2 = N$
Thus, $M = 5 - 2(2) = 5 - 4 = 1$ and the solution is $M = 1$, $N = 2$.

13. Given
$\frac{s}{2} = \frac{7}{6} - \frac{t}{3}$ and
$\frac{s}{4} = \frac{3}{4} - \frac{t}{4}$
multiply the first equation by 6 and the second equation by 4:
$3s = 7 - 2t$ (1)
$s = 3 - t$ (2)
Put equation (2) into equation (1):
$3(3 - t) = 7 - 2t$
$9 - 3t = 7 - 2t$
$2 - 3t = -2t$
$2 = t$
Put this value into (2):
$s = 3 - 2 = 1$.
The solution is $s = 1$, $t = 2$.

15. Given
$4.8x - 3.5y = 5.44$ and
$2.7x + 1.3y = 8.29$
first multiply both equations by 10 to better see which variable to eliminate.
$48x - 35y = 54.4$ (1)
$27x + 13y = 82.9$ (2)
Multiply (1) by −9 and (2) by 16:
$-432x + 315y = -489.6$
$432x + 208y = 1326.4$
Add these equations:
$523y = 836.8$
$y = 1.6$

71

Chapter 3, Section 3.2

Put this value into the first equation given:
$$4.8x - 3.5(1.6) = 5.44$$
$$4.8x - 5.6 = 5.44$$
$$4.8x = 11.04$$
$$x = 2.3$$
The solution is $x = 2.3$, $y = 1.6$.

17. Given
$$0.9x = 25.78 + 1.03y$$
$$0.25x + 0.3y = 85.7$$
Put the first equation into standard form by subtracting $1.03y$ from both sides:
$$0.9x - 1.03y = 25.78 \quad (1)$$
$$0.25x + 0.3y = 85.7 \quad (2)$$
Multiply (1) by -0.25 and (2) by 0.9:
$$-0.225x + 0.2575y = -6.445$$
$$0.225x + 0.27y = 77.13$$
Add these equations:
$$0.5275y = 70.685$$
$$y = 134$$
Put this value into the first equation given:
$$0.9x = 25.78 + 1.03(134)$$
$$0.9x = 25.78 + 138.02$$
$$0.9x = 163.8$$
$$x = 182$$
The solution is $x = 182$, $y = 134$.

19. Since the second equation is twice the first equation, the system is dependent.

21. Solving both equations for m:
$$m = \frac{1}{2}n + \frac{1}{2}$$
$$m = \frac{1}{2}n + \frac{3}{8}$$
These equations represent two parallel lines, so the system is inconsistent.

23. Solving both equations for r:
$$r = 3s + 4$$
$$r = \frac{-1}{2}s + 3$$
These equations represent intersecting lines, so the system is consistent.

25. Solving both equations for L:
$$L = \frac{5}{2}W + 3$$
$$L = \frac{5}{2}W + 3$$
Since these equations are identical, the system is dependent.

27. Let x = the number of votes that the winner received and y = the number of votes the loser received. Then $x + y = 7179$. The other information gives the equation
$x - 6 + 1 = y + 6$ or $x = y + 11$.
Substitute this into the first equation:
$$(y + 11) + y = 7179$$
$$2y + 11 = 7179$$
$$2y = 7168$$
$$y = 3584$$
From this value, $x = 3584 + 11 = 3595$.
The winner received 3595 votes and the loser received 3584 votes.

29. Let $x =$ the amount Francine has invested in bonds and $y =$ the amount she has in the certificate account. Then $x + y = 2000$, so $x = 2000 - y$. The income from the bonds is $0.10x$ while the income from the certificate is $0.08y$, thus $0.1x + 0.08y = 184$. Substitute the equation for x from above:
$$0.1(2000 - y) + 0.08y = 184$$
$$200 - 0.1y + 0.08y = 184$$
$$200 - 0.02y = 184$$
$$-0.02y = -16$$
$$y = 800$$
Thus Francine has $800 invested at 8% and $2000 - $800 = $1200 invested at 10%.

31. Let $x =$ the amount of 45% alloy and $y =$ the amount of 60% alloy that Paul uses. Then, to end up with 40 pounds of alloy, $x + y = 40$ or $y = 40 - x$. The amount of silver in the x pounds of 45% alloy is $0.45x$, while the amount of silver in the y pounds of 60% alloy is $0.60y$. The resulting 40 pounds of alloy should contain $0.48(40) = 19.2$ pounds of silver. Thus $0.45x + 0.60y = 19.2$. Put the equation above into this one:
$$0.45x + 0.60(40 - x) = 19.2$$
$$0.45x + 24 - 0.60x = 19.2$$
$$-0.15x + 24 = 19.2$$
$$-0.15x = -4.8$$
$$x = 32$$
Thus, Paul needs to use 32 pounds of the 45% alloy.

33. Let $l =$ Leon's speed and $m =$ Marlene's speed. Using (rate) × (time) = distance, with t standing for the amount of time that Leon and Marlene traveled,
$$lt = 1260$$
$$mt = 420$$
or $t = \dfrac{1260}{l} = \dfrac{420}{m}$.
Cross-multiplying, $1260m = 420l$ or $3m = l$. We also know that Leon flies 120 miles per hour faster than Marlene drives, or $m + 120 = l$. Combine the two equations:
$$m + 120 = 3m$$
$$120 = 2m$$
$$60 = m$$
Thus, Marlene is driving 60 miles per hour and Leon is flying $3(60) = 180$ miles per hour.

35. Let $p =$ the speed of the airplane and $w =$ the speed of the wind. The speed on the flight to Denver is $p - w$ and the speed on the return to Detroit is $p + w$. Using (rate) × (time) = distance:
$$(p - w)4 = 1120$$
$$(p + w)3.5 = 1120$$
or
$$4p - 4w = 1120 \quad (1)$$
$$3.5p + 3.5w = 1120 \quad (2)$$
Divide (1) by 4 and move the w term to the other side:
$p = 280 + w$.
Put this into (2):
$$3.5(280 + w) + 3.5w = 1120$$
$$980 + 3.5w + 3.5w = 1120$$
$$980 + 7w = 1120$$
$$7w = 140$$
$$w = 20$$

Chapter 3, Section 3.3

The speed of the wind is 20 mph, while the speed of the airplane is 280 + 20 = 300 mph.

37. Let o = the amount of oats and w = the amount of wheat flakes in one cup of the new cereal. Thus, $o + w = 1$ or $o = 1 - w$. The number of calories in o cups of oats is $310o$ and the number of calories in w cups of wheat flakes is $290w$. The new cereal has 302 calories per cup, so $310o + 290w = 302$. Put the expression above for o in terms of w into this equation:
$$310(1-w) + 290w = 302$$
$$310 - 310w + 290w = 302$$
$$310 - 20w = 302$$
$$-20w = -8$$
$$w = 0.4$$
Thus, the new cereal uses 0.4 cups of wheat flakes and $1 - 0.4 = 0.6$ cups of oats.

39. The equilibrium price occurs when $35x = 1700 - 15x$. Solving this equation:
$$35x = 1700 - 15x$$
$$50x = 1700$$
$$x = 34$$
The equilibrium price is $34 per pair, and Sanaz will produce and sell $35(34) = 1190$ pairs at that price.

Section 3.3

1. Solve the third equation for y to get $y = 2$. Put this value into the second equation:
$$3(2) + z = 5$$
$$6 + z = 5$$
$$z = -1$$
Now put the values for y and z into the first equation:
$$x + 2 + (-1) = 2$$
$$x + 1 = 2$$
$$x = 1$$
The solution is $(1, 2, -1)$.

3. Solve the third equation for y to get $y = -1$. Put this value into the second equation:
$$5(-1) + 3z = -8$$
$$-5 + 3z = -8$$
$$3z = -3$$
$$z = -1$$
Now put the values for y and z into the first equation:
$$2x - (-1) - (-1) = 6$$
$$2x + 2 = 6$$
$$2x = 4$$
$$x = 2$$
The solution is $(2, -1, -1)$.

5. Solve the third equation for x to get $x = 4$. Put this value for x into the first equation:
$$2(4) + z = 5$$
$$8 + z = 5$$
$$z = -3$$

Now put the value for z into the second equation:
$$3y + 2(-3) = 6$$
$$3y - 6 = 6$$
$$3y = 12$$
$$y = 4$$
The solution is $(4, 4, -3)$.

7. $x + y + z = 0$ (1)
    $2x - 2y + z = 8$ (2)
    $3x + 2y + z = 2$ (3)
    Multiply (1) by 2 and add the result to (2) to obtain (4)
    $2x + 2y + 2z = 0$ (1a)
    $2x - 2y + z = 8$ (2)
    $4x + 3z = 8$ (4)
    Now add (2) and (3):
    $2x - 2y + z = 8$ (2)
    $3x + 2y + z = 2$ (3)
    $5x + 2z = 10$ (5)
    Multiply (4) by -2 and (5) by 3 and add the result:
    $-8x - 6z = -16$ (4a)
    $15x + 6z = 30$ (5a)
    $7x = 14$
    $x = 2$ (6)
    We form a new system consisting of (1), (4), and (6):
    $x + y + z = 0$ (1)
    $4x + 3z = 8$ (4)
    $x = 2$ (6)
    Put the value for x from (6) into (4):
    $4(2) + 3z = 8$
    $8 + 3z = 8$
    $3z = 0$
    $z = 0$

Put these values for x and z into (1):
$$2 + y + 0 = 0$$
$$y = -2$$
The solution is $(2, -2, 0)$.

9. $x - 2y + z = -1$ (1)
    $2x + y - 3z = 3$ (2)
    $3x + 3y - 2z = 10$ (3)
    Multiply (1) by -2, add the result to (2), and simplify to get (4):
    $-2x + 4y - 2z = 2$ (1a)
    $2x + y - 3z = 3$ (2)
    $5y - 5z = 5$
    $y - z = 1$ (4)
    Now multiply (1) by -3, add the result to (3) and simplify to get (5):
    $-3x + 6y - 3z = 3$ (1b)
    $3x + 3y - 2z = 10$ (3)
    $9y - 5z = 13$ (5)
    Multiply (4) by -5, add the result to (5) and simplify to get (6):
    $-5y + 5z = -5$ (4a)
    $9y - 5z = 13$ (5)
    $4y = 8$
    $y = 2$ (6)
    Put the value for y from (6) into (4):
    $2 - z = 1$
    $-z = -1$
    $z = 1$
    Put these values for y and z into (1):
    $x - 2(2) + 1 = -1$
    $x - 3 = -1$
    $x = 2$
    The solution is $(2, 2, 1)$.

Chapter 3, Section 3.3

11. $4x + z = 3$ (1)
$2x - y = 2$ (2)
$3y + 2z = 0$ (3)

Multiply (2) by -2 and add the result to (1). The result will be an equation in y and z that can be combined with (3):

$4x + z = 3$ (1)
$-4x + 2y = -4$ (2a)
$2y + z = -1$ (4)

Multiply (4) by -2, add the result to (3) and simplify:

$-4y - 2z = 2$ (4a)
$3y + 2z = 0$ (3)
$-y = 2$
$y = -2$

Put this value for y into (4):

$2(-2) + z = -1$
$-4 + z = -1$
$z = 3$

Now put this value for z into (1):

$4x + 3 = 3$
$4x = 0$
$x = 0$

The solution is $(0, -2, 3)$.

13. $2x + 3y - 2z = 5$ (1)
$3x - 2y - 5z = 5$ (2)
$5x + 2y + 3z = -9$ (3)

Multiply (1) by 2, (2) by 3, and add the results:

$4x + 6y - 4z = 10$ (1a)
$9x - 6y - 15z = 15$ (2a)
$13x - 19z = 25$ (4)

Now, add (2) and (3), then simplifying the result:

$3x - 2y - 5z = 5$ (2)
$5x + 2y + 3z = -9$ (3)
$8x - 2z = -4$
$4x - z = -2$ (5)

Multiply (5) by -19, add the result to (4), and simplify:

$13x - 19z = 25$ (4)
$-76x + 19z = 38$ (5a)
$-63x = 63$
$x = -1$

Put this value for x into (5):

$4(-1) - z = -2$
$-z = 2$
$z = -2$

Put the values for x and z into (1):

$2(-1) + 3y - 2(-2) = 5$
$-2 + 3y + 4 = 5$
$3y = 3$
$y = 1$

The solution is $(-1, 1, -2)$.

15. $4x + 6y + 3z = -3$ (1)
$2x - 3y - 2z = 5$ (2)
$-6x + 6y + 2z = -5$ (3)

Multiply (2) by -2 and add the result to (1):

$4x + 6y + 3z = -3$ (1)
$-4x + 6y + 4z = -10$ (2a)
$12y + 7z = -13$ (4)

Now multiply (2) by 3 and add the result to (3):

$6x - 9y - 6z = 15$ (2b)
$-6x + 6y + 2z = -5$ (3)
$-3y - 4z = 10$ (5)

Multiply (5) by 4, add the result to (4) and simplify:
$$12y + 7z = -13 \quad (4)$$
$$-12y - 16z = 40 \quad (5a)$$
$$-9z = 27$$
$$z = -3$$
Put this value for z into (5):
$$-3y - 4(-3) = 10$$
$$-3y + 12 = 10$$
$$-3y = -2$$
$$y = \frac{2}{3}$$
Put the values for y and z into (1):
$$4x + 6\left(\frac{2}{3}\right) + 3(-3) = -3$$
$$4x + 4 - 9 = -3$$
$$4x = 2$$
$$x = \frac{1}{2}$$
The solution is $\left(\frac{1}{2}, \frac{2}{3}, -3\right)$.

17. $x - \frac{1}{2}y - \frac{1}{2}z = 4$
$x - \frac{3}{2}y - 2z = 3$
$\frac{1}{4}x + \frac{1}{4}y - \frac{1}{4}z = 0$

First, clear all the denominators by multiplying the first and second equations by 2, and multiplying the third equation by 4.
$$2x - y - z = 8 \quad (1)$$
$$2x - 3y - 4z = 6 \quad (2)$$
$$x + y - z = 0 \quad (3)$$

Multiply (1) by -1 and add the result to (2):
$$-2x + y + z = -8 \quad (1a)$$
$$2x - 3y - 4z = 6 \quad (2)$$
$$-2y - 3z = -2 \quad (4)$$
Multiply (3) by -2 and add the result to (2):
$$2x - 3y - 4z = 6 \quad (2)$$
$$-2x - 2y + 2z = 0 \quad (3a)$$
$$-5y - 2z = 6 \quad (5)$$
Now, multiply (4) by -2, (5) by 3, add and simplify the result.
$$4y + 6z = 4 \quad (4a)$$
$$-15y - 6z = 18 \quad (5a)$$
$$-11y = 22$$
$$y = -2$$
Put this value for y into (5):
$$-5(-2) - 2z = 6$$
$$10 - 2z = 6$$
$$-2z = -4$$
$$z = 2$$
Put the values for y and z into (3):
$$x + (-2) - 2 = 0$$
$$x - 4 = 0$$
$$x = 4$$
The solution is $(4, -2, 2)$.

19. $x + y - z = 2$
$\frac{1}{2}x - y + \frac{1}{2}z = \frac{-1}{2}$
$x + \frac{1}{3}y - \frac{2}{3}z = \frac{4}{3}$

Chapter 3, Section 3.3

First, clear all denominators by multiplying the second equation by 2 and the third equation by 3.

$x + y - z = 2$ (1)
$x - 2y + z = -1$ (2)
$3x + y - 2z = 4$ (3)

Multiply (2) by −1 and add the result to (1):

$x + y - z = 2$ (1)
$-x + 2y - z = 1$ (2a)
$3y - 2z = 3$ (4)

Multiply (2) by −3 and add the result to (3):

$-3x + 6y - 3z = 3$ (2b)
$3x + y - 2z = 4$ (3)
$7y - 5z = 7$ (5)

Multiply (4) by −5, (5) by 2, add and simplify the result:

$-15y + 10z = -15$ (4a)
$14y - 10z = 14$ (5a)
$-y = -1$
$y = 1$

Put this value for y into (4):
$3(1) - 2z = 3$
$3 - 2z = 3$
$-2z = 0$
$z = 0$

Put the values for y and z into (1):
$x + 1 - 0 = 2$
$x = 1$

The solution is (1, 1, 0).

21. $x = -y$
$x + z = \dfrac{5}{6}$
$y - 2z = \dfrac{-7}{6}$

First, put the system into standard form by moving the −y term to the left in the first equation and multiplying the second and third equations by 6.

$x + y = 0$ (1)
$6x + 6z = 5$ (2)
$6y - 12z = -7$ (3)

Multiply (1) by −6 and add the result to (2):

$-6x - 6y = 0$ (1a)
$6x + 6z = 5$ (2)
$-6y + 6z = 5$ (4)

Add (3) and (4), then simplify the result:

$6y - 12z = -7$ (3)
$-6y + 6z = 5$ (4)
$-6z = -2$
$z = \dfrac{1}{3}$

Put this value for z into (4) and simplify the result:

$-6y + 6\left(\dfrac{1}{3}\right) = 5$
$-6y + 2 = 5$
$-6y = 3$
$y = \dfrac{-1}{2}$

Put this value for y into (1):
$$x + \left(\frac{-1}{2}\right) = 0$$
$$x = \frac{1}{2}$$
The solution is $\left(\frac{1}{2}, \frac{-1}{2}, \frac{1}{3}\right)$.

23. $3x - 2y + z = 6$ (1)
$2x + y - z = 2$ (2)
$4x + 2y - 2z = 3$ (3)
Multiply (2) by –2 and add the result to (3):
$-4x - 2y + 2z = -4$ (2a)
$4x + 2y - 2z = 3$ (3)
$0 = -1$ (4)
Equation (4) can also be written as $0x + 0y + 0z = -1$, so the system is inconsistent.

25. $2x + 3y - z = -2$
$x - y + \frac{1}{2}z = 2$
$4x - \frac{1}{3}y + 2z = 8$
Write the system in standard form by multiplying the second equation by 2 and the third equation by 3:
$2x + 3y - z = -2$ (1)
$2x - 2y + z = 4$ (2)
$12x - y + 6z = 24$ (3)
Multiply (2) by –1 and add the result to (1):
$2x + 3y - z = -2$ (1)
$-2x + 2y - z = -4$ (2a)
$5y - 2z = -6$ (4)

Multiply (2) by –6 and add the result to (3):
$-12x + 12y - 6z = -24$ (2b)
$12x - y + 6z = 24$ (3)
$11y = 0$
$y = 0$
Put this value for y into (4):
$5(0) - 2z = -6$
$-2z = -6$
$z = 3$
Put the values for y and z into (1)
$2x + 3(0) - 3 = -2$
$2x = 1$
$x = \frac{1}{2}$
The solution is $\left(\frac{1}{2}, 0, 3\right)$.

27. $2x + y = 6$ (1)
$x - z = 4$ (2)
$3x + y - z = 10$ (3)
Add (1) and (2):
$2x + y = 6$ (1)
$x - z = 4$ (2)
$3x + y - z = 10$ (4)
Multiply (4) by –1 and add the result to (3):
$3x + y - z = 10$ (3)
$-3x - y + z = -10$ (4a)
$0 = 0$ (5)
Equation (5) can also be written $0x + 0y + 0z = 0$, so the system is dependent.

79

Chapter 3, Section 3.3

29. $x = 2y - 7$
$y = 4z + 3$
$z = 3x + y$

The first step is to write the system in standard form by moving the variable terms to the left sides of the equations.

$x - 2y = -7$ (1)
$y - 4z = 3$ (2)
$-3x - y + z = 0$ (3)

Multiply (1) by 3 and add the result to (3):

$3x - 6y = -21$ (1a)
$-3x - y + z = 0$ (3)
$-7y + z = -21$ (4)

Multiply (2) by 7, add the result to (4) and simplify:

$7y - 28z = 21$ (2a)
$-7y + z = -21$ (4)
$-27z = 0$
$z = 0$

Put this value for z into (2) and simplify:

$y - 4(0) = 3$
$y = 3$

Put this value for y into (1):

$x - 2(3) = -7$
$x - 6 = -7$
$x = -1$

The solution is $(-1, 3, 0)$.

31. $\frac{1}{2}x + y = \frac{1}{2}z$
$x - y = -z - 2$
$-x - 2y = -z + \frac{4}{3}$

Multiply the first equation by 2, the third equation by 3, and move the variable terms to the left sides of the equations:

$x + 2y - z = 0$ (1)
$x - y + z = -2$ (2)
$-3x - 6y + 3z = 4$ (3)

Multiply (1) by 3 and add the result to (3):

$3x + 6y - 3z = 0$ (1a)
$-3x - 6y + 3z = 4$ (3)
$0 = 4$ (4)

Equation (4) can also be written $0x + 0y + 0z = 4$, so the system is inconsistent.

33. $x - y = 0$
$2x + 2y + z = 5$
$2x + y - \frac{1}{2}z = 0$

We put this system into standard form by multiplying the third equation by 2.

$x - y = 0$ (1)
$2x + 2y + z = 5$ (2)
$4x + 2y - z = 0$ (3)

Multiply (1) by -2 and add the result to (2):

$-2x + 2y = 0$ (1a)
$2x + 2y + z = 5$ (2)
$4y + z = 5$ (4)

Multiply (2) by -2 and add the result to (3):

$-4x - 4y - 2z = -10$ (2a)
$4x + 2y - z = 0$ (3)
$-2y - 3z = -10$ (5)

Multiply (5) by 2, add the result to (4), and simplify:
$$-4y - 6z = -20 \quad (5a)$$
$$4y + z = 5 \quad (4)$$
$$-5z = -15$$
$$z = 3$$
Put this value for z into (4) and simplify:
$$4y + 3 = 5$$
$$4y = 2$$
$$y = \frac{1}{2}$$
Put this value for y into (1):
$$x - \frac{1}{2} = 0$$
$$x = \frac{1}{2}$$
The solution is $\left(\frac{1}{2}, \frac{1}{2}, 3\right)$.

35. Let n = the number of nickels, d = the number of dimes, and q = the number or quarters in the box. Since there are 85 coins, $n + d + q = 85$.
The value of the coins is $6.25, so $5n + 10d + 25q = 625$ (converting the values to cents). Three times as many nickels as dimes means that $n = 3d$ or $n - 3d = 0$
Thus we have the system:
$$n + d + q = 85 \quad (1)$$
$$5n + 10d + 25q = 625 \quad (2)$$
$$n - 3d = 0 \quad (3)$$

Multiply (1) by -25, add the result to (2), and simplify:
$$-25n - 25d - 25q = -2125 \quad (1a)$$
$$5n + 10d + 25q = 625 \quad (2)$$
$$-20n - 15d = -1500$$
$$4n + 3d = 300 \quad (4)$$
Add (3) to (4) and simplify the result:
$$n - 3d = 0 \quad (3)$$
$$4n + 3d = 300 \quad (4)$$
$$5n = 300$$
$$n = 60$$
Put this value for n into (3):
$$60 - 3d = 0$$
$$60 = 3d$$
$$20 = d$$
Put the values for n and d into (1):
$$60 + 20 + q = 85$$
$$80 + q = 85$$
$$q = 5$$
The box contains 60 nickels, 20 dimes, and 5 quarters.

37. From the perimeter, $x + y + z = 155$. The relationship between sides x and y is $x + 20 = y$ or $x - y = -20$.
The relationship between sides y and z is $y - 5 = z$ or $y - z = 5$.
The system is
$$x + y + z = 155 \quad (1)$$
$$x - y = -20 \quad (2)$$
$$y - z = 5 \quad (3)$$

Chapter 3, Section 3.3

Add (1) and (3):
$x + y + z = 155$ (1)
$y - z = 5$ (3)
$x + 2y = 160$ (4)

Multiply (2) by 2, add the result to (4), and simplify:
$2x - 2y = -40$ (2a)
$x + 2y = 160$ (4)
$3x = 120$
$x = 40$

Put this value for x into (2):
$40 - y = -20$
$-y = -60$
$y = 60$

Put this value for y into (3):
$60 - z = 5$
$-z = -55$
$z = 55$

The sides of the triangle are:
$x = 40$ in., $y = 60$ in., and $z = 55$ in.

39. Let c = the amount of carrots, b = the amount of green beans, and f = the amount of cauliflower in 1 cup of Vegetable Medley. So $c + b + f = 1$.
The information on Vitamin C gives the equation:
$9c + 15b + 69f = 29.4$.
The information on calcium gives the equation:
$48c + 63b + 26f = 47.4$.
The system is:
$c + b + f = 1$ (1)
$9c + 15b + 69f = 29.4$ (2)
$48c + 63b + 26f = 47.4$ (3)

Multiply (1) by -9, add the result to (2), and simplify:
$-9c - 9b - 9f = -9$ (1a)
$9c + 15b + 69f = 29.4$ (2)
$6b + 60f = 20.4$
$b + 10f = 3.4$ (4)

Multiply (1) by -48 and add the result to (3):
$-48c - 48b - 48f = -48$ (1b)
$48c + 63b + 26f = 47.4$ (3)
$15b - 22f = -0.6$ (5)

Multiply (4) by -15, add the result to (5), and simplify the result:
$-15b - 150f = -51$ (4a)
$15b - 22f = -0.6$ (5)
$-172f = -51.6$
$f = 0.3$

Put this value for f into (5):
$15b - 22(0.3) = -0.6$
$15b - 6.6 = -0.6$
$15b = 6$
$b = 0.4$

Put the values for f and b into (1):
$c + 0.4 + 0.3 = 1$
$c + 0.7 = 1$
$c = 0.3$

Each cup of Vegetable Medley contains 0.3 cups of carrots, 0.4 cups of green beans, and 0.3 cups of cauliflower.

41. Let s = the number of score only, e = the number of evaluation, and n = the number of narrative reports processed each day.
Converting all times to minutes, we have, for the optical scanner:
$3s + 3e + 3n = 420$.
For the analysis:
$0s + 4e + 5n = 480$.
For the printer:
$1s + 2e + 8n = 720$.
After reducing the equation which corresponds to the scanner, the system is:

$$s + e + n = 140 \quad (1)$$
$$4e + 5n = 480 \quad (2)$$
$$s + 2e + 8n = 720 \quad (3)$$

Multiply (1) by -1 and add the result to (3):

$$-s - e - n = -140 \quad (1a)$$
$$s + 2e + 8n = 720 \quad (3)$$
$$e + 7n = 580 \quad (4)$$

Multiply (4) by -4, add the result to (2), and simplify:

$$4e + 5n = 480 \quad (2)$$
$$-4e - 28n = -2320 \quad (4a)$$
$$-23n = -1840$$
$$n = 80$$

Put this value for n into (2):
$$4e + 5(80) = 480$$
$$4e + 400 = 480$$
$$4e = 80$$
$$e = 20$$

Put the values for e and n into (1):
$$s + 20 + 80 = 140$$
$$s + 100 = 140$$
$$s = 40$$

Thus, when using all of its resources, ABC can complete 40 score only, 20 evaluation, and 80 evaluation reports.

43. Let t = the number of tennis rackets, p = the number of Ping-Pong paddles, and s = the number of squash rackets that Ace makes each day. The time available for gluing gives the equation:
$3t + 1p + 2s = 95$.
The time available for sanding gives the equation:
$2t + 1p + 2s = 75$.
The time available for finishing gives the equation:
$3t + 1p + 2.5s = 100$.
The system is

$$3t + p + 2s = 95 \quad (1)$$
$$2t + p + 2s = 75 \quad (2)$$
$$3t + p + 2.5s = 100 \quad (3)$$

Multiply (1) by -1, add the result to (3), and simplify:

$$-3t - p - 2s = -95 \quad (1a)$$
$$3t + p + 2.5s = 100 \quad (3)$$
$$0.5s = 5$$
$$s = 10 \quad (4)$$

Multiply (2) by -1 and add the result to (1):

$$3t + p + 2s = 95 \quad (1)$$
$$-2t - p - 2s = -75 \quad (2a)$$
$$t = 20 \quad (5)$$

Put the values for t and s into (1):
$$3(20) + p + 2(10) = 95$$
$$60 + p + 20 = 95$$
$$p = 15$$

Ace can make 20 tennis rackets, 15 Ping-Pong paddles, and 10 squash rackets every day.

Chapter 3, Section 3.4

Section 3.4

1. Using the points (1, 81) and (5, 73),
$$m = \frac{73-81}{5-1} = -2.$$
$$y - 81 = -2(x-1)$$
$$y = -2x + 83$$
At 3:00 A.M., $x = 3$, so $y = -2(3) + 83 = 77$.
At 4:00 A.M., $x = 4$, so $y = -2(4) + 83 = 75$.
Therefore, the temperature is 77° at 3:00 A.M. and 75° at 4:00 A.M.

3. Using the points (1, 1) and (6, 8),
$$m = \frac{8-1}{6-1} = \frac{7}{5}.$$
$$y - 1 = \frac{7}{5}(x-1)$$
$$y = \frac{7}{5}x - \frac{2}{5}$$
On the third day, $x = 3$, so
$$y = \frac{7}{5}(3) - \frac{2}{5} = 3\frac{4}{5}.$$
On the fifth day, $x = 5$, so
$$y = \frac{7}{5}(5) - \frac{2}{5} = 6\frac{3}{5}.$$
Therefore, its height is $3\frac{4}{5}$ cm on the third day and $6\frac{3}{5}$ cm on the fifth day.

5. Using the points (0, 0) and (6, 60),
$$m = \frac{60-0}{6-0} = 10.$$
$$y - 0 = 10(x-0)$$
$$y = 10x$$
After 2 seconds, $x = 2$, so
$y = 10(2) = 20$.
After 4 seconds, $x = 4$, so
$y = 10(4) = 40$.
Therefore, the car's speed is 20 mph after 2 seconds and 40 mph after 4 seconds.

7. Using the points (0, 23) and (8, 13),
$$m = \frac{13-23}{8-0} = -\frac{5}{4}.$$
$$C - 23 = -\frac{5}{4}(t-0)$$
$$C = -\frac{5}{4}t + 23$$
At midnight, $t = 2$, so
$$C = -\frac{5}{4}(2) + 23 = 20.5.$$
At 2:00 A.M., $t = 4$, so
$$C = -\frac{5}{4}(4) + 23 = 18.$$
Therefore, the temperature is 20.5° at midnight and 18° at 2:00 A.M.

9. Using the points (0, 9) and (7, 51),
$$m = \frac{51-9}{7-0} = 6.$$
$$y - 9 = 6(x-0)$$
$$y = 6x + 9$$
After two minutes, $x = 2$, so
$y = 6(2) + 9 = 21$.
After two hours, $x = 120$, so
$y = 6(120) + 9 = 729$.
Therefore, the engine temperature is 21°C two minutes after it started and 729°C two hours after it started. The last prediction is not reasonable. The engine temperature should not continue to rise linearly.

Chapter 3, Section 3.4

11. Using the points (0, 8) and (1, 20),
$m = \dfrac{20-8}{1-0} = 12$.
$y - 8 = 12(x - 0)$
$y = 12x + 8$
At age 10, $x = 10$, so
$y = 12(10) + 8 = 128$.
Therefore, his weight will be 128 lb.

13. a.

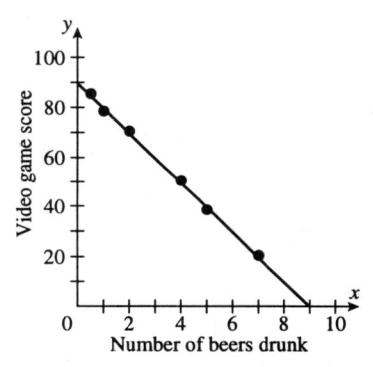

b. At $x = 3$, $y = 60$. At $x = 8$, $y = 10$. Therefore, predict the score to be 60 after 3 beers and 10 after 8 beers.

c. Using the points (3, 60) and (8, 10),
$m = \dfrac{10-60}{8-3} = -10$.
$y - 60 = -10(x - 3)$
$y = -10x + 90$

d. Substitute $x = 4.5$ into the equation.
$y = -10(4.5) + 90 = 45$.
Predict the score to be 45.

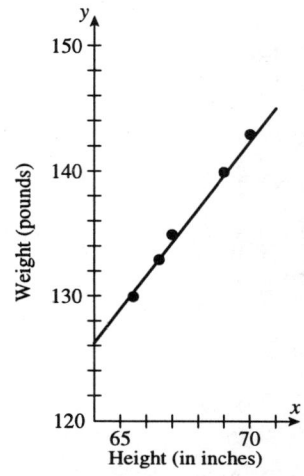

15. a.

b. At $x = 65$, $y = 129$. At $x = 71$, $y = 145$. Therefore, predict the weight to be 129 pounds for a 65-inch-tall distance runner and 145 pounds for a 71-inch-tall distance runner.

c. Using the points (65, 129) and (71, 145),
$m = \dfrac{145-129}{71-65} = \dfrac{8}{3}$.
$y - 129 = \dfrac{8}{3}(x - 65)$
$y = \dfrac{8}{3}x - \dfrac{133}{3}$

d. Substitute $x = 68$ into the equation.
$y = \dfrac{8}{3}(68) - \dfrac{133}{3} = 137$.
Predict the weight to be 137 pounds.

Chapter 3, Section 3.4

17. a.

[Graph: Lawyers per 100,000 residents vs Physicians per 100,000 residents, with scattered data points and a fitted line]

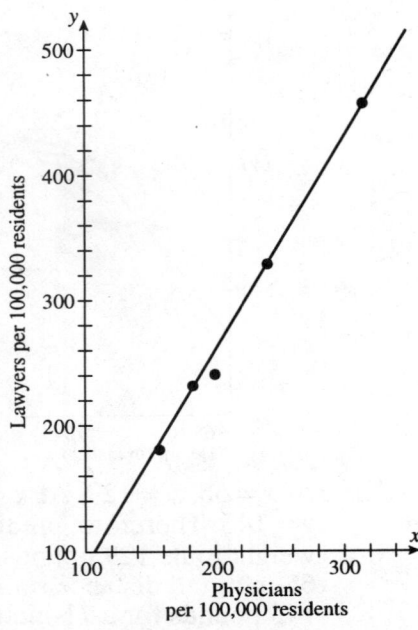

b. At $x = 210$, $y = 275$. At $x = 310$, $y = 450$. Therefore, predict the number of lawyers to be 275 (per 100,000 residents) in a state with 210 physicians (per 100,000 residents) and 450 (per 100,000 residents) in a state with 310 physicians (per 100,000 residents).

c. Using the points (210, 275) and (310, 450),
$$m = \frac{450 - 275}{310 - 210} = \frac{7}{4}.$$
$$y - 275 = \frac{7}{4}(x - 210)$$
$$y = \frac{7}{4}x - \frac{185}{2}$$

d. Substitute $x = 220$ into the equation.
$$y = \frac{7}{4}(220) - \frac{185}{2} = 292.5.$$
Predict the number to be 293 (per 100,000 residents).

19. Enter the data and compute the regression line:

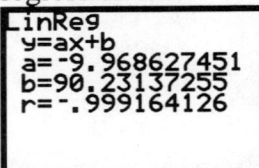

```
LinReg
y=ax+b
a=-9.968627451
b=90.23137255
r=-.999164126
```

The regression line is approximately $y = -9.97x + 90.23$.
Predict the score to be approximately 45.4.

21. Enter the data and compute the regression line:

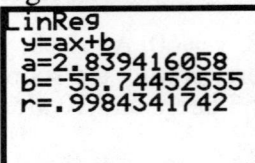

```
LinReg
y=ax+b
a=2.839416058
b=-55.74452555
r=.9984341742
```

The regression line is approximately $y = 2.84x - 55.74$.
Predict the weight to be approximately 137.3 pounds.

23. Enter the data and compute the regression line:

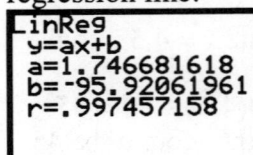

```
LinReg
y=ax+b
a=1.746681618
b=-95.92061961
r=.997457158
```

The regression line is approximately $y = 1.75x - 95.92$.
Predict the number to be approximately 288 (per 100,000 residents).

Section 3.5

1. Refer to the graph in the back of the textbook. Graph the line $y = 2x + 4$. Since the inequality is $y > 2x + 4$, the region above the line is shaded and the line is dashed.

3. Refer to the graph in the back of the textbook. Graph the line $3x - 2y = 12$. Since the inequality can be written as $y \geq -6 + \frac{3}{2}x$, the region above the line is shaded and the line is solid.

5. Refer to the graph in the back of the textbook. Graph the line $x + 4y = -6$. Since the inequality can be written as $y \geq -\frac{3}{2} - \frac{x}{4}$, the region above the line is shaded and the line is solid.

7. Refer to the graph in the back of the textbook. Graph the line $x = -3y + 1$. Since the inequality can be written as $y > -\frac{x}{3} + \frac{1}{3}$, the region above the line is shaded and the line is dashed.

9. Refer to the graph in the back of the textbook. Graph the line $x = -3$. Since the inequality is $x \geq -3$, the region to the right of the line is shaded and the line is solid.

11. Refer to the graph in the back of the textbook. Graph the line $y = \frac{1}{2}$. Since The inequality is $y > \frac{1}{2}$, the region above the line is shaded and the line is dashed.

13. Refer to the graph in the back of the textbook. Graph the line $0 = x - y$. Since the inequality can be written as $y \geq x$, the region above the line is shaded and the line is solid.

15. Refer to the graph in the back of the textbook. Graph the lines $-1 = y$ and $y = 4$. The line $-1 = y$ should be dashed, $y = 4$ should be solid. Since the inequality is $-1 < y \leq 4$, the region between the two lines is shaded.

17. Refer to the graph in the back of the textbook.
For $y > 2$, graph the line $y = 2$ with a dashed line and shade the region above it.
For $x \geq -2$, graph the line $x = -2$ with a solid line and shade the region to the right of the line.

19. Refer to the graph in the back of the textbook.
For $y < x$, graph the line $y = x$ with a dashed line and shade the region below the line.
For $y \geq -3$, graph the line $y = -3$ with a solid line and shade the region above the line.

21. Refer to the graph in the back of the textbook.
For $x + y \leq 6$, graph the line $x + y = 6$ with a solid line. Shade the region below this line, since the inequality can be written as $y \leq 6 - x$.
For $x + y \geq 4$, graph the line $x + y = 4$ with a solid line. Shade the region above this line, since the inequality can be written as $y \geq 4 - x$.

Chapter 3, Section 3.5

23. Refer to the graph in the back of the textbook.
For $2x - y \leq 4$, graph the line $2x - y = 4$ with a solid line. Shade the region above this line since the inequality can be written as $y \geq -4 + 2x$.
For $x + 2y > 6$, graph the line $x + 2y = 6$ with a dashed line. Shade the region above this line, since the inequality can be written as
$y > 3 - \dfrac{x}{2}$.

25. Refer to the graph in the back of the textbook.
For $3y - 2x < 2$, graph the line $3y - 2x = 2$ with a dashed line. Shade the region below this line since the inequality can be written as $y < \dfrac{2}{3}x + \dfrac{2}{3}$.
For $y > x - 1$, graph the line $y = x - 1$ with a dashed line and shade the region above this line.

27. Refer to the graph in the back of the textbook.
For $2x + 3y - 6 < 0$, graph the line $2x + 3y - 6 = 0$ with a dashed line. Shade the region below this line, since the inequality can be written as $y < 2 - \dfrac{2}{3}x$.
For $x \geq 0$, graph the line $x = 0$ (the y-axis) with a solid line and shade the region to the right of this line.
For $y \geq 0$, graph the line $y = 0$ (the x-axis) with a solid line and shade the region above this line.
The vertices are the points of intersection of the lines $2x + 3y - 6 = 0$, $x = 0$, and $y = 0$.
The lines $2x + 3y - 6 = 0$ and $x = 0$ intersect at the point $(0, 2)$,
$2x + 3y - 6 = 0$ and $y = 0$ intersect at the point $(3, 0)$, while $x = 0$ and $y = 0$ intersect at the origin $(0, 0)$.

29. Refer to the graph in the back of the textbook.
For $5y - 3x \leq 15$, graph the line $5y - 3x = 15$ with a solid line. Shade the region below this line since the inequality can be written as
$y \leq 3 + \dfrac{3}{5}x$. For $x + y \leq 11$, graph the line $x + y = 11$ with a solid line. Shade the region below this line since the inequality can be written as $y \leq 11 - x$. For graphing $x = 0$ and $y = 0$, see 27. The vertices are $(0, 0)$, the intersection of the lines $x = 0$ and $y = 0$; $(0, 3)$, the intersection of the lines $x = 0$ and $5y - 3x = 15$; $(5, 6)$, the intersection of the lines $5y - 3x = 15$ and $x + y = 11$; and $(11, 0)$, the intersection of $y = 0$ and $x + y = 11$.

31. Refer to the graph in the back of the textbook. For $2y \leq x$, graph the line $2y = x$ with a solid line. Shade the region below this line since the inequality can be written as $y \leq \dfrac{x}{2}$.
For $2x \leq y + 12$, graph the line $2x = y + 12$ with a solid line. Shade the region above this line since the inequality can be written as $y \geq 2x - 12$.
For graphing $x \geq 0$ and $y \geq 0$, see 27. The vertices are $(0, 0)$; $(8, 4)$, the intersection of the lines $2y = x$ and $2x = y + 12$; and $(6, 0)$, the intersection of the lines $2x = y + 12$ and $y = 0$.

Chapter 3, Section 3.5

33. Refer to the graph in the back of the textbook.
For $x + y \geq 3$, graph the line $x + y = 3$ with a solid line. Shade the region above this line since the inequality can be written as $y \geq 3 - x$.
For $2y \leq x + 8$, graph the line $2y = x + 8$ with a solid line. Shade the region below this line since the inequality can be written as
$y \leq \dfrac{x}{2} + 4$.
For $2y + 3x \leq 24$, graph the line $2y + 3x = 24$ with a solid line. Shade the region below this line since the inequality can be written as
$y \leq 12 - \dfrac{3x}{2}$.
For graphing $x \geq 0$ and $y \geq 0$, see 27.
The vertices are
(0, 3), the intersection of the lines $x = 0$ and $x + y = 3$;
(0, 4), the intersection of the lines $x = 0$ and $2y = x + 8$;
(4, 6), the intersection of the lines $2y = x + 8$ and $2y + 3x = 24$;
(8, 0), the intersection of the lines $2y + 3x = 24$ and $y = 0$; and
(3, 0) the intersection of the lines $y = 0$ and $x + y = 3$.

35. Refer to the graph in the back of the textbook.
For $3y - x \geq 3$, graph the line $3y - x = 3$ with a solid line. Shade the region above this line since the inequality can be written as
$y \geq 1 + \dfrac{x}{3}$. For $y - 4x \geq -10$, graph the line $y - 4x = -10$ with a solid line. Shade the region above this line since the inequality can be written as $y \geq 4x - 10$.
For $y - 2 \leq x$, graph the line $y - 2 = x$ with a solid line. Shade the region below this line since the inequality can be written as $y \leq x + 2$.
For graphing $x \geq 0$ and $y \geq 0$, see 27.
The vertices are (0, 1), the intersection of the lines $x = 0$ and $3y - x = 3$; (0, 2), the intersection of the lines $x = 0$ and $y - 2 = x$; (4,6), the intersection of the lines $y - 2 = x$ and $y - 4x = -10$; and (3, 2), the intersection of the lines $y - 4x = -10$ and $3y - x = 3$.

37. Let x represent the number of student tickets sold, and y represent the number of faculty tickets sold. The information that student tickets cost \$1, faculty tickets cost \$2, and the receipts must be at least \$250, can be stated in the inequality $x + 2y \geq 250$. The fact that only positive numbers of tickets will be sold can be stated in the inequalities $x \geq 0$ and $y \geq 0$.
The system of inequalities is
$x + 2y \geq 250$
$x \geq 0, y \geq 0$
Refer to the graph in the back of the textbook.

39. Let x represent the amount that Vassilis invests at 6% interest and let y represent the amount that he invests at 5%. Vassilis will invest at most \$10,000, so $x + y \leq 10,000$. Earning at least \$540 from these investments requires that $0.06x + 0.05y \geq 540$. It is impossible for Vassilis to invest a negative amount with either bank, so $x \geq 0$ and $y \geq 0$.

Chapter 3, Section 3.6

The system of inequalities is
$x + y \leq 10{,}000$
$0.06x + 0.05y \geq 540$
$x \geq 0,\ y \geq 0$
Refer to the graph in the back of the textbook.

41. Let x represent the amount of corn meal and let y represent the amount of whole wheat flour that Gary uses. Since he cannot use more than 3 cups of the two ingredients combined, $x + y \leq 3$. The amount of linoleic acid in x cups of corn meal is $2.4x$ grams, while the amount of linoleic acid in y cups of whole wheat flour is $0.8y$ grams. To provide at least 3.2 grams in the combined mixture requires that $2.4x + 0.8y \geq 3.2$. Similarly, to makes sure that the mixture contains at least 10 milligrams of niacin, $2.5x + 5y \geq 10$. Since using a negative amount of either ingredient does not make sense, $x \geq 0$ and $y \geq 0$. The system of inequalities is
$x + y \leq 3$
$2.4x + 0.8y \geq 3.2$
$2.5x + 5y \geq 10$
$x \geq 0,\ y \geq 0$
Refer to the graph in the back of the textbook.

Section 3.6

1. From the graph below, the line $12 = 3x + 4y$ does not intersect the set of feasible solutions. Since $3x + 4y$ is the cost, there is no point in the set of all feasible solutions where the cost can be $12.

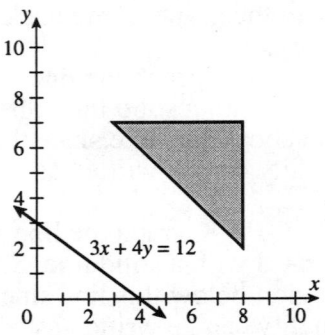

3. The lines where the cost is constant will all be parallel to the line $12 = 3x + 4y$ shown in the graph above for 1. Higher costs will be represented by lines further from the origin. The line closest to the origin and parallel to $12 = 3x + 4y$ intersecting the set of feasible solutions will intersect the solution set at the point $(8, 2)$. The cost at this point will be
$3(8) + 4(2) = 24 + 8 = 32$ dollars.

5.

Any point on the line $8 = 4x - 2y$ will correspond to a profit of \$8. One point on the line which is also in the set of feasible solutions is $(2, 0)$.

Chapter 3, Section 3.6

7. a. The line $8 = 4x - 2y$ is closer to the origin than the line $22 = 4x - 2y$, so the line corresponding to profit of $22 is farther from the origin.

b. From a., lines corresponding to higher profit are farther from the origin, so the vertex of the set of feasible solutions corresponding to maximum profit will be the vertex farthest from the origin. This vertex is (8, 0).

c. Using b., the maximum profit is $4(8) - 2(0) = 32 - 0 = 32$ dollars.

9. a. The lines representing $C = 3$ and $C = 12$ are shown below with the graph of the feasible solutions. Increasing the values of C from $C = 3$, the first vertex of the set of feasible solutions will be (1, 4). Thus, the minimum value of C will occur at (1, 4).

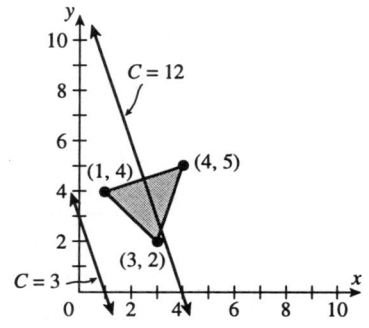

b. $3(1) + 4 = 3 + 4 = 7$
The minimum value of C is 7.

c. As C increases, the last vertex of the graph of the feasible solutions will be (4, 5). Thus, the maximum value of C will occur at (4, 5)

d. $3(4) + 5 = 12 + 5 = 17$
The maximum value of C is 17.

11. a. The lines representing $C = 0$ and $C = 10$ are shown below with the graph of the feasible solutions. Decreasing the values of C from $C = 0$, the last vertex of the set of feasible solutions will be (0, 5). Thus, the minimum value of C will occur at (0, 5).

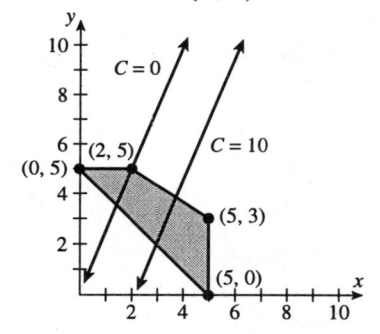

b. $5(0) - 2(5) = 0 - 10 = -10$
The minimum value of C is -10.

c. Increasing C from 10, the last vertex of the graph of feasible solutions will be (5, 0). Thus, the maximum value of C will occur at (5, 0).

d. $5(5) - 2(0) = 25 - 0 = 25$.
The maximum value of C is 25.

91

Chapter 3, Section 3.6

13. a.

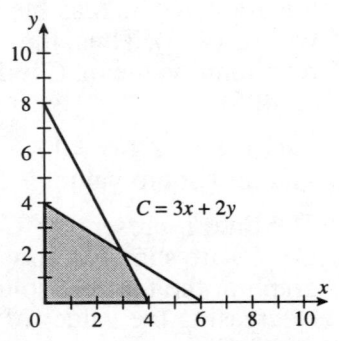

[Graph showing feasible region with line labeled $C = 3x + 2y$, axes from 0 to 10]

b. The vertices of the set of feasible solutions are $(0, 0)$, $(0, 4)$, $(3, 2)$, and $(4, 0)$. Since the objective function involves only addition of multiples of x and y, the minimum value for C is 0. This occurs at $(0, 0)$.

c. The maximum value will occur at either $(0, 4)$, $(3, 2)$ or $(4, 0)$.
At $(0, 4)$, $C = 3(0) + 2(4) = 8$.
At $(3, 2)$, $C = 3(3) + 2(2) = 13$.
At $(4, 0)$, $C = 3(4) + 2(0) = 12$.
The maximum value for C is 13 and it occurs at $(3, 2)$

15. a.

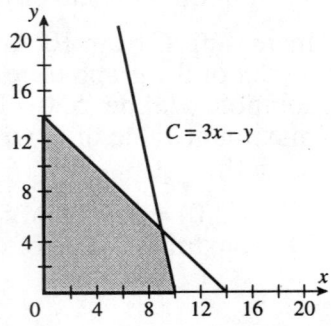

[Graph showing feasible region with line labeled $C = 3x - y$, axes from 0 to 20]

b. The vertices of the set of feasible solutions are $(0, 0)$, $(0, 14)$, $(9, 5)$, and $(10, 0)$. The lines representing constant values of the objective function will all have slope 3. Thus the minimum value of C will occur at either $(0, 14)$ or $(10, 0)$. Looking at the objective function, $C = 3x - y$, note that if $x = 0$ and $y \geq 0$, then $C \leq 0$, while if $y = 0$ and $x \geq 0$, then $C \geq 0$. Thus the minimum will occur at $(0, 14)$. The minimum value of C is $3(0) - 14 = -14$.

c. Following the reasoning in b., the maximum will occur at $(10, 0)$. The maximum value of C is $3(10) - 0 = 30$.

17. a.

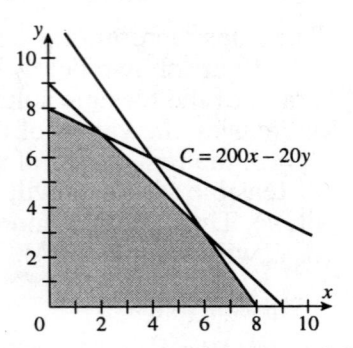

[Graph showing feasible region with line labeled $C = 200x - 20y$, axes from 0 to 10]

b. The vertices of the set of feasible solutions are $(0, 0)$, $(0, 8)$, $(2, 7)$, $(6, 3)$ and $(8, 0)$. The lines representing constant values of the objective function will all have slope 10. Thus the minimum value of C will occur at either $(0, 8)$ or $(8, 0)$. By reasoning similar to that in 15., the minimum value will occur at $(0, 8)$. The minimum value of C is $200(0) - 20(8) = -160$.

c. The maximum value of C will occur at $(8, 0)$. The maximum value of C is $200(8) - 20(0) = 1600$.

Chapter 3, Section 3.6

19. a.

Graph showing feasible region with C = 18x + 48y

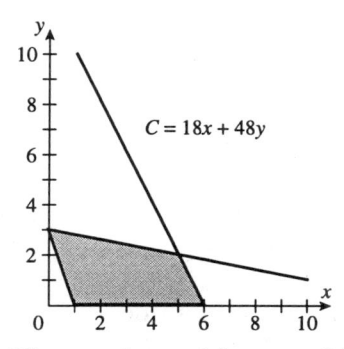

b. The vertices of the set of feasible solutions are (0, 3), (5, 2), (6, 0), and (1, 0). The lines representing constant values of the objective function will all have slope $-\frac{3}{8}$. Thus the minimum value of C will occur at either (1, 0) or (5, 2). Looking at the objective function, $C = 18x + 48y$, the value of C will be higher when both x and y are greater than 0. Thus, the minimum value of C will occur at (1, 0). The minimum value of C is $18(1) + 48(0) = 18$.

c. The maximum value of C will occur at (5, 2). The maximum value of C is $18(5) + 48(2) = 186$.

21. Let x = the number of student tickets sold and y = the number of faculty tickets sold. The requirement for $250 in ticket receipts can be expressed as $x + 2y \geq 250$. Since the number of tickets sold cannot be negative, $x \geq 0$ and $y \geq 0$. The system of inequalities is
$x + 2y \geq 250$
$x \geq 0, y \geq 0$

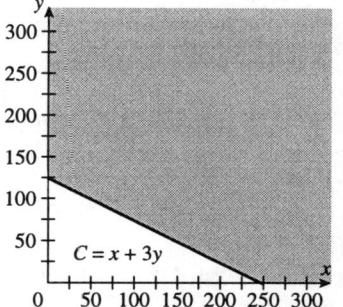

Graph showing feasible region with C = x + 3y

The vertices of the set of feasible solutions are (0, 125) and (250, 0). The objective function represents the number of calculators that the alumnus will donate, $C = x + 3y$. At (0, 125), $C = 0 + 3(125) = 375$. At (250, 0), $C = 250 + 3(0) = 250$. The alumnus will donate at least 250 calculators.

23. Let x = the number of acres of wheat and y = the number of acres of soybeans that Jeannette will plant. Since she has 180 acres available for planting, $x + y \leq 180$. The requirements of time for harvesting, and the number of hours available, yield $2x + y \leq 240$. Since she cannot plant a negative number of acres, $x \geq 0$ and $y \geq 0$. The system of inequalities is
$x + y \leq 180$
$2x + y \leq 240$
$x \geq 0, y \geq 0$

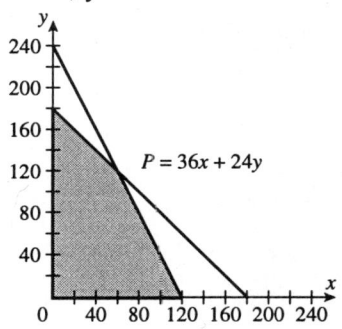

Graph showing feasible region with P = 36x + 24y

93

Chapter 3, Section 3.6

The vertices of the set of feasible solutions are (0, 0), (0, 180), (60, 120) and (120, 0). The objective function represents the total profit of planting x acres of wheat and y acres of soybeans. Thus, $P = 36x + 24y$.
At (0, 0), $P = 36(0) + 24(0) = 0$.
At (0, 180),
$P = 36(0) + 24(180) = 4320$.
At (60, 120),
$P = 36(60) + 24(120) = 5040$.
At (120, 0),
$P = 36(120) + 24(0) = 4320$.
The maximum profit that Jeannette can make is $5040, which she will receive if she plants 60 acres of wheat and 120 acres of soybeans.

25. Let x = the amount of corn meal and y = the amount of whole wheat flour that Gary uses. Since these ingredients combine for at most 3 cups, $x + y \leq 3$. To make sure that there's enough linoleic acid in the pancakes, $2.4x + 0.8y \geq 3.2$. For the niacin, the recipe requires $2.5x + 5y \geq 10$. Since it is impossible for Gary to put negative amounts of either ingredient in the pancakes, $x \geq 0$ and $y \geq 0$.
The system of inequalities is
$x + y \leq 3$
$2.4x + 0.8y \geq 3.2$
$2.5x + 5y \geq 10$
$x \geq 0, y \geq 0$

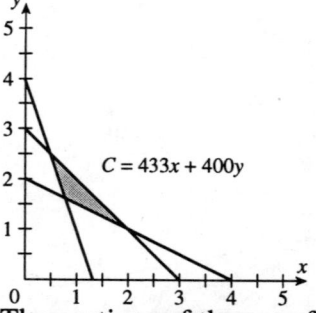

The vertices of the set of feasible solutions are (0.5, 2.5), (2, 1), and (0.8, 1.6). The objective function represents the number of calories in the recipe from the given ingredients. Thus, $C = 433x + 400y$.
At (0.5, 2.5),
$C = 433(0.5) + 400(2.5) = 1216.5$.
At (2, 1),
$C = 433(2) + 400(1)$
$= 1266$.
At (0.8, 1.6),
$C = 433(0.8) + 400(1.6)$
$= 986.4$.
The minimum number of calories in Gary's pancakes is 986.4, which occurs when he uses 0.8 cups of corn meal and 1.6 cups of whole wheat flour.

27.

Window: Xmin = 0, Xmax = 3, Ymin = 0, and Ymax = 3.
The set of feasible solutions is the box in the lower left hand corner, since, when solved for y, both constraints have y less than the expression in x.

The minimum value of C is −8.4, which occurs at (0, 2). The maximum value of C is 17.4, which occurs at (2, 0). You will probably need to [ZOOM] to get the bug to this point.

29.

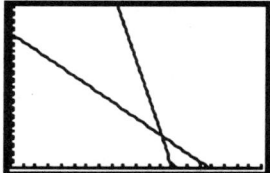

Window: Xmin = 0, Xmax = 20, Ymin = 0, and Ymax = 20.
The set of feasible solutions is the box in the lower left hand corner, since when both constraints are solved for y, they have y less than the expression in x.
The minimum value of C is 0, which occurs at (0, 0).
The maximum value of C is approximately 4112, which occurs at (12, 4).

31.

Window: Xmin = 0, Xmax = 8, Ymin = 0, and Ymax = 8.
The set of feasible solutions is the box in the lower left hand corner, since when all of the constraints are solved for y, they have y less than the expression in x.
The minimum value of C is 0, which occurs at (0, 0).
The maximum value is 1908, which occurs at (4, 5).

Chapter 3 Review

1. Refer to the graph in the back of the textbook.
   The solution is $x = -1, y = 2$.

2. Refer to the graph in the back of the textbook.
   The solution is $x = 1.9, y = -0.8$.

3. $x + 5y = 18$ (1)
   $x - y = -3$ (2)
   Subtract (2) from (1) to eliminate x:
   $x + 5y = 18$ (1)
   $-x + y = 3$ (2a)
   $6y = 21$
   $y = \dfrac{21}{6}$
   $y = \dfrac{7}{2}$
   Put this value for y into (2):
   $x - \dfrac{7}{2} = -3$
   $x = -3 + \dfrac{7}{2}$
   $x = \dfrac{1}{2}$
   The solution is $x = \dfrac{1}{2}, y = \dfrac{7}{2}$.

4. $x + 5y = 11$ (1)
   $2x + 3y = 8$ (2)
   Subtract twice (1) from (2):
   $-2x - 10y = -22$ (1a)
   $2x + 3y = 8$ (2)
   $-7y = -14$
   $y = 2$

Chapter 3 Review

Put this value for y into (1):
$$x + 5(2) = 11$$
$$x + 10 = 11$$
$$x = 1$$
The solution is $x = 1$, $y = 2$.

5. $\frac{2}{3}x - 3y = 8$ (1)

$x + \frac{3}{4}y = 12$ (2)

Add 4 times (2) to (1):
$\frac{2}{3}x - 3y = 8$ (1)
$4x + 3y = 48$ (2a)
$$\frac{14}{3}x = 56$$
$$x = 12$$
Put this value for x into (1):
$$\frac{2}{3}(12) - 3y = 8$$
$$8 - 3y = 8$$
$$-3y = 0$$
$$y = 0$$
The solution is $x = 12$, $y = 0$.

6. $3x = 5y - 6$

$3y = 10 - 11x$

Put the equations in standard form:
$3x - 5y = -6$ (1)
$11x + 3y = 10$ (2)

Multiply (1) by 3, (2) by 5, and add the results:
$9x - 15y = -18$ (1a)
$55x + 15y = 50$ (2a)
$$64x = 32$$
$$x = \frac{1}{2}$$

Put this value for x into (1):
$$3\left(\frac{1}{2}\right) - 5y = -6$$
$$\frac{3}{2} - 5y = -6$$
$$-5y = -\frac{15}{2}$$
$$y = \frac{3}{2}$$
The solution is $x = \frac{1}{2}$, $y = \frac{3}{2}$.

7. $2x - 3y = 4$ (1)

$x + 2y = 7$ (2)

Multiply (2) by 2:
$2x - 3y = 4$ (1)
$2x + 4y = 14$ (2a)

These lines are neither the same nor parallel, so the system is consistent.

8. $2x - 3y = 4$ (1)

$6x - 9y = 4$ (2)

Multiply (1) by −3 and add the result to (2):
$-6x + 9y = -12$ (1a)
$6x - 9y = 4$ (2)
$$0 = -8$$

The last equation can be written as $0x + 0y = -8$, which shows that the system is inconsistent.

9. $2x - 3y = 4$ (1)

$6x - 9y = 12$ (2)

Multiply (1) by 3:
$6x - 9y = 12$ (1a)
$6x - 9y = 12$ (2)

These equations are identical, so the system is dependent.

10. $x - y = 6$ (1)

$x + y = 6$ (2)

Add these equations:

$2x = 12$

$x = 6$

Put this value for x into (1):

$6 - y = 6$

$y = 0$

The system has a solution, therefore it is consistent.

11. $x + 3y - z = 3$ (1)

$2x - y + 3z = 1$ (2)

$3x + 2y + z = 5$ (3)

To eliminate x, multiply (1) by -2 and add the result to (2):

$-2x - 6y + 2z = -6$ (1a)

$2x - y + 3z = 1$ (2)

$-7y + 5z = -5$ (4)

Multiply (1) by -3 and add the result to (3):

$-3x - 9y + 3z = -9$ (1b)

$3x + 2y + z = 5$ (3)

$-7y + 4z = -4$ (5)

Now consider the system formed by (4) and (5):

$-7y + 5z = -5$ (4)

$-7y + 4z = -4$ (5)

Multiply (5) by -1 and add the result to (4):

$-7y + 5z = -5$ (4)

$7y - 4z = 4$ (5a)

$z = -1$

Put this value for z into (4):

$-7y + 5(-1) = -5$

$-7y - 5 = -5$

$-7y = 0$

$y = 0$

Put the values for y and z into (1):

$x + 3(0) - (-1) = 3$

$x + 1 = 3$

$x = 2$

The solution is $x = 2$, $y = 0$, $z = -1$.

12. $x + y + z = 2$ (1)

$3x - y + z = 4$ (2)

$2x + y + 2z = 3$ (3)

To eliminate y, add (1) and (2):

$x + y + z = 2$ (1)

$3x - y + z = 4$ (2)

$4x + 2z = 6$ (4)

Now add (2) and (3):

$3x - y + z = 4$ (2)

$2x + y + 2z = 3$ (3)

$5x + 3z = 7$ (5)

Consider the system formed by (4) and (5):

$4x + 2z = 6$ (4)

$5x + 3z = 7$ (5)

To eliminate z, multiply (4) by -3, (5) by 2, and add the result:

$-12x - 6z = -18$ (4a)

$10x + 6z = 14$ (5a)

$-2x = -4$

$x = 2$

Chapter 3 Review

Put this value for x into (4):
$4(2) + 2z = 6$
$8 + 2z = 6$
$2z = -2$
$z = -1$
Put the values for x and z into (1):
$2 + y - 1 = 2$
$y + 1 = 2$
$y = 1$
The solution is $x = 2$, $y = 1$, $z = -1$.

13. $x + z = 5$ (1)
$y - z = -8$ (2)
$2x + z = 7$ (3)
To eliminate z, multiply (1) by -1 and add the result to (3):
$-x - z = -5$ (1a)
$2x + z = 7$ (3)
$x = 2$
Put this value for x into (1):
$2 + z = 5$
$z = 3$
Put this value for z into (2):
$y - 3 = -8$
$y = -5$
The solution is $x = 2$, $y = -5$, $z = 3$.

14. $x + 4y + 4z = 0$ (1)
$3x - 2y + z = -10$ (2)
$2x - 4y + z = -11$ (3)
To eliminate y, multiply (2) by 2 and add the result to (1):
$x + 4y + 4z = 0$ (1)
$6x - 4y + 2z = -20$ (2a)
$7x + 6z = -20$ (4)

Now add (1) and (3):
$x + 4y + 4z = 0$ (1)
$2x - 4y + z = -11$ (3)
$3x + 5z = -11$ (5)
Consider the system formed by (4) and (5):
$7x + 6z = -20$ (4)
$3x + 5z = -11$ (5)
Multiply (4) by -3, (5) by 7, and add the results:
$-21x - 18z = 60$ (4a)
$21x + 35z = -77$ (5a)
$17z = -17$
$z = -1$
Put this value for z into (4):
$7x + 6(-1) = -20$
$7x - 6 = -20$
$7x = -14$
$x = -2$
Put these values for x and z into (1):
$-2 + 4y + 4(-1) = 0$
$-6 + 4y = 0$
$4y = 6$
$y = \dfrac{3}{2}$
The solution is $x = -2$, $y = \dfrac{3}{2}$, $z = -1$.

15. $\dfrac{1}{2}x + y + z = 3$

$x - 2y - \dfrac{1}{3}z = -5$

$\dfrac{1}{2}x - 3y - \dfrac{2}{3}z = -6$

Put the system into standard form by multiplying the first equation by 2, the second equation by 3, and the third equation by 6.

$x + 2y + 2z = 6$ (1)

$3x - 6y - z = -15$ (2)

$3x - 18y - 4z = -36$ (3)

To eliminate x, multiply (1) by -3 and add the result to (2):

$-3x - 6y - 6z = -18$ (1a)

$3x - 6y - z = -15$ (2)

$-12y - 7z = -33$ (4)

Now multiply (2) by -1 and add the result to (3):

$-3x + 6y + z = 15$ (2a)

$3x - 18y - 4z = -36$ (3)

$-12y - 3z = -21$ (5)

Consider the system formed by (4) and (5):

$-12y - 7z = -33$ (4)

$-12y - 3z = -21$ (5)

Multiply (4) by -1 and add the result to (5):

$12y + 7z = 33$ (4a)

$-12y - 3z = -21$ (5)

$4z = 12$

$z = 3$

Put this value for z into (4):

$-12y - 7(3) = -33$

$-12y - 21 = -33$

$-12y = -12$

$y = 1$

Put these values for y and z into (1):

$x + 2(1) + 2(3) = 6$

$x + 8 = 6$

$x = -2$

The solution is $x = -2$, $y = 1$, $z = 3$.

16. $\dfrac{3}{4}x - \dfrac{1}{2}y + 6z = 2$

$\dfrac{1}{2}x + y - \dfrac{3}{4}z = 0$

$\dfrac{1}{4}x + \dfrac{1}{2}y - \dfrac{1}{2}z = 0$

Put the system into standard form by multiplying all three equations by 4.

$3x - 2y + 24z = 8$ (1)

$2x + 4y - 3z = 0$ (2)

$x + 2y - 2z = 0$ (3)

To eliminate x, multiply (3) by -3 and add the result to (1):

$3x - 2y + 24z = 8$ (1)

$-3x - 6y + 6z = 0$ (3a)

$-8y + 30z = 8$ (4)

Now multiply (3) by -2 and add the result to (2):

$2x + 4y - 3z = 0$ (2)

$-2x - 4y + 4z = 0$ (3b)

$z = 0$

Put this value for z into (4):

$-8y + 30(0) = 8$

$-8y = 8$

$y = -1$

Put these values for y and z into (3):

$x + 2(-1) - 2(0) = 0$

$x - 2 = 0$

$x = 2$

The solution is $x = 2$, $y = -1$, $z = 0$.

Chapter 3 Review

17. Let x = the number of questions Lupe answered correctly and y = the number of questions she answered incorrectly. Since there are 40 questions on the exam, $x + y = 40$. The scoring system is $5x - 2y$. Since Lupe got a score of 102, $5x - 2y = 102$.
The system of equations is
$\quad x + y = 40 \quad (1)$
$\quad 5x - 2y = 102 \quad (2)$
We want to know how many questions Lupe answered correctly, i.e., we want to find x. To eliminate y, multiply (1) by 2 and add the result to (2):
$\quad 2x + 2y = 80 \quad (1a)$
$\quad 5x - 2y = 102 \quad (2)$
$\quad\quad\quad 7x = 182$
$\quad\quad\quad\ \ x = 26$
Lupe answered 26 questions correctly.

18. Let x = the number of questions that Roger answered correctly and y = the number answered incorrectly. Since Roger answered 24 questions, $x + y = 24$. The way that the game show pays is $25x - 10y$, so in Roger's case, $25x - 10y = 355$.
The system of equations is
$\quad x + y = 24 \quad (1)$
$\quad 25x - 10y = 355 \quad (2)$
To find x, eliminate y. Multiply (1) by 10 and add the result to (2):
$\quad 10x + 10y = 240 \quad (1a)$
$\quad 25x - 10y = 355 \quad (2)$
$\quad\quad\quad 35x = 595$
$\quad\quad\quad\ \ x = 17$
Roger got 17 answers right.

19. Let x = the amount that Barbara invests at 8% and y = the amount that she invests at 13.5%. Since she is investing $5000, $x + y = 5000$. To earn $500 per year in interest, she needs $0.08x + 0.135y = 500$.
The system of equations is
$\quad x + y = 5000 \quad (1)$
$\quad 0.08x + 0.135y = 500 \quad (2)$
To eliminate x, multiply (1) by -0.08 and add the result to (2).
$\quad -0.08x - 0.08y = -400 \quad (1a)$
$\quad 0.08x + 0.135y = 500 \quad (2)$
$\quad\quad\quad 0.055y = 100$
$\quad\quad\quad\ \ y = 1818.\overline{18}$
Put the value of $y = 1818.18$ into (1):
$\quad x + 1818.18 = 5000$
$\quad\quad\quad x = 3181.82$
Barbara needs to invest $3181.82 at 8% and $1818.18 at 13.5%.

20. Let x = the amount the broker needs to invest at 15%. The total amount he invests will be $3000 + x$. A 12% return on this amount is given by $0.12(3000 + x)$. The amount of interest is also given by $3000(0.08) + x(0.15) = 240 + 0.15x$. Thus to find x, solve
$\quad 0.12(3000 + x) = 240 + 0.15x$
$\quad 360 + 0.12x = 240 + 0.15x$
$\quad\quad 120 = 0.03x$
$\quad\quad 4000 = x$
The broker needs to invest $4000 at 15%.

21. Let x, y, and z represent the lengths of the sides of the triangle. Since the perimeter of the triangle is 30 cm, $x + y + z = 30$. One side is 7 cm shorter than the second side, say the side of length x is 7 cm shorter than the side of length y, or $x = y - 7$. The third side is 1 cm longer than the second side, so $z = y + 1$.
The system of equations is
$$x + y + z = 30 \quad (1)$$
$$x = y - 7 \quad (2)$$
$$z = y + 1 \quad (3)$$
(2) and (3) give x and z in terms of y, so substitute these into (1):
$$y - 7 + y + y + 1 = 30$$
$$3y - 6 = 30$$
$$3y = 36$$
$$y = 12$$
Put this value for y into (2):
$x = 12 - 7 = 5$
Put the value for y into (3):
$z = 12 + 1 = 13$
The sides of the triangle are 5 cm, 12 cm, and 13 cm long.

22. Let x = the number of crates shipped to Boston, y = the number of crates shipped to Chicago, and z = the number of crates shipped to Los Angeles. There are 55 crates to ship, so $x + y + z = 55$. From the budgeting constraints, $10x + 5y + 12z = 445$. Sending twice as many crates to Boston as to Los Angeles is expressed by $x = 2z$ or $x - 2z = 0$.
The system of equations is
$$x + y + z = 55 \quad (1)$$
$$10x + 5y + 12z = 445 \quad (2)$$
$$x - 2z = 0 \quad (3)$$

To eliminate x, multiply (1) by -10 and add the result to (2).
$$-10x - 10y - 10z = -550 \quad (1a)$$
$$10x + 5y + 12z = 445 \quad (2)$$
$$-5y + 2z = -105 \quad (4)$$
Now multiply (3) by -1 and add the result to (1):
$$x + y + z = 55 \quad (1)$$
$$-x + 2z = 0 \quad (3a)$$
$$y + 3z = 55 \quad (5)$$
Consider the system formed by (4) and (5).
$$-5y + 2z = -105 \quad (4)$$
$$y + 3z = 55 \quad (5)$$
Multiply (5) by 5 and add the result to (4):
$$-5y + 2z = -105 \quad (4)$$
$$5y + 15z = 275 \quad (5a)$$
$$17z = 170$$
$$z = 10$$
Put this value for z into (5):
$$y + 3(10) = 55$$
$$y + 30 = 55$$
$$y = 25$$
Put the value for z into (3):
$$x - 2(10) = 0$$
$$x = 20$$
The company should send 20 crates to Boston, 25 crates to Chicago, and 10 crates to Los Angeles.

23. Graph the line $3x - 4y = 12$ with a dashed line. The inequality can be written as $y > \frac{3}{4}x - 3$, so the region above the line is shaded. Refer to the graph in the back of the textbook.

Chapter 3 Review

24. Graph the line $x = 3y - 6$ with a dashed line. The inequality can be written as $y < \dfrac{x}{3} + 2$, so the region below the line is shaded. Refer to the graph in the back of the textbook.

25. Graph the line $y = -\dfrac{1}{2}$ with a dashed line. Because the inequality is $y < -\dfrac{1}{2}$, the region below the line is shaded. Refer to the graph in the back of the textbook.

26. Graph the line $-4 = x$ with a solid line, and the line $x = 2$ with a dashed line. Because the inequality has x between -4 and 2, the region between the two lines is shaded. Refer to the graph in the back of the textbook.

27. Graph the line $y = 3$ with a dashed line and the line $x = 2$ with a solid line. The region above the line $y = 3$ and the region to the left of $x = 2$ are shaded. The solution set is where the shaded regions overlap. Refer to the graph in the back of the textbook.

28. Graph the line $y = x$ with a solid line and the line $x = 2$ with a dashed line. The region above the line $y = x$ and the region to the right of the line $x = 2$ are shaded. The solution set is where the shaded regions overlap. Refer to the graph in the back of the textbook.

29. Graph the lines $3x - y = 6$ and $x + 2y = 6$ with dashed lines. The first inequality can be written as $y > 3x - 6$, so the region above the line $3x - y = 6$ is shaded. The second inequality can be written as $y > -\dfrac{x}{2} + 3$, so the region above the line $x + 2y = 6$ is shaded. The solution set is where the shaded regions overlap. Refer to the graph in the back of the textbook.

30. Graph the lines $x - 3y = 3$ and $y = x + 2$ with dashed lines. The first inequality can be written as $y < \dfrac{x}{3} - 1$, so the region below the line $x - 3y = 3$ is shaded. Since the second inequality is $y < x + 2$, the region below the line $y = x + 2$ is shaded. The solution set is where the shaded regions overlap. Refer to the graph in the back of the textbook.

31. Graph the lines $3x - 4y = 12$, $x = 0$ (the y-axis), and $y = 0$ (the x-axis). The first inequality can be written as $y \geq \dfrac{3}{4}x - 3$, so the region that is shaded is above the line $3x - 4y = 12$, to the right of the y-axis, and below the x-axis. The vertices are $(0, 0)$, the intersection of the lines $x = 0$ and $y = 0$; $(4, 0)$, the intersection of $y = 0$ and $3x - 4y = 12$; and $(0, -3)$, the intersection of $x = 0$ and $3x - 4y = 12$. Refer to the graph in the back of the textbook.

Chapter 3 Review

32. Graph the lines $x - 2y = 6$, $y = x$, $x = 0$, and $y = 0$. The first inequality can be written as $y \geq \dfrac{x}{2} - 3$, so the solution set is the region above the line $x - 2y = 6$, below the line $y = x$, to the right of the y-axis, and above the x-axis. The vertices are
(0, 0), the intersection of the lines $x = 0$, $y = 0$, and $x = y$ and
(6, 0), the intersection of $y = 0$ and $x - 2y = 6$.
Refer to the graph in the back of the textbook.

33. Graph the lines $x + y = 5$, $y = x$, $y = 2$, and $x = 0$. The first inequality can be written as $y \leq 5 - x$, so the solution set is the region below the line $x + y = 5$, above the line $y = x$, above the line $y = 2$, and to the right of the y-axis. The vertices are
(0, 2), the intersection of the lines $x = 0$ and $y = 2$;
(0, 5), the intersection of $x = 0$ and $x + y = 5$;
$\left(\dfrac{5}{2}, \dfrac{5}{2}\right)$, the intersection of $y = x$ and $x + y = 5$; and
(2, 2), the intersection of $y = x$ and $y = 2$.
Refer to the graph in the back of the textbook.

34. Graph the lines $x - y = -3$, $x + y = 6$, $x = 4$, $x = 0$, and $y = 0$. The first inequality can be written $y \geq x + 3$ and the second inequality can be written as $y \leq 6 - x$, so the solution set is the region above the line $x - y = -3$, below the line $x + y = 6$, to the left of the line $x = 4$, above the x-axis, and to the right of the y-axis. The vertices are
(0, 3), the intersection of the lines $x = 0$ and $x - y = -3$;
(0, 6), the intersection of $x = 0$ and $x + y = 6$; and
$\left(\dfrac{3}{2}, \dfrac{9}{2}\right)$, the intersection of $x - y = -3$ and $x + y = 6$.
Refer to the graph in the back of the textbook.

35. Let p = the number of batches of peanut butter cookies and g = the number of batches of granola cookies that Ruth can make. From the mixing considerations, $20p + 8g \leq 120$, or $5p + 2g \leq 30$, with time in minutes. From the baking considerations, $10p + 10g \leq 120$, or $p + g \leq 12$. Since Ruth cannot make a negative number of either type of cookie, $p \geq 0$ and $g \geq 0$.
The system of inequalities is
$5p + 2g \leq 30$
$p + g \leq 12$
$p \geq 0$, $g \geq 0$
Graph the lines $5p + 2g = 30$, $p + g = 12$, $p = 0$, and $g = 0$, with p along the horizontal axis and g along the vertical axis. The first inequality can be written as $g \leq 15 - \dfrac{5}{2}p$ and the second inequality can be written as $g \leq 12 - p$, so the solution set is the region below the lines $5p + 2g = 30$ and $p + g = 12$, above the p-axis, and to the right of the g-axis.
Refer to the graph in the back of the textbook.

103

Chapter 3 Review

36. Let t = the number of ounces of tofu and b = the number of ounces of brown rice in the recipe. Since the total amount of tofu and brown rice cannot exceed 32 ounces, $t + b \le 32$. To have the protein in the recipe be at least 56 grams, $2t + 1.6b \ge 56$. It makes no sense to have a negative amount of either ingredient, so $t \ge 0$ and $b \ge 0$.
The system of inequalities is
$t + b \le 32$
$2t + 1.6b \ge 56$
$t \ge 0$, $b \ge 0$
Graph the lines $t + b = 32$, $2t + 1.6b = 56$, $t = 0$, and $b = 0$, with t along the horizontal axis and b along the vertical axis. The first inequality can be written as $b \le 32 - t$, and the second inequality can be written as $b \ge 35 - \frac{5}{4}t$, so the solution set is the region below the line $t + b = 32$, above the line $2t + 1.6b = 56$, above the t-axis, and to the right of the b-axis.
Refer to the graph in the back of the textbook.

37. Since each batch contains 50 cookies, the income from selling a batch of peanut butter cookies is $50 \times 0.25 = \$12.50$, while the income from selling a batch of granola cookies is $50 \times 0.20 = \$10.00$. Thus, the constraint to be maximized is $C = 12.5p + 10g$. The vertices of the solution set in Problem 35 are $(0, 0)$, the intersection of $p = 0$ and $g = 0$; $(0, 12)$, the intersection of $p = 0$ and $p + g = 12$; $(2, 10)$, the intersection of $p + g = 12$ and $5p + 2g = 30$; and $(6, 0)$, the intersection of $g = 0$ and $5p + 2g = 30$. At $(0, 0)$,
$C = 12.5(0) + 10(0) = 0$.
At $(0, 12)$,
$C = 12.5(0) + 10(12) = 120$.
At $(2, 10)$,
$C = 12.5(2) + 10(10)$
$= 25 + 100 = 125$.
At $(6, 0)$,
$C = 12.5(6) + 10(0) = 75$.
To maximize her income, Ruth should bake 2 batches of peanut butter cookies and 10 batches of granola cookies.

38. Since tofu costs 12 cents per ounce and brown rice costs 16 cents per ounce, the constraint to be minimized is $C = 12t + 16b$. The vertices of the solution set in Problem 36 are $(12, 20)$, the intersection of $t + b = 32$ and $2t + 1.6b = 56$; $(32, 0)$, the intersection of $b = 0$ and $t + b = 32$; and $(28, 0)$, the intersection of $b = 0$ and $2t + 1.6b = 56$. At $(12, 20)$,
$C = 12(12) + 16(20)$
$= 144 + 320 = 464$.
At $(32, 0)$,
$C = 12(32) + 16(0) = 384$.
At $(28, 0)$,
$C = 12(28) + 16(0) = 336$.
The least expensive mixture for the recipe is 28 ounces of tofu and no brown rice.

39. Using the points $(70, 1)$ and $(54, 3)$,
$m = \dfrac{3-1}{54-70} = \dfrac{2}{-16} = -\dfrac{1}{8}$
$y - 1 = -\dfrac{1}{8}(x - 70)$
$y = -\dfrac{1}{8}x + \dfrac{78}{8}$

Chapter 3 Review

When the temperature is 30°F, $x = 30$:

$$y = -\frac{1}{8}(30) + \frac{78}{8}$$

$$= -\frac{30}{8} + \frac{78}{8}$$

$$= \frac{48}{8} = 6$$

Estimate 6 incidents of O-ring thermal distress when the temperature is 30°F.

40. Using the points (19, 40) and (8, 18), $m = \dfrac{40-18}{19-8} = \dfrac{22}{11} = 2$

$y - 40 = 2(x - 19)$

$y = 2x + 2$

For a 12-page technical report, $x = 12$.
$y = 2(12) + 2 = 24 + 2 = 26$.
Estimate 26 minutes for Thelma to type a 12-page technical report.

41. a. Drawing a line close to the data points gives a line that goes through the point (30, 850). The prediction is that a woman swimmer with body mass of 30 kilograms has a metabolic rate of 850.

  b. The line from a. goes through the point (50, 1450). The prediction is that a woman swimmer with body mass of 50 kilograms has a metabolic rate of 1450.

  c. Using the points (30, 850) and (50, 1450),
  $m = \dfrac{1450 - 850}{50 - 30} = \dfrac{600}{20} = 30$
  $y - 850 = 30(x - 30)$
  $y = 30x - 50$

  d. For a body mass of 45 kilograms, $x = 45$.
  $y = 30(45) - 50$
  $= 1350 - 50 = 1300$
  The prediction is that a woman swimmer with body mass of 45 kilograms has a metabolic rate of 1300.

  e. Put the data into your calculator and use linear regression. The result is $a = 31.1025641$, and $b = -93.90769231$. Thus, the least squares regression line is $y = 31.10x - 93.91$ after rounding to two decimal places.
  Using $x = 45$, as in d.,
  $y = 31.10(45) - 93.91$
  $= 1399.5 - 93.91 = 1305.59$
  This line predicts that the metabolic rate of a woman swimmer of body mass 45 kilograms is 1305.59.

42. a. Drawing a line close to the data points gives a line that goes through the point (40, 45). The prediction is that an archaeopteryx whose femur is 40 cm long will have a humerus 45 cm long.

Chapter 3 Review

   b. Drawing a line close to the data points gives a line that goes through the point (75, 87). The prediction is that an archaeopteryx whose femur is 75 cm long will have a humerus 87 cm long.

   c. Using the points (40, 45) and (75, 87),

$$m = \frac{87-45}{75-40} = \frac{42}{35} = 1.2$$

$$y - 45 = 1.2(x - 40)$$

$$y = 1.2x - 3$$

   d. For a femur length of 60 cm, $x = 60$

$$y = 1.2(60) - 3 = 72 - 3$$

$$= 69$$

The prediction is that an archaeopteryx whose femur is 60 cm long will have a humerus 69 cm long.

   e. Put the data into your calculator and use linear regression. The result is $a = 1.196900115$ and $b = -3.659586682$. Thus, the least squares regression line is $y = 1.197x - 3.660$ after rounding to three decimal places.
Using $x = 60$, as in d.,

$$y = 1.197(60) - 3.660$$

$$= 71.82 - 3.660 = 68.16$$

This line predicts the humerus to be 68.16 cm long on an archaeopteryx whose femur is 60 cm long.

Chapter 4

Section 4.0

1. $3x(x-5) = 3x(x) + 3x(-5) = 3x^2 - 15x$

3. $(b+6)(2b-3) = b(2b) + b(-3) + 6(2b) + 6(-3) = 2b^2 - 3b + 12b - 18$
   $= 2b^2 + 9b - 18$

5. $(4w-3)^2 = (4w-3)(4w-3) = 4w(4w) + 4w(-3) - 3(4w) - 3(-3)$
   $= 16w^2 - 12w - 12w + 9 = 16w^2 - 24w + 9$

7. $3p(2p-5)(p-3) = 3p[2p(p) + 2p(-3) - 5(p) - 5(-3)]$
   $= 3p[2p^2 - 6p - 5p + 15] = 3p[2p^2 - 11p + 15] = 3p[2p^2] + 3p[-11p] + 3p[15]$
   $= 6p^3 - 33p^2 + 45p$

9. $-50(1+r)^2 = -50(1+r)(1+r) = -50[1(1) + 1(r) + r(1) + r(r)]$
   $= -50[1 + r + r + r^2] = -50[1 + 2r + r^2] = -50[1] - 50[2r] - 50[r^2]$
   $= -50 - 100r - 50r^2$

11. $3q^2(2q-3)^2 = 3q^2(2q-3)(2q-3) = 3q^2[2q(2q) + 2q(-3) - 3(2q) - 3(-3)]$
    $= 3q^2[4q^2 - 6q - 6q + 9] = 3q^2[4q^2 - 12q + 9] = 3q^2[4q^2] + 3q^2[-12q] + 3q^2[9]$
    $= 12q^4 - 36q^3 + 27q^2$

13. $x^2 - 7x + 10 = x^2 - 2x - 5x + 10 = x(x-2) - 5(x-2)$
    $= (x-5)(x-2)$

15. $x^2 - 225 = (x)^2 - (15)^2 = (x-15)(x+15)$

17. $w^2 - 4w - 32 = w^2 - 8w + 4w - 32 = w(w-8) + 4(w-8) = (w+4)(w-8)$

19. $2z^2 + 11z - 40 = 2z^2 + 16z - 5z - 40 = 2z(z+8) - 5(z+8) = (2z-5)(z+8)$

21. $9n^2 + 24n + 16 = 9n^2 + 12n + 12n + 16 = 3n(3n+4) + 4(3n+4) = (3n+4)(3n+4)$
    $= (3n+4)^2$

23. $3a^4 + 6a^3 + 3a^2 = 3a^2(a^2 + 2a + 1) = 3a^2(a^2 + a + a + 1)$
    $= 3a^2[a(a+1) + (1)(a+1)] = 3a^2(a+1)(a+1) = 3a^2(a+1)^2$

25. $4h^4 - 36h^2 = 4h^2(h^2 - 9) = 4h^2[(h)^2 - (3)^2] = 4h^2(h-3)(h+3)$

Chapter 4, Section 4.1

27. $-10u^2 - 100u + 390 = -10(u^2 + 10u - 39) = -10(u^2 + 13u - 3u - 39)$
$= -10[u(u+13) - 3(u+13)] = -10(u-3)(u+13)$

29. $24t^4 + 6t^2 = 6t^2(4t^2 + 1)$

Section 4.1

1. a.

Height	Base	Perimeter	Area
1	34	70	34
2	32	68	64
3	30	66	90
4	28	64	112
5	26	62	130
6	24	60	144
7	22	58	154
8	20	56	160
9	18	54	162
10	16	52	160
11	14	50	154
12	12	48	144
13	10	46	130
14	8	44	112
15	6	42	90
16	4	40	64
17	2	38	34

Chapter 4, Section 4.1

b.

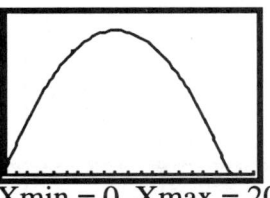

Xmin = 0, Xmax = 20, Ymin = 0, and Ymax = 180.

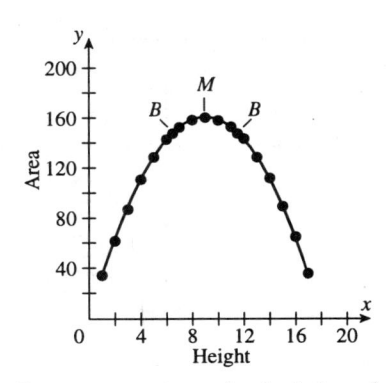

c. Let x represent the height of the region and b the length of the base. Since $b + 2x = 36$, then $b = 36 - 2x$. The perimeter is $2b + 2x = 2(36 - 2x) + 2x = 72 - 4x + 2x = 72 - 2x$. The area of the region is $bx = (36 - 2x)x = 36x - 2x^2$

d. To verify the expressions in c., insert the appropriate value for the height into each expression. For example, when $x = 10$:
For the base: $b = 36 - 2(10) = 36 - 20 = 16$
For the perimeter: $72 - 2(10) = 72 - 20 = 52$
For the area:
$36(10) - 2(10)^2 = 360 - 2(100) = 360 - 200 = 160$
The table agrees with the algebraic expressions.

e. The graphing calculator graph agrees with the graph in b. wherever x is at least 1 and less than or equal to 17.

f. The largest area that can be enclosed is 162 in.^{2} This comes from a rectangle with height 9 in. and base 18 in. This point is labeled M on the graph in part b.

g. If the area of the rectangle is 149.5 in., we can find the height by solving $36h - 2h^2 = 149.5$ for h. The solutions are $h = 6.5$ in. and $h = 11.5$ in. These points are labeled B on the graph in part b.

3. a. A good picture is obtained by using Xmin = –5, Xmax = 5, Ymin = 220, and Ymax = 310.

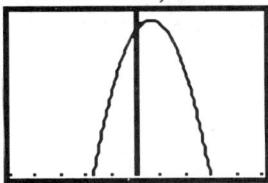

b. The highest altitude reached is about 306.25 feet after about 0.625 seconds.

c. His book passes him after about 1.25 seconds.

d. The book will hit the ground after 5 seconds.

Chapter 4, Section 4.1

5. a.

Number of price increases	Price of room	Number of rooms rented	Total revenue
0	20	60	1200
1	22	57	1254
2	24	54	1296
3	26	51	1326
4	28	48	1344
5	30	45	1350
6	32	42	1344
7	34	39	1326
8	36	36	1296
9	38	33	1254
10	40	30	1200
11	42	27	1134
12	44	24	1056
13	46	21	966
14	48	18	864
15	50	15	750
16	52	12	624
17	54	9	486
18	56	6	336
19	58	3	174
20	60	0	0

b. The price of a room is $20 plus an additional $2 for each price increase or $20 + 2x$. The number of rooms rented is 60 less 3 rooms for each price increase or $60 - 3x$.
Total revenue = (price of a room) × (number of rooms rented)
$= (20 + 2x)(60 - 3x) = 1200 - 60x + 120x - 6x^2 = 1200 + 60x - 6x^2$

c. Use $Y_1 = 20 + 2x$, $Y_2 = 60 - 3x$, and $Y_3 = 1200 + 60x - 6x^2$.

d. The total revenue is 0 when $x = 20$ (after 20 price increases).

e. Use Xmin = 0, Xmax = 20, Ymin = 0, and Ymax = 1355.

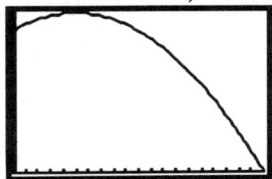

f. The revenue is equal to $1296 when the price is $24 and also at $36 per night, so the lowest price she could charge to have the revenues exceed $1296 is $26 per night, and the highest price she could charge is $34.

g. The maximum revenue in one night is $1350, which would occur if she charged $30 per night. At that price, 45 rooms would be rented each night.

Section 4.2

1. Use Xmin = –5, Xmax = 5, Ymin = –15, and Ymax = 15.

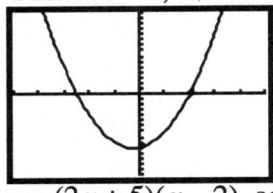

$y = (2x + 5)(x - 2)$, so solve $(2x + 5)(x - 2) = 0$.
$2x + 5 = 0$ or $x - 2 = 0$
$2x = -5$ $x = 2$
$x = \frac{-5}{2}$

The solutions are $x = \frac{-5}{2}$ and $x = 2$.

3. Use Xmin = –5, Xmax = 5, Ymin = –10, and Ymax = 10.

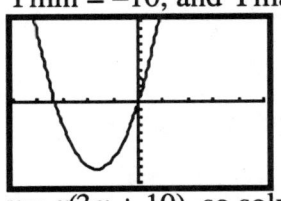

$y = x(3x + 10)$, so solve $x(3x + 10) = 0$.
$x = 0$ or $3x + 10 = 0$
$\phantom{x = 0 \text{ or } }3x = -10$
$\phantom{x = 0 \text{ or } }x = \frac{-10}{3}$

The solutions are $x = 0$ and $x = \frac{-10}{3}$.

Chapter 4, Section 4.2

5. Use Xmin = –5, Xmax = 5, Ymin = –15, and Ymax = 15.

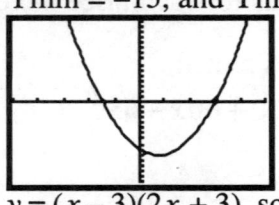

$y = (x-3)(2x+3)$, so solve
$(x-3)(2x+3) = 0$.
$x - 3 = 0$ or $2x + 3 = 0$
$x = 3$ $\qquad 2x = -3$
$\qquad\qquad x = \frac{-3}{2}$

The solutions are $x = 3$ and $x = \frac{-3}{2}$.

7. Use Xmin = –10, Xmax = 10, Ymin = –50 and Ymax = 50.

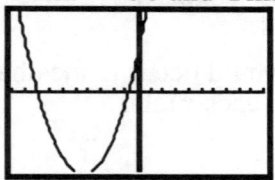

$y = (4x+3)(x+8)$, so solve
$(4x+3)(x+8) = 0$.
$4x + 3 = 0$ or $x + 8 = 0$
$4x = -3$ $\qquad x = -8$
$x = \frac{-3}{4}$

The solutions are $x = \frac{-3}{4}$ and $x = -8$.

9. Use Xmin = 0, Xmax = 8, Ymin = –5, and Ymax = 5.

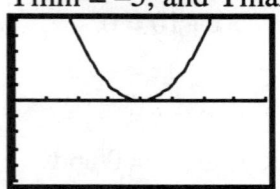

$y = (x-4)^2$, so solve
$(x-4)^2 = 0$
$x - 4 = 0$
$x = 4$

The solution is $x = 4$.

11.
$2a^2 + 5a - 3 = 0$
$2a^2 + 6a - a - 3 = 0$
$2a(a+3) - 1(a+3) = 0$
$(2a-1)(a+3) = 0$
$2a - 1 = 0$ or $a + 3 = 0$
$2a = 1$ $\qquad a = -3$
$a = \frac{1}{2}$

The solutions are $a = -3$ and $a = \frac{1}{2}$.

13.
$2x^2 = 6x$
$2x^2 - 6x = 0$
$2x(x-3) = 0$
$2x = 0$ or $x - 3 = 0$
$x = 0$ $\qquad x = 3$

The solutions are $x = 0$ and $x = 3$.

15.
$3y^2 - 6y = -3$
$3y^2 - 6y + 3 = 0$
$3y^2 - 3y - 3y + 3 = 0$
$3y(y-1) - 3(y-1) = 0$
$(3y-3)(y-1) = 0$
$3(y-1)(y-1) = 0$
$3(y-1)^2 = 0$
$(y-1)^2 = 0$
$y - 1 = 0$
$y = 1$

The solution is $y = 1$.

112

17.
$$x(2x-3) = -1$$
$$2x^2 - 3x = -1$$
$$2x^2 - 3x + 1 = 0$$
$$2x^2 - 2x - x + 1 = 0$$
$$2x(x-1) - 1(x-1) = 0$$
$$(2x-1)(x-1) = 0$$
$$2x - 1 = 0 \quad \text{or} \quad x - 1 = 0$$
$$2x = 1 \qquad \qquad x = 1$$
$$x = \frac{1}{2}$$

The solutions are $x = 1$ and $x = \frac{1}{2}$.

19.
$$t(t-3) = 2(t-3)$$
$$t(t-3) - 2(t-3) = 0$$
$$(t-2)(t-3) = 0$$
$$t - 2 = 0 \quad \text{or} \quad t - 3 = 0$$
$$t = 2 \qquad \qquad t = 3$$

The solutions are $t = 3$ and $t = 2$.

21.
$$z(3z+2) = (z+2)^2$$
$$3z^2 + 2z = z^2 + 4z + 4$$
$$3z^2 + 2z - z^2 - 4z - 4 = 0$$
$$2z^2 - 2z - 4 = 0$$
$$2(z^2 - z - 2) = 0$$
$$2(z^2 - 2z + z - 2) = 0$$
$$2[z(z-2) + 1(z-2)] = 0$$
$$2(z+1)(z-2) = 0$$
$$z + 1 = 0 \quad \text{or} \quad z - 2 = 0$$
$$z = -1 \qquad \qquad z = 2$$

The solutions are $z = 2$ and $z = -1$.

23.
$$(v+2)(v-5) = 8$$
$$v^2 + 2v - 5v - 10 = 8$$
$$v^2 - 3v - 10 = 8$$
$$v^2 - 3v - 18 = 0$$
$$v^2 - 6v + 3v - 18 = 0$$
$$v(v-6) + 3(v-6) = 0$$
$$(v+3)(v-6) = 0$$
$$v + 3 = 0 \quad \text{or} \quad v - 6 = 0$$
$$v = -3 \qquad \qquad v = 6$$

The solutions are $v = 6$ and $v = -3$

25.

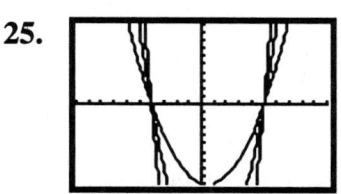

The x-intercepts are the same.

27.

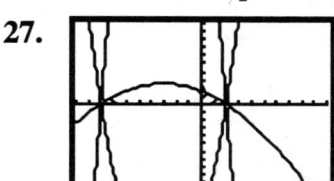

The x-intercepts are the same.

29.
$$[x - (-2)](x - 1) = 0$$
$$(x+2)(x-1) = 0$$
$$x^2 - x + 2x - 2 = 0$$
$$x^2 + x - 2 = 0$$

31.
$$(x - 0)[x - (-5)] = 0$$
$$x(x+5) = 0$$
$$x^2 + 5x = 0$$

33.
$$[x-(-3)]\left(x - \frac{1}{2}\right) = 0$$
$$(x+3)\left(x - \frac{1}{2}\right) = 0$$
$$x^2 - \frac{1}{2}x + 3x - \frac{3}{2} = 0$$
$$x^2 + \frac{5}{2}x - \frac{3}{2} = 0$$
$$2\left(x^2 + \frac{5}{2}x - \frac{3}{2}\right) = 2(0)$$
$$2x^2 + 5x - 3 = 0$$

Chapter 4, Section 4.2

35. $\left[x-\left(\frac{-1}{4}\right)\right]\left(x-\frac{3}{2}\right)=0$
$\left(x+\frac{1}{4}\right)\left(x-\frac{3}{2}\right)=0$
$x^2-\frac{3}{2}x+\frac{1}{4}x-\frac{3}{8}=0$
$x^2-\frac{5}{4}x-\frac{3}{8}=0$
$8\left(x^2-\frac{5}{4}x-\frac{3}{8}\right)=8(0)$
$8x^2-10x-3=0$

37.

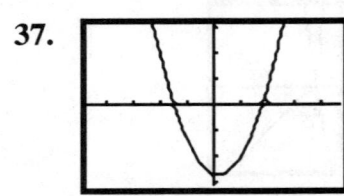

x-intercepts: –15 and 18. So the equation is $y = 0.1(x - 18)(x + 15)$.

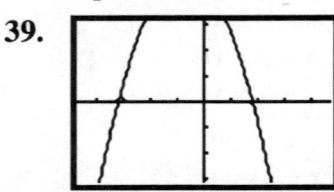

39.

x-intercepts: –32 and 18. So the equation is $y = -0.08(x + 32)(x - 18)$

41. $(x-2)^2=9$
$x-2=\pm\sqrt{9}$
$x-2=\pm 3$
$x-2=3$ or $x-2=-3$
$x=5 \qquad\qquad x=-1$
The solutions are $x = 5$ and $x = -1$.

43. $(2x-1)^2=16$
$2x-1=\pm\sqrt{16}$
$2x-1=\pm 4$
$2x-1=4$ or $2x-1=-4$
$2x=5 \qquad\qquad 2x=-3$
$x=\frac{5}{2} \qquad\qquad x=\frac{-3}{2}$
The solutions are $x=\frac{5}{2}$ and $x=\frac{-3}{2}$.

45. $(x+2)^2=3$
$x+2=\pm\sqrt{3}$
$x=-2\pm\sqrt{3}$
The solutions are $x=-2+\sqrt{3}$ and $x=-2-\sqrt{3}$.

47. $\left(x-\frac{1}{2}\right)^2=\frac{3}{4}$
$x-\frac{1}{2}=\pm\sqrt{\frac{3}{4}}$
$x=\frac{1}{2}\pm\frac{\sqrt{3}}{\sqrt{4}}$
$x=\frac{1}{2}\pm\frac{\sqrt{3}}{2}$
The solutions are $x=\frac{1}{2}+\frac{\sqrt{3}}{2}$ and $x=\frac{1}{2}-\frac{\sqrt{3}}{2}$.

49. $\left(x+\frac{1}{3}\right)^2=\frac{1}{81}$
$x+\frac{1}{3}=\pm\sqrt{\frac{1}{81}}$
$x+\frac{1}{3}=\pm\frac{1}{9}$
$x+\frac{1}{3}=\frac{1}{9}$ or $x+\frac{1}{3}=\frac{-1}{9}$
$x=\frac{-2}{9} \qquad\qquad x=\frac{-4}{9}$
The solutions are $x=\frac{-2}{9}$ and $x=\frac{-4}{9}$.

51. $(8x-7)^2=8$
$8x-7=\pm\sqrt{8}$
$8x=7\pm\sqrt{8}$
$x=\frac{7}{8}\pm\frac{\sqrt{8}}{8}$
The solutions are $x=\frac{7}{8}+\frac{\sqrt{8}}{8}$ and $x=\frac{7}{8}-\frac{\sqrt{8}}{8}$

Chapter 4, Section 4.2

53. a. Use Xmin = 0, Xmax = 10, Ymin = 0, and Ymax = 10.

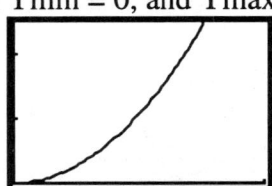

A point (s, A) represents the relation of the length of the sides of an equilateral triangle to the area A.

b. $\frac{\sqrt{3}}{4}s^2 = 12$

$s^2 = 12 \cdot \frac{4}{\sqrt{3}}$

$s^2 \approx 27.713$

$s \approx 5.26$

Each side is approximately 5.26 cm long.

55. a. Let x represent the height of the window. Then the length of the ladder is $x + 2$. By the Pythagorean theorem:

$(x)^2 + (10)^2 = (x+2)^2$

$x^2 + 100 = x^2 + 4x + 4$

b. $x^2 + 100 = x^2 + 4x + 4$

$100 = 4x + 4$

$96 = 4x$

$24 = x$

The window is 24 feet high.

57. a. $h = \frac{-1}{2}(32)t^2 + 16t + 8$

$h = -16t^2 + 16t + 8$

b. When $t = \frac{1}{2}$,

$h = -16\left(\frac{1}{2}\right)^2 + 16\left(\frac{1}{2}\right) + 8$

$h = -16\left(\frac{1}{4}\right) + 8 + 8$

$h = -4 + 16$

$h = 12$

The tennis ball is 12 feet in the air when $t = \frac{1}{2}$ sec.

When $t = 1$,

$h = -16(1)^2 + 16(1) + 8$

$h = -16 + 16 + 8$

$h = 8$

The tennis ball is 8 feet in the air when $t = 1$ sec.

c. To find when the ball is 11 feet in the air, solve:

$11 = -16t^2 + 16t + 8$

$16t^2 - 16t + 3 = 0$

$(4t - 1)(4t - 3) = 0$

$4t - 1 = 0$ or $4t - 3 = 0$

$4t = 1$ $4t = 3$

$t = \frac{1}{4}$ $t = \frac{3}{4}$

The tennis ball is 11 feet in the air after $\frac{1}{4}$ sec and after $\frac{3}{4}$ sec.

d. Use Xmin = 0, Xmax = 5, Ymin = 0, and Ymax = 15.

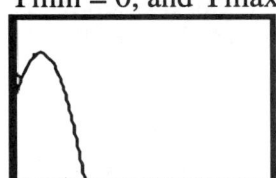

Chapter 4, Section 4.2

59. a. Since there are 360 feet of fence, the perimeter of the pasture is 360. Let l and w represent the length and width of the pasture, respectively. Then $2l + 2w = 360$, or $l + w = 180$.

Width	Length	Area
50	130	6500
55	125	6875
60	120	7200
65	115	7475
70	110	7700
75	105	7875
80	100	8000
85	95	8075
90	90	8100
95	85	8075
100	80	8000
105	75	7875
110	70	7700
115	65	7475
120	60	7200

The pasture will contain 8000 square yards when its dimensions are 80 yards by 100 yards.

b. If $x =$ the width, then from a., $l + x = 180$, $l = 180 - x$, and area $= A = lx = (180 - x)x$
$= 180x - x^2$
Use Xmin = 0, Xmax = 200, Ymin = 0, and Ymax = 8500.

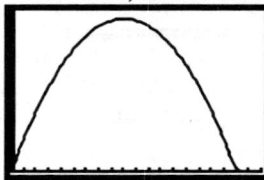

c. From b., we have $A = 180x - x^2$, so we solve $180x - x^2 = 8000$.
$180x - x^2 = 8000$
$0 = x^2 - 180x + 8000$
$0 = (x - 80)(x - 100)$
So the width is either 80 yards or 100 yards. When the width is 80 yards, the length is $180 - 80 = 100$ yards and vice versa, so the pasture has dimensions 80 yards by 100 yards.

61. a. If the cardboard is x inches long, then the sides (length and width) of the box will be $x - 4$ inches long, and the height of the box is 2 inches. $V = lwh$, so
$V = (x - 4)(x - 4)2$
$= 2(x - 4)^2$

b. Use Xmin = 0, Xmax = 10, Ymin = 0, and Ymax = 100.

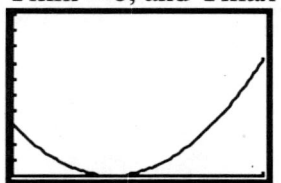

V increases as x increases for $x > 4$.

c. If $V = 50$ in.^{3}, solve
$$2(x-4)^2 = 50$$
$$(x-4)^2 = 25$$
$$x - 4 = \pm\sqrt{25}$$
$$x - 4 = \pm 5$$
$$x - 4 = 5 \quad \text{or} \quad x - 4 = -5$$
$$x = 9 \qquad\qquad x = -1$$
Since the length can never be negative, the piece of cardboard must be 9 inches on a side.

63. a. The size of the group $= 20 + x$; the price per person $= 600 - 10x$.

b. The total income is the size of the group times the price per person, or
$$(20 + x)(600 - 10x)$$
$$= 12{,}000 + 400x - 10x^2.$$

c. When 25 people go on the trip, it means that $x = 5$, so
$$12{,}000 + 400(5) - 10(5)^2$$
$$= 12{,}000 + 2000 - 250$$
$$= 13{,}750$$
The income is $13,750.
When 30 people go on the trip, $x = 10$, so
$$12{,}000 + 400(10) - 10(10)^2$$
$$= 12{,}000 + 4000 - 1000$$
$$= 15{,}000$$
and the income is $15,000.

d. Let the income be $15,750, and solve
$$12{,}000 + 400x - 10x^2 = 15750$$
$$0 = 10x^2 - 400x + 3750$$
$$0 = 10(x^2 - 40x + 375)$$
$$0 = 10(x - 15)(x - 25)$$
$$x - 15 = 0 \quad \text{or} \quad x - 25 = 0$$
$$x = 15 \qquad\qquad x = 25$$

If $x = 15$, it means that $20 + 15 = 35$ people go on the trip; if $x = 25$, it means that $20 + 25 = 45$ people go on the trip.

e. Use Xmin $= 0$, Xmax $= 50$, Ymin $= 0$, and Ymax $= 16{,}000$.

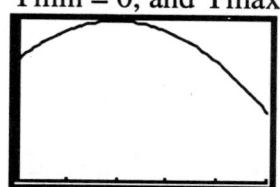

Section 4.3

1. In $x^2 + 8x$, since one-half of 8 is 4, add $4^2 = 16$ to get
$$x^2 + 8x + 16 = (x + 4)^2.$$

3. In $x^2 - 7x$, since one-half of -7 is $-\frac{7}{2}$, add $\left(\frac{-7}{2}\right)^2 = \frac{49}{4}$ to get
$$x^2 - 7x + \frac{49}{4} = \left(x - \frac{7}{2}\right)^2.$$

5. In $x^2 + \frac{3}{2}x$, since one-half of $\frac{3}{2}$ is $\frac{3}{4}$, add $\left(\frac{3}{4}\right)^2 = \frac{9}{16}$ to get
$$x^2 + \frac{3}{2}x + \frac{9}{16} = \left(x + \frac{3}{4}\right)^2.$$

7. In $x^2 - \frac{4}{5}x$, since one-half of $-\frac{4}{5}$ is $-\frac{2}{5}$, add $\left(-\frac{2}{5}\right)^2 = \frac{4}{25}$ to get
$$x^2 - \frac{4}{5}x + \frac{4}{25} = \left(x - \frac{2}{5}\right)^2.$$

Chapter 4, Section 4.3

9. Note that $1 = (-1)^2 = \left[\frac{1}{2}(-2)\right]^2$, so this is already a perfect square.
$$x^2 - 2x + 1 = 0$$
$$(x-1)^2 = 0$$
$$x - 1 = 0$$
$$x = 1$$

11. $x^2 + 9x + 20 = 0$
$$x^2 + 9x = -20$$
One-half of 9 is $\frac{9}{2}$, so add $\left(\frac{9}{2}\right)^2 = \frac{81}{4}$ to both sides to get
$$x^2 + 9x + \frac{81}{4} = -20 + \frac{81}{4}$$
$$\left(x + \frac{9}{2}\right)^2 = \frac{1}{4}$$
$$x + \frac{9}{2} = \pm\sqrt{\frac{1}{4}}$$
$$x + \frac{9}{2} = \pm\frac{1}{2}$$
$$x + \frac{9}{2} = \frac{1}{2} \quad \text{or} \quad x + \frac{9}{2} = -\frac{1}{2}$$
$$x = -\frac{8}{2} \qquad\qquad x = -\frac{10}{2}$$
$$x = -4 \qquad\qquad x = -5$$
The solutions are $x = -4$ and $x = -5$.

13. $x^2 = 3 - 3x$
$$x^2 + 3x = 3$$
Since one-half of 3 is $\frac{3}{2}$, add $\left(\frac{3}{2}\right)^2 = \frac{9}{4}$ to both sides.

$$x^2 + 3x + \frac{9}{4} = 3 + \frac{9}{4}$$
$$\left(x + \frac{3}{2}\right)^2 = \frac{21}{4}$$
$$x + \frac{3}{2} = \pm\sqrt{\frac{21}{4}}$$
$$x + \frac{3}{2} = \pm\frac{\sqrt{21}}{2}$$
$$x = -\frac{3}{2} \pm \frac{\sqrt{21}}{2}$$
The solutions are $x = -\frac{3}{2} + \frac{\sqrt{21}}{2}$ and $x = -\frac{3}{2} - \frac{\sqrt{21}}{2}$.

15. $2x^2 + 4x - 3 = 0$
$$\frac{1}{2}\left(2x^2 + 4x - 3\right) = \frac{1}{2}(0)$$
$$x^2 + 2x - \frac{3}{2} = 0$$
$$x^2 + 2x = \frac{3}{2}$$
Since one-half of 2 is 1, add $1^2 = 1$ to both sides
$$x^2 + 2x + 1 = \frac{3}{2} + 1$$
$$(x+1)^2 = \frac{5}{2}$$
$$x + 1 = \pm\sqrt{\frac{5}{2}}$$
$$x = -1 \pm \sqrt{\frac{5}{2}}$$
The solutions are $x = -1 + \sqrt{\frac{5}{2}}$ and $x = -1 - \sqrt{\frac{5}{2}}$.

Chapter 4, Section 4.3

17.
$$4x^2 - 3 = 2x$$
$$4x^2 - 2x - 3 = 0$$
$$\tfrac{1}{4}(4x^2 - 2x - 3) = \tfrac{1}{4}(0)$$
$$x^2 - \tfrac{1}{2}x - \tfrac{3}{4} = 0$$
$$x^2 - \tfrac{1}{2}x = \tfrac{3}{4}$$

Since one-half of $-\tfrac{1}{2}$ is $-\tfrac{1}{4}$, add $\left(-\tfrac{1}{4}\right)^2 = \tfrac{1}{16}$ to both sides.

$$x^2 - \tfrac{1}{2}x + \tfrac{1}{16} = \tfrac{3}{4} + \tfrac{1}{16}$$
$$\left(x - \tfrac{1}{4}\right)^2 = \tfrac{13}{16}$$
$$x - \tfrac{1}{4} = \pm\sqrt{\tfrac{13}{16}}$$
$$x - \tfrac{1}{4} = \pm\tfrac{\sqrt{13}}{4}$$
$$x = \tfrac{1}{4} \pm \tfrac{\sqrt{13}}{4}$$

The solutions are $x = \tfrac{1}{4} + \tfrac{\sqrt{13}}{4}$ and $x = \tfrac{1}{4} - \tfrac{\sqrt{13}}{4}$.

19.
$$3x^2 - x - 4 = 0$$
$$\tfrac{1}{3}(3x^2 - x - 4) = \tfrac{1}{3}(0)$$
$$x^2 - \tfrac{1}{3}x - \tfrac{4}{3} = 0$$
$$x^2 - \tfrac{1}{3}x = \tfrac{4}{3}$$

Since one-half of $-\tfrac{1}{3}$ is $-\tfrac{1}{6}$, add $\left(-\tfrac{1}{6}\right)^2 = \tfrac{1}{36}$ to both sides.

$$x^2 - \tfrac{1}{3}x + \tfrac{1}{36} = \tfrac{4}{3} + \tfrac{1}{36}$$
$$\left(x - \tfrac{1}{6}\right)^2 = \tfrac{49}{36}$$
$$x - \tfrac{1}{6} = \pm\sqrt{\tfrac{49}{36}}$$
$$x - \tfrac{1}{6} = \pm\tfrac{7}{6}$$
$$x - \tfrac{1}{6} = \tfrac{7}{6} \quad \text{or} \quad x - \tfrac{1}{6} = \tfrac{-7}{6}$$
$$x = \tfrac{4}{3} \qquad\qquad x = -1$$

The solutions are $x = \tfrac{4}{3}$ and $x = -1$.

21. In $x^2 - x - 1 = 0$, $a = 1$, $b = -1$, and $c = -1$, so

$$x = \frac{-(-1) \pm \sqrt{(-1)^2 - 4(1)(-1)}}{2(1)}$$
$$= \frac{1 \pm \sqrt{1+4}}{2}$$
$$= \frac{1 \pm \sqrt{5}}{2}$$
$$= \tfrac{1}{2} \pm \tfrac{\sqrt{5}}{2}$$

The solutions are

$$x = \tfrac{1}{2} + \tfrac{\sqrt{5}}{2} \approx 1.618 \text{ and}$$
$$x = \tfrac{1}{2} - \tfrac{\sqrt{5}}{2} \approx -0.618.$$

Chapter 4, Section 4.3

23. $y^2 + 2y = 5$
$y^2 + 2y - 5 = 0$
Thus, $a = 1$, $b = 2$, and $c = -5$, so

$$x = \frac{-2 \pm \sqrt{(2)^2 - 4(1)(-5)}}{2(1)}$$
$$= \frac{-2 \pm \sqrt{4 + 20}}{2}$$
$$= \frac{-2 \pm \sqrt{24}}{2}$$
$$= \frac{-2}{2} \pm \frac{\sqrt{24}}{2}$$
$$= -1 \pm \frac{\sqrt{24}}{2}$$

The solutions are $x = -1 + \frac{\sqrt{24}}{2} \approx 1.449$ and $x = -1 - \frac{\sqrt{24}}{2} \approx -3.449$.

25. $3z^2 = 4.2z + 1.5$
$3z^2 - 4.2z - 1.5 = 0$
Thus, $a = 3$, $b = -4.2$, and $c = -1.5$.

$$z = \frac{-(-4.2) \pm \sqrt{(-4.2)^2 - 4(3)(-1.5)}}{2(3)}$$
$$= \frac{4.2 \pm \sqrt{17.64 + 18}}{6}$$
$$= \frac{4.2 \pm \sqrt{35.64}}{6}$$

The solutions are $z = \frac{4.2 + \sqrt{35.64}}{6} \approx 1.695$ and $z = \frac{4.2 - \sqrt{35.64}}{6} \approx -0.295$.

27. In $0 = x^2 - \frac{5}{3}x + \frac{1}{3}$, $a = 1$, $b = \frac{-5}{3}$, and $c = \frac{1}{3}$, so

$$x = \frac{-\left(\frac{-5}{3}\right) \pm \sqrt{\left(\frac{-5}{3}\right)^2 - 4(1)\left(\frac{1}{3}\right)}}{2(1)}$$

$$= \frac{\frac{5}{3} \pm \sqrt{\frac{25}{9} - \frac{4}{3}}}{2}$$

$$= \frac{\frac{5}{3} \pm \sqrt{\frac{13}{9}}}{2}$$

$$= \frac{\frac{5}{3} \pm \frac{\sqrt{13}}{3}}{2}$$

$$= \frac{\frac{1}{3}(5 \pm \sqrt{13})}{2}$$

$$= \frac{5 \pm \sqrt{13}}{6}$$

The solutions are $x = \frac{5 + \sqrt{13}}{6} \approx 1.434$ and $x = \frac{5 - \sqrt{13}}{6} \approx 0.232$

29. In $-5.2z^2 + 176z + 1218 = 0$, $a = -5.2$, $b = 176$, and $c = 1218$, so

$$z = \frac{-176 \pm \sqrt{(176)^2 - 4(-5.2)(1218)}}{2(-5.2)}$$

$$= \frac{-176 \pm \sqrt{30,976 + 25,334.4}}{-10.4}$$

$$= \frac{-176 \pm \sqrt{56,310.4}}{-10.4}$$

The solutions are $z = \frac{-176 + \sqrt{56,310.4}}{-10.4} \approx -5.894$ and

$z = \frac{-176 - \sqrt{56,310.4}}{-10.4} \approx 39.740$.

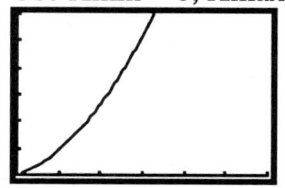

31. a. Use Xmin = 0, Xmax = 60, Ymin = 0, and Ymax = 60.

Chapter 4, Section 4.3

b. For the car to stop in 50 feet, solve $50 = \frac{s^2}{24} + \frac{s}{2}$ for s.

$$50 = \frac{s^2}{24} + \frac{s}{2}$$
$$24(50) = 24\left(\frac{s^2}{24} + \frac{s}{2}\right)$$
$$1200 = s^2 + 12s$$
$$0 = s^2 + 12s - 1200$$

So $a = 1$, $b = 12$, and $c = -1200$.

$$s = \frac{-12 \pm \sqrt{(12)^2 - 4(1)(-1200)}}{2(1)}$$
$$= \frac{-12 \pm \sqrt{144 + 4800}}{2}$$
$$= \frac{-12 \pm \sqrt{4944}}{2}$$

Since the answer is the speed of a car, use the positive square root (a negative speed makes no sense). $s = \frac{-12 + \sqrt{4944}}{2} \approx 29.157$ mph.

33. a. To find when her altitude is 1000 feet, solve $1000 = -16t^2 - 16t + 11{,}000$ or $16t^2 + 16t - 10{,}000 = 0$. So $a = 16$, $b = 16$, and $c = -10{,}000$.

$$t = \frac{-16 \pm \sqrt{(16)^2 - 4(16)(-10{,}000)}}{2(16)}$$
$$= \frac{-16 \pm \sqrt{256 + 640{,}000}}{32}$$
$$= \frac{-16 \pm \sqrt{640{,}256}}{32}$$

Since the answer must be positive, use the positive square root. Thus,

$$t = \frac{-16 + \sqrt{640{,}256}}{32} \approx 24.5 \text{ sec.}$$

Chapter 4, Section 4.3

b. The altitude of the marker is still given by the equation in the textbook, so solve $0 = -16t^2 - 16t + 11{,}000$. Here $a = -16$, $b = -16$, and $c = 11{,}000$.

$$t = \frac{-(-16) \pm \sqrt{(-16)^2 - 4(-16)(11{,}000)}}{2(-16)}$$

$$= \frac{16 \pm \sqrt{256 + 704{,}000}}{-32}$$

$$= \frac{16 \pm \sqrt{704{,}256}}{-32}$$

$$= \frac{-16 \pm \sqrt{704{,}256}}{32}$$

Again, t must be positive, so $t = \dfrac{-16 + \sqrt{704{,}256}}{32} \approx 25.7$ sec. This is the time from when the marker and the skydiver leave the plane, so the time it takes the marker to reach the ground from when the skydiver releases it is $25.7 - 24.5 = 1.2$ seconds.

c. Use Xmin = 0, Xmax = 30, Ymin = 0, and Ymax = 11,000.

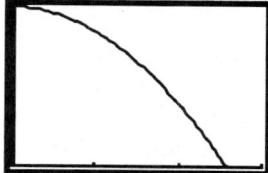

35. $A = P(1+r)^n$ We are given that $P = 5000$, $n = 2$, and that Cyril wants A to equal 6250.

$$6250 = 5000(1+r)^2$$
$$1.25 = (1+r)^2$$
$$\pm\sqrt{1.25} = 1 + r$$
$$-1 \pm \sqrt{1.25} = r$$

The interest rate must be positive, so $r = -1 + \sqrt{1.25} \approx 0.118$ or 11.8%.

37. Let w represent the width of a pen and l the length of the enclosure in feet, as shown below.

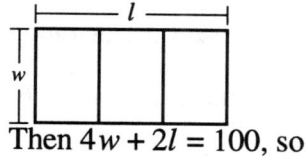

Then $4w + 2l = 100$, so $l = \dfrac{100 - 4w}{2} = 50 - 2w$. The area enclosed is

$A = wl = w(50 - 2w) = 50w - 2w^2$. The area is 250 feet, so

$$50w - 2w^2 = 250$$
$$0 = 2w^2 - 50w + 250$$
$$\tfrac{1}{2}(0) = \tfrac{1}{2}(2w^2 - 50w + 250)$$
$$0 = w^2 - 25w + 125$$

123

Chapter 4, Section 4.3

Thus $a = 1$, $b = -25$, and $c = 125$.
$$w = \frac{-(-25) \pm \sqrt{(-25)^2 - 4(1)(125)}}{2(1)}$$
$$= \frac{25 \pm \sqrt{625 - 500}}{2}$$
$$= \frac{25 \pm \sqrt{125}}{2}$$

The solutions are
$w = \frac{25 + \sqrt{125}}{2} \approx 18.09$ feet and
$w = \frac{25 - \sqrt{125}}{2} \approx 6.91$ feet. When $w = 18.09$ feet, $l = 50 - 2(18.09) = 50 - 36.18 = 13.82$. The length of each pen is one-third the length of the whole enclosure, so the length of one pen is $\frac{13.82}{3} \approx 4.61$ feet and the dimensions of each pen are 18.09 feet by 4.61 feet.
When $w = 6.91$ feet, $l = 50 - 2(6.91) = 50 - 13.82 = 36.18$ feet, so the length of each pen is
$\frac{36.18}{3} = 12.06$ feet, and the dimensions of each pen are 6.91 feet by 12.06 feet.

39. a. Let x represent the distance that you can see from the top of a building that is h miles tall, as shown below.

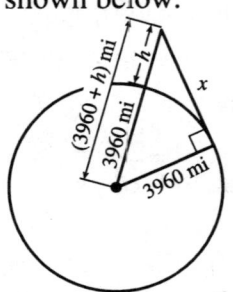

By the Pythagorean theorem,
$x^2 + (3960)^2 = (3960 + h)^2$.
Since there are 5280 feet in a mile, the World Trade Center is $\frac{1350}{5280} \approx 0.2557$ miles tall. From above:
$$x^2 + (3960)^2 = (3960.2557)^2$$
$$x^2 + 15{,}681{,}600 = 15{,}683{,}625$$
$$x^2 = 2025$$
$$x = \pm 45$$
So you can see for 45 miles on a clear day.

b. To see for 100 miles, solve for h in
$$(100)^2 + (3960)^2 = (h + 3960)^2$$
$$15{,}691{,}600 = (h + 3960)^2$$
$$\pm\sqrt{15{,}691{,}600} = h + 3960$$
$$\pm 3961.2624 = h + 3960$$
$$-3960 \pm 3961.2624 = h$$
Since the height of the building must be positive, use the positive square root and get that
$h = -3960 + 3961.2624 = 1.2624$, so the building must be 1.26 miles tall.

41. $x^2 + bx + c = 0$
$x^2 + bx = -c$

Since one-half of b is $\frac{b}{2}$, add $\left(\frac{b}{2}\right)^2 = \frac{b^2}{4}$ to both sides.

$$x^2 + bx + \frac{b^2}{4} = -c + \frac{b^2}{4}$$
$$\left(x + \frac{b}{2}\right)^2 = \frac{-4c + b^2}{4}$$
$$x + \frac{b}{2} = \pm\sqrt{\frac{-4c + b^2}{4}}$$
$$x = \frac{-b}{2} \pm \frac{\sqrt{b^2 - 4c}}{2}$$
$$x = \frac{-b \pm \sqrt{b^2 - 4c}}{2}$$

43. If the quadratic equation is given by $ax^2 + bx + c = 0$, then the solutions are $x_1 = \frac{-b + \sqrt{b^2 - 4ac}}{2a}$ and $x_2 = \frac{-b - \sqrt{b^2 - 4ac}}{2a}$ which we add together to get
$$\frac{-b + \sqrt{b^2 - 4ac}}{2a} + \frac{-b - \sqrt{b^2 - 4ac}}{2a}$$
$$= \frac{-b + \sqrt{b^2 - 4ac} - b - \sqrt{b^2 - 4ac}}{2a}$$
$$= \frac{-2b}{2a}$$
$$= \frac{-b}{a}$$

Section 4.4

1. Refer to the graph in the back of the textbook. The parabola will open upward and be narrower than the standard parabola.

3. Refer to the graph in the back of the textbook. The parabola will open upward and be wider than the standard parabola.

5. Refer to the graph in the back of the textbook. The parabola will open downward, but will be no wider or narrower than the standard parabola.

7. Refer to the graph in the back of the textbook. The parabola will open downward and be wider than the standard parabola.

9. Refer to the graph in the back of the textbook. This is the standard parabola shifted 2 units upward. The vertex is at (0, 2). There are no x-intercepts.

11. Refer to the graph in the back of the textbook. This is the standard parabola shifted 1 unit downward. The vertex is at (0, −1). The x-intercepts are at (1, 0) and (−1, 0).

13. Refer to the graph in the back of the textbook. This is the standard parabola shifted 5 units downward. The vertex is at (0, −5). The x-intercepts are at $(\sqrt{5}, 0)$ and $(-\sqrt{5}, 0)$.

15. Refer to the graph in the back of the textbook. This is the standard parabola reflected about the x-axis and shifted 100 units upward. The vertex is at (0, 100). The x-intercepts are at (10, 0) and (−10, 0).

17. Refer to the graph in the back of the textbook. In $y = x^2 - 4x$, $a = 1$, $b = -4$, and $c = 0$. The vertex is where $x = \frac{-b}{2a}$, so in this case, $x = \frac{-(-4)}{2(1)} = \frac{4}{2} = 2$. When $x = 2$, $y = (2)^2 - 4(2) = 4 - 8 = -4$, so the vertex is at (2, −4). The x-intercepts occur when $y = 0$, so solve $x^2 - 4x = 0$ which is $x(x - 4) = 0$.

Chapter 4, Section 4.4

$x = 0$ or $x - 4 = 0$
$x = 4$
The x-intercepts are (0, 0) and (4, 0).

19. Refer to the graph in the back of the textbook. In $y = x^2 + 2x$, $a = 1$, $b = 2$, and $c = 0$. The vertex is where $x = \frac{-2}{2(1)} = -\frac{2}{2} = -1$. When $x = -1$,
$y = (-1)^2 + 2(-1) = 1 - 2 = -1$, so the vertex is at (–1, –1). For the x-intercepts, solve $0 = x^2 + 2x$ which is $0 = x(x + 2)$.
$x = 0$ or $x + 2 = 0$
$x = -2$
The x-intercepts are at (0, 0) and (–2, 0).

21. Refer to the graph in the back of the textbook. In $y = 3x^2 + 6x$, $a = 3$, $b = 6$, and $c = 0$. The vertex is where $x = \frac{-6}{2(3)} = \frac{-6}{6} = -1$. When $x = -1$,
$y = 3(-1)^2 + 6(-1) = 3 - 6 = -3$, so the vertex is at (–1, –3). For the x-intercepts, solve $3x^2 + 6x = 0$ which is $3x(x + 2) = 0$.
$3x = 0$ or $x + 2 = 0$
$x = 0$ $\qquad$ $x = -2$
The x-intercepts are at (0, 0) and (–2, 0).

23. Refer to the graph in the back of the textbook. In $y = -2x^2 + 5x$, $a = -2$, $b = 5$, and $c = 0$. The vertex is where $x = \frac{-5}{2(-2)} = \frac{-5}{-4} = \frac{5}{4}$. When $x = \frac{5}{4}$,

$y = -2\left(\frac{5}{4}\right)^2 + 5\left(\frac{5}{4}\right)$
$= -2\left(\frac{25}{16}\right) + \frac{25}{4}$
$= \frac{-25}{8} + \frac{25}{4}$
$= \frac{25}{8}$

so the vertex is at $\left(\frac{5}{4}, \frac{25}{8}\right)$. For the x-intercepts, solve $0 = -2x^2 + 5x$ which is $0 = x(-2x + 5)$.
$x = 0$ or $-2x + 5 = 0$
$-2x = -5$
$x = \frac{5}{2}$
The x-intercepts are at (0, 0) and $\left(\frac{5}{2}, 0\right)$.

25. In $y = 3x^2 - 6x + 4$, $a = 3$, $b = -6$, and $c = 4$, so the vertex is where $x = \frac{-(-6)}{2(3)} = \frac{6}{6} = 1$. When $x = 1$,
$y = 3(1)^2 - 6(1) + 4$
$= 3 - 6 + 4$
$= 1$
so the vertex is at (1, 1).

27. In $y = 2 + 3x - x^2$, $a = -1$, $b = 3$, and $c = 2$. The vertex is where $x = \frac{-3}{2(-1)} = \frac{-3}{-2} = \frac{3}{2}$. When $x = \frac{3}{2}$,
$y = 2 + 3\left(\frac{3}{2}\right) - \left(\frac{3}{2}\right)^2$
$= 2 + \frac{9}{2} - \frac{9}{4}$
$= \frac{8}{4} + \frac{18}{4} - \frac{9}{4}$
$= \frac{17}{4}$
so the vertex is at $\left(\frac{3}{2}, \frac{17}{4}\right)$.

Chapter 4, Section 4.4

29. In $y = \frac{1}{2}x^2 - \frac{2}{3}x + \frac{1}{3}$, $a = \frac{1}{2}$, $b = \frac{-2}{3}$, and $c = \frac{1}{3}$. The vertex is where $x = \frac{-\left(\frac{-2}{3}\right)}{2\left(\frac{1}{2}\right)} = \frac{\frac{2}{3}}{1} = \frac{2}{3}$. When $x = \frac{2}{3}$,

$$y = \frac{1}{2}\left(\frac{2}{3}\right)^2 - \frac{2}{3}\left(\frac{2}{3}\right) + \frac{1}{3}$$
$$= \frac{1}{2}\left(\frac{4}{9}\right) - \frac{4}{9} + \frac{1}{3}$$
$$= \frac{2}{9} - \frac{4}{9} + \frac{3}{9}$$
$$= \frac{1}{9}$$

so the vertex is at $\left(\frac{2}{3}, \frac{1}{9}\right)$.

31. In $y = 2.3 - 7.2x - 0.8x^2$, $a = -0.8$, $b = -7.2$, and $c = 2.3$. The vertex is where $x = \frac{-(-7.2)}{2(-0.8)} = \frac{7.2}{-1.6} = -4.5$.
When $x = -4.5$,
$y = 2.3 - 7.2(-4.5) - 0.8(4.5)^2$
$= 2.3 + 32.4 - 16.2$
$= 18.5$
so the vertex is at $(-4.5, 18.5)$.

33. Refer to the graph in the back of the textbook. In $y = x^2 - 5x + 4$, $a = 1$, $b = -5$, and $c = 4$. The vertex is where $x = \frac{-(-5)}{2(1)} = \frac{5}{2}$. When $x = \frac{5}{2}$,

$$y = \left(\frac{5}{2}\right)^2 - 5\left(\frac{5}{2}\right) + 4$$
$$= \frac{25}{4} - \frac{25}{2} + 4$$
$$= \frac{25}{4} - \frac{50}{4} + \frac{16}{4}$$
$$= \frac{-9}{4}$$

so the vertex is at $\left(\frac{5}{2}, \frac{-9}{4}\right)$. The y-intercept is at
$y = (0)^2 - 5(0) + 4 = 4$ or $(0, 4)$.
For the x-intercepts, solve
$0 = x^2 - 5x + 4$
$= (x - 4)(x - 1)$
$x - 4 = 0$ or $x - 1 = 0$
$x = 4$ $x = 1$
The x-intercepts are at $(4, 0)$, and $(1, 0)$.

35. Refer to the graph in the back of the textbook. In $y = -2x^2 + 7x + 4$, $a = -2$, $b = 7$, and $c = 4$. The vertex is where $x = \frac{-7}{2(-2)} = \frac{-7}{-4} = \frac{7}{4}$.
When $x = \frac{7}{4}$,

$$y = -2\left(\frac{7}{4}\right)^2 + 7\left(\frac{7}{4}\right) + 4$$
$$= -2\left(\frac{49}{16}\right) + \frac{49}{4} + 4$$
$$= \frac{-49}{8} + \frac{49}{4} + 4$$
$$= \frac{-49}{8} + \frac{98}{8} + \frac{32}{8}$$
$$= \frac{81}{8}$$

so the vertex is at $\left(\frac{7}{4}, \frac{81}{8}\right)$. The y-intercept is at $(0, 4)$. (Note that the y-intercept is always at $(0, c)$.) For the x-intercepts, solve
$0 = -2x^2 + 7x + 4$ or
$2x^2 - 7x - 4 = 0$, which is
$(2x + 1)(x - 4) = 0$.
$2x + 1 = 0$ or $x - 4 = 0$
$2x = -1$ $x = 4$
$x = \frac{-1}{2}$

127

Chapter 4, Section 4.4

The x-intercepts are at $\left(\frac{-1}{2}, 0\right)$ and $(4, 0)$.

37. In $y = 0.6x^2 + 0.6x - 1.2$, $a = 0.6$, $b = 0.6$, and $c = -1.2$. The vertex is where $x = \frac{-0.6}{2(0.6)} = \frac{-1}{2} = -0.5$.
When $x = -0.5$,
$$y = 0.6(-0.5)^2 + 0.6(-0.5) - 1.2$$
$$= 0.6(0.25) - 0.3 - 1.2$$
$$= 0.15 - 1.5$$
$$= -1.35$$
so the vertex is at $(-0.5, -1.35)$. The y-intercept is at $(0, -1.2)$, while to find the x-intercepts, solve
$$0 = 0.6x^2 + 0.6x - 1.2$$
$$0 = 0.6(x^2 + x - 2)$$
$$0 = 0.6(x - 1)(x + 2)$$
$$x - 1 = 0 \quad \text{or} \quad x + 2 = 0$$
$$x = 1 \qquad\qquad x = -2$$
The x-intercepts are at $(1, 0)$ and $(-2, 0)$.

39. Refer to the graph in the back of the textbook. In $y = x^2 + 4x + 7$, $a = 1$, $b = 4$, and $c = 7$. The vertex is where $x = \frac{-4}{2(1)} = -2$. When $x = -2$,
$$y = (-2)^2 + 4(-2) + 7$$
$$= 4 - 8 + 7$$
$$= 3$$
so the vertex is at $(-2, 3)$. The y-intercept is at $(0, 7)$. For the x-intercepts, solve $0 = x^2 + 4x + 7$. Since this does not obviously factor "nicely", use the quadratic formula, with a, b, and c as above.

$$x = \frac{-4 \pm \sqrt{(4)^2 - 4(1)(7)}}{2(1)}$$
$$= \frac{-4 \pm \sqrt{16 - 28}}{2}$$
$$= \frac{-4 \pm \sqrt{-12}}{2}$$

Since there is a negative number under the radical, there are no x-intercepts.

41. Refer to the graph in the back of the textbook. In $y = x^2 + 2x - 1$, $a = 1$, $b = 2$, and $c = -1$. The vertex is where $x = \frac{-2}{2(1)} = -1$. When $x = -1$,
$$y = (-1)^2 + 2(-1) - 1$$
$$= 1 - 2 - 1$$
$$= -2$$
so the vertex is at $(-1, -2)$. The y-intercept is at $(0, -1)$. For the x-intercepts, solve $0 = x^2 + 2x - 1$ by using the quadratic formula.

$$x = \frac{-2 \pm \sqrt{(2)^2 - 4(1)(-1)}}{2(1)}$$
$$= \frac{-2 \pm \sqrt{4 + 4}}{2}$$
$$= \frac{-2 \pm \sqrt{8}}{2}$$
$$= -1 \pm \frac{\sqrt{8}}{2}$$
$$= -1 \pm \frac{2\sqrt{2}}{2}$$
$$= -1 \pm \sqrt{2}$$

The x-intercepts are at $\left(-1 + \sqrt{2}, 0\right) \approx (0.41, 0)$ and $\left(-1 - \sqrt{2}, 0\right) \approx (-2.41, 0)$.

Chapter 4, Section 4.4

43. Refer to the graph in the back of the textbook. In $y = -2x^2 + 6x - 3$, $a = -2$, $b = 6$, and $c = -3$. The vertex is where $x = \dfrac{-6}{2(-2)} = \dfrac{-6}{-4} = \dfrac{3}{2}$.

When $x = \dfrac{3}{2}$,

$$y = -2\left(\dfrac{3}{2}\right)^2 + 6\left(\dfrac{3}{2}\right) - 3$$
$$= -2\left(\dfrac{9}{4}\right) + 9 - 3$$
$$= \dfrac{-9}{2} + 6$$
$$= \dfrac{3}{2}$$

so the vertex is at $\left(\dfrac{3}{2}, \dfrac{3}{2}\right)$. The y-intercept is at $(0, -3)$, while for the x-intercepts, solve $0 = -2x^2 + 6x - 3$ by using the quadratic formula.

$$x = \dfrac{-6 \pm \sqrt{(6)^2 - 4(-2)(-3)}}{2(-2)}$$
$$= \dfrac{-6 \pm \sqrt{36 - 24}}{-4}$$
$$= \dfrac{-6 \pm \sqrt{12}}{-4}$$
$$= \dfrac{-6 \pm 2\sqrt{3}}{-4}$$
$$= \dfrac{6 \pm 2\sqrt{3}}{4}$$
$$= \dfrac{3}{2} \pm \dfrac{\sqrt{3}}{2}$$

The x-intercepts are at $\left(\dfrac{3}{2} - \dfrac{\sqrt{3}}{2}, 0\right) \approx (0.63, 0)$ and $\left(\dfrac{3}{2} + \dfrac{\sqrt{3}}{2}, 0\right) \approx (2.37, 0)$.

45. a. If the x-intercepts are at $x = 2$, and $x = -3$, the parabola can have the equation $y = (x - 2)(x + 3)$ or $y = x^2 + x - 6$.

b. Another parabola with the same x-intercepts is $y = 2x^2 + 2x - 12$, or, in general $y = kx^2 + kx - 6k$ where k is any real number.

47. We can use that the y-intercept of the parabola $y = ax^2 + bx + c$ is $(0, c)$. So to find two parabolas with y-intercept of $(0, 2)$, we pick some values for a and b. For example, $y = x^2 + 2$ and $y = -5x^2 + 2x + 2$ have y-intercept of $(0, 2)$.

49. a. Simplify the equation given:
$$y = 2(x - 3)^2 + 4$$
$$y = 2\left(x^2 - 6x + 9\right) + 4$$
$$y = 2x^2 - 12x + 18 + 4$$
so $a = 2$ and $b = -12$. The vertex is where $x = \dfrac{-(-12)}{2(2)} = \dfrac{12}{4} = 3$.

When $x = 3$,
$$y = 2(3 - 3)^2 + 4$$
$$= 2(0) + 4$$
$$= 4$$

so the vertex is at $(3, 4)$. The x-coordinate is the number that is subtracted from x within the parentheses and the y-coordinate is the number that is added outside of the parentheses.

Chapter 4, Section 4.5

b. From the work in a., the parabola has the equation
$$y = 2x^2 - 12x + 18 + 4 \text{ which is}$$
$$y = 2x^2 - 12x + 22 \text{ in standard form.}$$

51. a. Simplify the equation given:
$$y = \tfrac{-1}{2}(x+4)^2 - 3$$
$$y = \tfrac{-1}{2}(x^2 + 8x + 16) - 3$$
$$y = \tfrac{-1}{2}x^2 - 4x - 8 - 3$$
so $a = \tfrac{-1}{2}$ and $b = -4$. The vertex is where $x = \dfrac{-(-4)}{2\left(\tfrac{-1}{2}\right)} = \dfrac{4}{-1} = -4$.

When $x = -4$,
$$y = \tfrac{-1}{2}(-4+4)^2 - 3$$
$$= \tfrac{-1}{2}(0)^2 - 3$$
$$= -3$$
so the vertex is at $(-4, -3)$. Since $(x + 4) = [x - (-4)]$, the x-coordinate of the vertex is the number that is subtracted from x within the parentheses, while the y-coordinate of the vertex is the number that is added outside of the parentheses.

b. From the work in a.,
$$y = \tfrac{-1}{2}x^2 - 4x - 8 - 3, \text{ which is}$$
$$y = \tfrac{-1}{2}x^2 - 4x - 11 \text{ in standard form.}$$

53. If we know the coordinates of the vertex, we can substitute them into the equation using the x-coordinate of the vertex for x_v and the y-coordinate of the vertex for y_v.

Then we only need to determine the value of a.

55. a. Using the formula from Problem 53,
$$y = a[x - (-2)]^2 + 6$$
$$= a(x + 2)^2 + 6$$
We can substitute any value of a that we want without changing the vertex. For example, using $a = 2$:
$$y = 2(x + 2)^2 + 6$$
$$y = 2(x^2 + 4x + 4) + 6$$
$$y = 2x^2 + 8x + 8 + 6$$
$$y = 2x^2 + 8x + 14$$

b. If the y-intercept is 18, then $(0, 18)$ is a point on the graph:
$$18 = a(0 + 2)^2 + 6$$
$$18 = a(2)^2 + 6$$
$$18 = 4a + 6$$
$$12 = 4a$$
$$3 = a$$

57. Using the formula from Problem 53, $y = a(x - 0)^2 - 3 = ax^2 - 3$. Since the vertex coincides with the y-intercept, we don't have any other points to use to find what a is, but any value of a will result in a parabola whose vertex and y-intercept coincide at $(0, -3)$.

Section 4.5

1. The parts of the graph with positive y-coordinate appear to be where $x < -12$ or $x > 15$.

3. The part of the graph with positive or zero y-coordinate appears to be where $0.3 \le x \le 0.5$.

Chapter 4, Section 4.5

5. The parts of the graph with positive y-coordinate appear to be where $x < -2$ or $x > 3$.

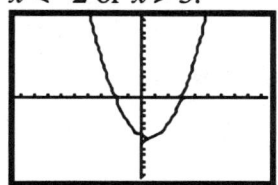

7. The part of the graph with positive or zero y-coordinate appears to be where $0 \le k \le 4$.

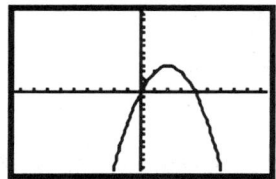

9. The parts of the graph with negative y-coordinate appear to be where $p < -1$ or $p > 6$.

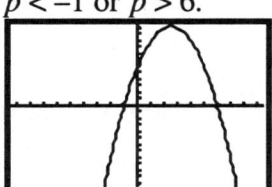

11. The solutions to the inequality $x^2 - 1.4x - 20 < 9.76$ are the solutions of $x^2 - 1.4x - 29.76 < 0$. The part of the graph of $y = x^2 - 1.4x - 29.76$ with negative y-coordinate appears to be where $-4.8 < x < 6.2$.

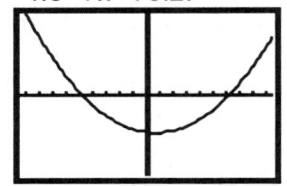

13. The solutions to the inequality $5x^2 + 39x + 27 \ge 5.4$ are the solutions of $5x^2 + 39x + 21.6 \ge 0$. The parts of the graph of $y = 5x^2 + 39x + 21.6$ with positive or zero y-coordinate appear to be where $x \ge -0.6$ or $x \le -7.2$.

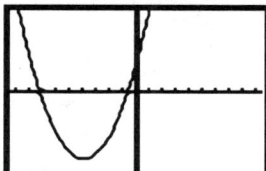

15. The solutions of the inequality $-8x^2 + 112x - 360 < 6.08$ are the solutions of $-8x^2 + 112x - 366.08 < 0$. The parts of the graph of $y = -8x^2 + 112x - 366.08$ with negative y-coordinate appear to be where $x < 5.2$ or $x > 8.8$.

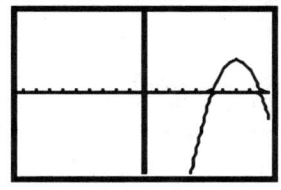

17. The solutions to the inequality $x^2 > 12.2$ are the solutions of $x^2 - 12.2 > 0$. Graph $y = x^2 - 12.2$ using Xmin = -9.4, Xmax = 9.4, Ymin = -12.8, and Ymax = 12.8.

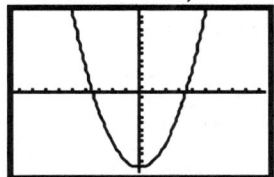

The parts of the graph with positive y-coordinate appear to be where $x < -3.5$ or $x > 3.5$.

131

Chapter 4, Section 4.5

19. Graph $y = -3x^2 + 7x - 25$ using Xmin = –9.4, Xmax = 9.4, Ymin = –64, and Ymax = 64.

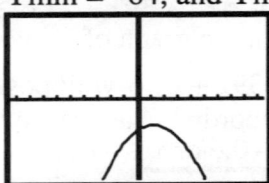

Since the parabola opens downward, there are no points with positive y-coordinate, so all values of x are solutions of the inequality.

21. The solutions of the inequality $0.4x^2 - 54x < 620$ are the solutions of $0.4x^2 - 54x - 620 < 0$. Graph $y = 0.4x^2 - 54x - 620$ using Xmin = –200, Xmax = 200, Ymin = –5000, and Ymax = 5000.

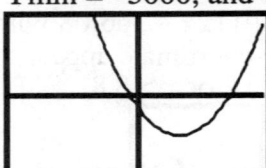

Then Zoom to find that the part of the graph with negative y-coordinate is where $-10.6 < x < 145.6$.

23. (–5, 3]

25. [–4, 0]

27. (–6, ∞)

29. $(-\infty, -3) \cup [-1, \infty)$

31. $[-6, -4) \cup (-2, 0]$

33. $2z^2 - 7z > 4$
$2z^2 - 7z - 4 > 0$

Solve $2z^2 - 7z - 4 = 0$ using the quadratic formula.

$$z = \frac{-(-7) \pm \sqrt{(-7)^2 - 4(2)(-4)}}{2(2)}$$
$$= \frac{7 \pm \sqrt{49 + 32}}{4}$$
$$= \frac{7 \pm \sqrt{81}}{4}$$
$$= \frac{7 \pm 9}{4}$$

The z-intercepts of the parabola are $z = \frac{7+9}{4} = \frac{16}{4} = 4$ and $z = \frac{7-9}{4} = \frac{-2}{4} = \frac{-1}{2}$. Since the parabola opens upward and we are looking for points with positive y-coordinate, the solution set is $\left(-\infty, \frac{-1}{2}\right) \cup (4, \infty)$.

35. $v^2 < 5$
$v^2 - 5 < 0$

Solve $v^2 - 5 = 0$.
$v^2 - 5 = 0$
$v^2 = 5$
$v = \pm\sqrt{5}$

Thus, $v = -\sqrt{5}$ and $v = \sqrt{5}$ are the v-intercepts of the parabola. Since the parabola opens upward and we are looking for points with negative y-coordinate the solution set is $(-\sqrt{5}, \sqrt{5})$ or (–2.24, 2.24).

Chapter 4, Section 4.5

37. Solve $5a^2 - 32a + 12 = 0$ using the quadratic formula.

$$a = \frac{-(-32) \pm \sqrt{(-32)^2 - 4(5)(12)}}{2(5)}$$

$$= \frac{32 \pm \sqrt{1024 - 240}}{10}$$

$$= \frac{32 \pm \sqrt{784}}{10}$$

$$= \frac{32 \pm 28}{10}$$

Thus, $a = \frac{32+28}{10} = \frac{60}{10} = 6$ and $a = \frac{32-28}{10} = \frac{4}{10} = \frac{2}{5}$ are the a-intercepts of the parabola. Since the parabola opens upward and we are looking for points with positive or zero y-coordinate, the solution set is $\left(-\infty, \frac{2}{5}\right] \cup [6, \infty)$.

39. a. To solve $320t - 16t^2 > 1024$ or $-16t^2 + 320t - 1024 > 0$, solve $-16t^2 + 320t - 1024 = 0$ using the quadratic formula.

$$t = \frac{-320 \pm \sqrt{(320)^2 - 4(-16)(-1024)}}{2(-16)}$$

$$= \frac{-320 \pm \sqrt{102,400 - 65,536}}{-32}$$

$$= \frac{-320 \pm \sqrt{36,864}}{-32}$$

$$= \frac{-320 \pm 192}{-32}$$

Thus, $t = \frac{-320+192}{-32} = \frac{-128}{-32} = 4$ and $t = \frac{-320-192}{-32} = \frac{-512}{-32} = 16$ are the t-intercepts of the parabola. Since the parabola opens downward and we are looking for points with positive y-coordinate, the solution is that the rocket is above 1024 feet from 4 to 16 seconds into its flight.

b. Graph $y = 320x - 16x^2$ with Xmin = 0, Xmax = 20, Ymin = 0, and Ymax = 2000. Then use [TRACE] and [ZOOM] to verify the solution in a.

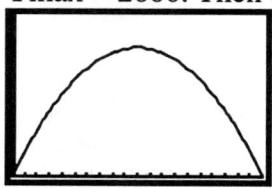

133

Chapter 4, Section 4.5

41. a. To solve $-0.02x^2 + 14x + 1600 < 2800$, or $-0.02x^2 + 14x - 1200 < 0$ with $0 \le x \le 700$, solve $-0.02x^2 + 14x - 1200 = 0$ by using the quadratic formula.

$$x = \frac{-14 \pm \sqrt{(14)^2 - 4(-0.02)(-1200)}}{2(-0.02)}$$
$$= \frac{-14 \pm \sqrt{196 - 96}}{-0.04}$$
$$= \frac{-14 \pm \sqrt{100}}{-0.04}$$
$$= \frac{-14 \pm 10}{-0.04}$$

Thus, $x = \frac{-14 + 10}{-0.04} = \frac{-4}{-0.04} = 100$ and $x = \frac{-14 - 10}{-0.04} = \frac{-24}{-0.04} = 600$ are the x-intercepts of the parabola. Since the parabola opens downward and we are looking for points with negative or zero y-coordinate, the solution is when $0 \le x < 100$ and when $600 < x \le 700$. So the company should produce fewer than 100 pairs of shears or more than 600 but less than 701 pairs of shears.

b. Graph $y = -0.02x^2 + 14x + 1600$ with Xmin = 0, Xmax = 800, Ymin = 0, and Ymax = 5000. Then use TRACE and ZOOM to verify the solution in a.

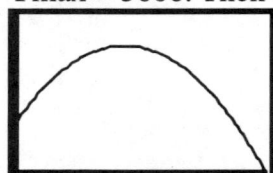

43. a. Revenue = $(1200 - 30p)p = 1200p - 30p^2$. To solve $1200p - 30p^2 > 9000$ or $-30p^2 + 1200p - 9000 > 0$, first solve $-30p^2 + 1200p - 9000 = 0$ using the quadratic formula.

$$p = \frac{-1200 \pm \sqrt{(1200)^2 - 4(-30)(-9000)}}{2(-30)}$$
$$= \frac{-1200 \pm \sqrt{1,440,000 - 1,080,000}}{-60}$$
$$= \frac{-1200 \pm \sqrt{360,000}}{-60}$$
$$= \frac{-1200 \pm 600}{-60}$$

Thus, $p = \frac{-1200 + 600}{-60} = \frac{-600}{-60} = 10$ and $p = \frac{-1200 - 600}{-60} = \frac{-1800}{-60} = 30$ are the p-intercepts of the parabola. Since the parabola opens downward and we are looking for points with positive y-coordinate, the solution set is $10 < p < 30$, so the price of the sweatshirts should be between $10 and $30.

b. Graph $y = 1200x - 30x^2$ with Xmin = 0, Xmax = 40, Ymin = 0, and Ymax = 15,000. Then use TRACE and ZOOM to verify the solution in a.

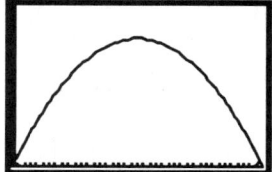

Section 4.6

1. The equation $d = 96t - 16t^2$ is a parabola which opens downward so the maximum value occurs at the vertex. Here, $a = -16$ and $b = 96$, so the vertex is where
$t = \dfrac{-96}{2(-16)} = \dfrac{-96}{-32} = 3$. The rocket reaches its greatest height 3 seconds after it is launched. The height that it reaches is
$d = 96(3) - 16(3)^2$
$= 288 - 16(9) = 288 - 144 = 144$
It reaches a height of 144 feet.

Graph $y = 96x - 16x^2$ using Xmin = 0, Xmax = 10, Ymin = -200, and Ymax = 200.

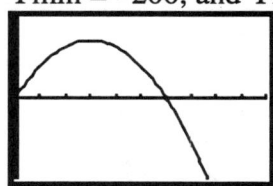

3. a. The perimeter of the rectangle must be 100 inches. If l represents the length of the rectangle, $2w + 2l = 100$, so $w + l = 50$ and $l = 50 - w$. The area is length times width or
$A = lw = (50 - w)w = 50w - w^2$.

b. The graph of the area is a parabola which opens downward, so the maximum value will occur at the vertex. Here $a = -1$ and $b = 50$ so the vertex is where
$w = \dfrac{-50}{2(-1)} = \dfrac{-50}{-2} = 25$. The maximum area is
$A = 50(25) - (25)^2$
$= 1250 - 625 = 625$
So Sheila can enclose 625 in.^{2}

Graph $y = 50x - x^2$ using Xmin = 0, Xmax = 50, Ymin = 0, and Ymax = 800.

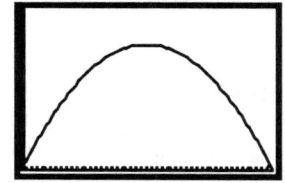

5. a. Let w represent the width of the rectangle (the length of the fence perpendicular to the river). Then there are two fenced sides of length w and one fenced side of length l. Thus $2w + l = 300$, since there are 300 yards of fence available, and $l = 300 - 2w$. The area is $A = lw = (300 - 2w)w$
$= 300w - 2w^2$.

Chapter 4, Section 4.6

b. The graph of the area is a parabola which opens downward, so the maximum value will occur at the vertex. Here $a = -2$ and $b = 300$, so the vertex is at $w = \dfrac{-300}{2(-2)} = \dfrac{-300}{-4} = 75$. When $w = 75$, the area is
$300(75) - 2(75)^2$
$= 22,500 - 11,250 = 11,250$.
The maximum area that the farmer can enclose is 11,250 yd^2.

Graph $y = 300x - 2x^2$ using Xmin = 0, Xmax = 200, Ymin = 0, and Ymax = 15,000.

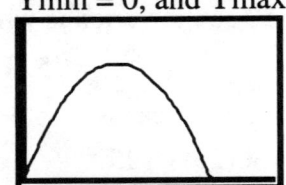

7. a. The total number of people signed up is $16 + x$, and the price per person is $2400 - 100x$, so the total revenue is
$(16 + x)(2400 - 100x)$
$= 38,400 + 800x - 100x^2$.

b. The graph of the revenue is a parabola which opens downward, so the maximum revenue occurs at the vertex. Here $a = -100$ and $b = 800$, so the vertex is at
$x = \dfrac{-800}{2(-100)} = \dfrac{-800}{-200} = 4$ and 4 additional people (20 people total) must sign up for the travel agent to maximize her revenue.

Graph $y = 38,400 + 800x - 100x^2$ using Xmin = 0, Xmax = 25, Ymin = 20,000, and Ymax = 50,000.

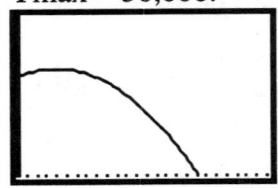

9. The graph of the cost is a parabola which opens upward, so the cost will be minimized at the vertex. Here $a = 0.1$ and $b = -20$, so the vertex is at
$x = \dfrac{-(-20)}{2(0.1)} = \dfrac{20}{0.2} = 100$. They should produce 100 baskets to minimize the cost. The cost of producing 100 baskets is
$0.1(100)^2 - 20(100) + 1800$
$= 1000 - 2000 + 1800$
$= 800$
Thus the cost for producing each basket is $\dfrac{800}{100} = 8$ dollars.

Graph $y = 0.1x^2 - 20x + 1800$ using Xmin = 0, Xmax = 200, Ymin = 500, and Ymax = 1500.

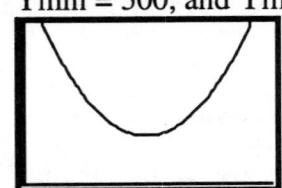

11. We want to minimize
$$V = 576a^2 + 5184(1-a)^2$$
$$= 576a^2 + 5184(1 - 2a + a^2)$$
$$= 576a^2 + 5184 - 10{,}368a + 5184a^2$$
$$= 5760a^2 - 10{,}368a + 5184$$

The graph of V is a parabola which opens upward so the minimum value occurs at the vertex. The vertex is where $a = \dfrac{-(-10{,}368)}{2(5760)} = \dfrac{10{,}368}{11{,}520} = 0.9$. Graph $y = 5760x^2 - 10{,}368x + 5184$ using Xmin = 0, Xmax = 2, Ymin = 0, and Ymax = 10,000.

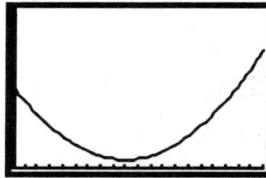

The refined estimate for total income is
$I = (0.9)860 + (1 - 0.9)918 = 774 + 91.8 = 865.8$, so the estimate of income is $865.80 per month.

13. To fit the points on a parabola, use the equations:
$$0 = a(-1)^2 + b(-1) + c$$
$$12 = a(2)^2 + b(2) + c$$
$$8 = a(-2)^2 + b(-2) + c$$
which simplify to
$\quad a - b + c = 0 \quad$ (1)
$\quad 4a + 2b + c = 12 \quad$ (2)
$\quad 4a - 2b + c = 8 \quad$ (3)
Subtract (3) from (2):
$\quad 4a + 2b + c = 12 \quad$ (2)
$\quad -4a + 2b - c = -8 \quad$ (3a)
$\quad\quad\quad\quad 4b = 4$
$\quad\quad\quad\quad\; b = 1 \quad$ (4)
Subtract (1) from (2):
$\quad 4a + 2b + c = 12 \quad$ (2)
$\quad -a + b - c = 0 \quad$ (1a)
$\quad\; 3a + 3b = 12 \quad$ (5)

Chapter 4, Section 4.6

Put the value for b from (4) into (5):
$$3a + 3(1) = 12$$
$$3a + 3 = 12$$
$$3a = 9$$
$$a = 3$$
Put the values for a and b into (1):
$$3 - 1 + c = 0$$
$$2 + c = 0$$
$$c = -2$$
The parabola is $y = 3x^2 + x - 2$

15. a. To fit the points (15, 4), (20, 13), and (30, 7), onto a parabola we have the equations:
$$(15)^2 a + 15b + c = 4$$
$$(20)^2 a + 20b + c = 13$$
$$(30)^2 a + 30b + c = 7$$
Which simplify to
$$225a + 15b + c = 4 \quad (1)$$
$$400a + 20b + c = 13 \quad (2)$$
$$900a + 30b + c = 7 \quad (3)$$
Subtract (2) from (3):
$$900a + 30b + c = 7 \quad (3)$$
$$-400a - 20b - c = -13 \quad (2a)$$
$$500a + 10b = -6 \quad (4)$$
Subtract (1) from (2):
$$400a + 20b + c = 13 \quad (2)$$
$$-225a - 15b - c = -4 \quad (1a)$$
$$175a + 5b = 9 \quad (5)$$
Subtract twice (5) from (4):
$$500a + 10b = -6 \quad (4)$$
$$-350a - 10b = -18 \quad (5a)$$
$$150a = -24$$
$$a = -0.16 \quad (6)$$

Put this value for a into (4):
$$500(-0.16) + 10b = -6$$
$$-80 + 10b = -6$$
$$10b = 74$$
$$b = 7.4$$
Substitute the values for a and b into (1):
$$225(-0.16) + 15(7.4) + c = 4$$
$$-36 + 111 + c = 4$$
$$c = -71$$
The parabola is
$$P = -0.16x^2 + 7.4x - 71.$$

b. Using $x = 25$:
$$-0.16(25)^2 + 7.4(25) - 71$$
$$= -0.16(625) + 185 - 71$$
$$= -100 + 114 = 14$$
The parabola predicts that 14% of 25-year-olds use marijuana regularly.

c. Refer to the graph in the back of the textbook.

17. a. Since t represents the years since 1985, the figure for chicken consumption in 1987 corresponds to $t = 2$. So, the points to be fit to a parabola are (2, 55.5), (3, 57.4), and (4, 60.8). The equations are:
$$(2)^2 a + 2b + c = 55.5$$
$$(3)^2 a + 3b + c = 57.4$$
$$(4)^2 a + 4b + c = 60.8$$
which simplify to
$$4a + 2b + c = 55.5 \quad (1)$$
$$9a + 3b + c = 57.4 \quad (2)$$
$$16a + 4b + c = 60.8 \quad (3)$$
Subtract (2) from (3):
$$16a + 4b + c = 60.8 \quad (3)$$
$$-9a - 3b - c = -57.4 \quad (2a)$$
$$7a + b = 3.4 \quad (4)$$

Chapter 4, Section 4.6

Subtract (1) from (2):
$9a + 3b + c = 57.4$ (2)
$-4a - 2b - c = -55.5$ (1a)
$5a + b = 1.9$ (5)
Now, subtract (5) from (4):
$7a + b = 3.4$ (4)
$-5a - b = -1.9$ (5a)
$2a = 1.5$
$a = 0.75$ (6)
Put the value for a into (4):
$7(0.75) + b = 3.4$
$5.25 + b = 3.4$
$b = -1.85$
Substitute these values into (1):
$4(0.75) + 2(-1.85) + c = 55.5$
$3 - 3.7 + c = 55.5$
$c = 56.2$
The parabola is
$C = 0.75t^2 - 1.85t + 56.2$.

b. Using $t = 5$:
$0.75(5)^2 - 1.85(5) + 56.2$
$= 18.75 - 9.25 + 56.2 = 65.7$
so the parabola predicts a per capita chicken consumption of 65.7 pounds.

c. Refer to the graph in the back of the textbook.

19. Let D represent the number of diagonals in a polygon of n sides. We are looking for a quadratic of the form $D = an^2 + bn + c$. Using the data given for polygons with 4, 5, and 6 sides:
$16a + 4b + c = 2$ (1)
$25a + 5b + c = 5$ (2)
$36a + 6b + c = 9$ (3)
Subtract (1) from (2):
$25a + 5b + c = 5$ (2)
$-16a - 4b - c = -2$ (1a)
$9a + b = 3$ (4)
Subtract (2) from (3):
$36a + 6b + c = 9$ (3)
$-25a - 5b - c = -5$ (2a)
$11a + b = 4$ (5)
Subtract (4) from (5):
$11a + b = 4$ (5)
$-9a - b = -3$ (4a)
$2a = 1$
$a = \frac{1}{2}$ (6)
Put the value from (6) into (4):
$9\left(\frac{1}{2}\right) + b = 3$
$\frac{9}{2} + b = 3$
$b = -\frac{3}{2}$
Substitute these values into (1):
$16\left(\frac{1}{2}\right) + 4\left(-\frac{3}{2}\right) + c = 2$
$8 - 6 + c = 2$
$c = 0$
The parabola is $D = \frac{1}{2}n^2 - \frac{3}{2}n$.
(This is the correct parabola even if you use different points from the data given.)

139

Chapter 4, Section 4.6

21. From the picture in the textbook, we have the three points (0, 500), (2000, 20), and (4000, 500). Put these into $y = ax^2 + bx + c$:
$$0a + 0b + c = 500 \quad (1)$$
$$4{,}000{,}000a + 2000b + c = 20 \quad (2)$$
$$16{,}000{,}000a + 4000b + c = 500 \quad (3)$$
From (1), $c = 500$. Put that into (2) and (3):
$$4{,}000{,}000a + 2000b = -480 \quad (4)$$
$$16{,}000{,}000a + 4000b = 0 \quad (5)$$
Subtract twice (4) from (5):
$$16{,}000{,}000a + 4000b = 0 \quad (5)$$
$$-8{,}000{,}000a - 4000b = 960 \quad (4a)$$
$$8{,}000{,}000a = 960$$
$$a = 0.00012 \quad (6)$$
Put the value from (6) into (5):
$$16{,}000{,}000(0.00012) + 4000b = 0$$
$$1920 + 4000b = 0$$
$$4000b = -1920$$
$$b = -0.48$$
An equation for the shape of the curve is $y = 0.00012x^2 - 0.48x + 500$.

23. Given $y = x^2 - 4x + 7$ and $y = 11 - x$, equate the expressions for y:
$$x^2 - 4x + 7 = 11 - x$$
$$x^2 - 3x - 4 = 0$$
$$(x - 4)(x + 1) = 0$$
The solutions are when $x = 4$, and $x = -1$. The points are (4, 7) and (–1, 12).
Graph the equations using Xmin = –10, Xmax = 10, Ymin = –20, and Ymax = 20.

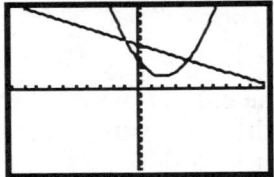

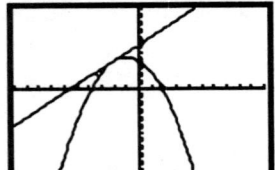

25. Given $y = -x^2 - 2x + 7$ and $y = 2x + 11$, equate the expressions for y:
$$-x^2 - 2x + 7 = 2x + 11$$
$$0 = x^2 + 4x + 4$$
$$0 = (x + 2)^2$$
The solution is when $x = -2$. The point is (–2, 7).
Graph the equations using Xmin = –10, Xmax = 10, Ymin = –20, and Ymax = 20.

Chapter 4, Section 4.6

27. Given $y = x^2 + 8x + 8$ and $3y + 2x = -36$, we solve the second equation for y
$$3y + 2x = -36$$
$$3y = -2x - 36$$
$$y = \left(-\frac{2}{3}\right)x - 12$$
and then equate the two expressions for y:
$$x^2 + 8x + 8 = \left(-\frac{2}{3}\right)x - 12$$
$$x^2 + \frac{26}{3}x + 20 = 0$$
Solve using the quadratic formula
$$x = \frac{-\frac{26}{3} \pm \sqrt{\left(\frac{26}{3}\right)^2 - 4(1)(20)}}{2(1)}$$
$$= \frac{-\frac{26}{3} \pm \sqrt{75.11 - 80}}{2}$$
$$= \frac{-26 \pm 3\sqrt{-4.89}}{6}$$
Since there is a negative under the radical, there is no solution to the system.
Graph the equations using Xmin = −10, Xmax = 10, Ymin = −20, and Ymax = 20.

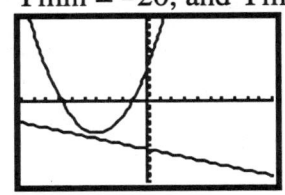

29. Given $y = x^2 - 9$ and $y = -2x^2 + 9x + 21$, equate the expressions for y:
$$x^2 - 9 = -2x^2 + 9x + 21$$
$$3x^2 - 9x - 30 = 0$$
$$3\left(x^2 - 3x - 10\right) = 0$$
$$3(x - 5)(x + 2) = 0$$
The solutions are when $x = 5$ and $x = -2$. The points are (5, 16) and (−2, −5).
Graph the equations using Xmin = −10, Xmax = 10, Ymin = −50, and Ymax = 50.

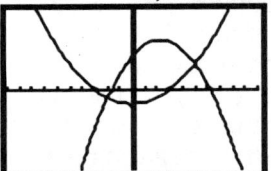

31. Given $y = x^2 - 0.5x + 3.5$ and $y = -x^2 + 3.5x + 1.5$, equate the expressions for y:
$$x^2 - 0.5x + 3.5 = -x^2 + 3.5x + 1.5$$
$$2x^2 - 4x + 2 = 0$$
$$2\left(x^2 - 2x + 1\right) = 0$$
$$2(x - 1)^2 = 0$$
The solution is when $x = 1$. When $x = 1$, $y = (1)^2 - 0.5(1) + 3.5$
$= 1 - 0.5 + 3.5 = 4$, so the point is (1, 4).
Graph the equations using Xmin = −10, Xmax = 10, Ymin = −10, and Ymax = 10.

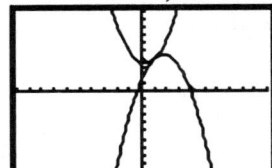

141

Chapter 4, Section 4.6

33. Given $y = x^2 - 4x + 4$ and $y = x^2 - 8x + 16$, equate the expressions for y:
$$x^2 - 4x + 4 = x^2 - 8x + 16$$
$$4x - 12 = 0$$
$$4x = 12$$
$$x = 3$$

The solution is when $x = 3$. When $x = 3$, $y = (3)^2 - 4(3) + 4$
$= 9 - 12 + 4 = 1$, so the point is (3, 1).
Graph the equations using Xmin = –10, Xmax = 10, Ymin = –10, and Ymax = 10.

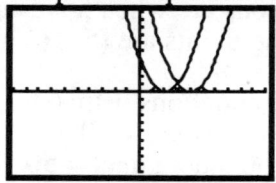

35. a. The equation for revenue is a parabola which opens downward, so the maximum revenue occurs at the vertex. From Section 4.5, the vertex is at (800, 12,800), so the maximum revenue occurs when 800 pens are sold and the maximum revenue is $12,800.
Graph the equations using Xmin = –1500, Xmax = 1500, Ymin = –13,000, and Ymax = 13,000.

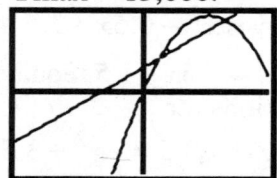

b. Equating the equations for C and R:
$$8x + 4000 = -0.02(x - 800)^2 + 12,800$$
$$8x + 4000 = -0.02(x^2 - 1600x + 640,000) + 12,800$$
$$8x + 4000 = -0.02x^2 + 32x - 12,800 + 12,800$$
$$8x + 4000 = -0.02x^2 + 32x$$
$$0 = -0.02x^2 + 24x - 4000$$
$$0 = -0.02(x^2 - 1200x + 200,000)$$
$$0 = -0.02(x - 200)(x - 1000)$$
so the break-even points are when Writewell sells either 200 or 1000 pens.

37. a. The equation for revenue is a parabola which opens downward, so the maximum revenue occurs at the vertex. From Section 4.5, the vertex is at (200, 70,000), so the maximum revenue occurs when 200 washing machines are sold, and the maximum revenue is $70,000.

Graph the equations using Xmin = –500, Xmax = 500, Ymin = –75,000, and Ymax = 75,000.

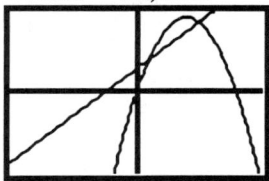

b. Equating the equations for C and R:
$$175x + 21,875 = -1.75(x - 200)^2 + 70,000$$
$$175x + 21,875 = -1.75\left(x^2 - 400x + 40,000\right) + 70,000$$
$$175x + 21,875 = -1.75x^2 + 700x - 70,000 + 70,000$$
$$175x + 21,875 = -1.75x^2 + 700x$$
$$0 = -1.75x^2 + 525x - 21,875$$
$$0 = -1.75\left(x^2 - 300x + 12,500\right)$$
$$0 = -1.75(x - 50)(x - 250)$$
so the break-even points are when either 50 or 250 washing machines are sold.

Chapter 4 Review

1. $$x(3x + 2) = (x + 2)^2$$
    $$3x^2 + 2x = x^2 + 4x + 4$$
    $$2x^2 - 2x - 4 = 0$$
    $$2\left(x^2 - x - 2\right) = 0$$
    $$2(x - 2)(x + 1) = 0$$
    The solutions are $x = 2$ and $x = -1$.

2. $$6y = (y + 1)^2 + 3$$
    $$6y = y^2 + 2y + 1 + 3$$
    $$0 = y^2 - 4y + 4$$
    $$0 = (y - 2)^2$$
    The solution is $y = 2$.

3. $$4x - (x + 1)(x + 2) = -8$$
    $$4x - \left(x^2 + 3x + 2\right) = -8$$
    $$4x - x^2 - 3x - 2 = -8$$
    $$-x^2 + x - 2 = -8$$
    $$0 = x^2 - x - 6$$
    $$0 = (x - 3)(x + 2)$$
    The solutions are $x = 3$ and $x = -2$.

Chapter 4 Review

4.
$$3(x+2)^2 = 15 + 12x$$
$$3(x^2 + 4x + 4) = 15 + 12x$$
$$3x^2 + 12x + 12 = 15 + 12x$$
$$3x^2 - 3 = 0$$
$$3(x^2 - 1) = 0$$
$$3(x-1)(x+1) = 0$$
The solutions are $x = 1$ and $x = -1$.

5.
$$\left(x - \left(\tfrac{-3}{4}\right)\right)(x-8) = 0$$
$$\left(x + \tfrac{3}{4}\right)(x-8) = 0$$
$$x^2 - \tfrac{29}{4}x - 6 = 0$$
$$4\left(x^2 - \tfrac{29}{4}x - 6\right) = 4(0)$$
$$4x^2 - 29x - 24 = 0$$

6.
$$\left(x - \tfrac{5}{3}\right)\left(x - \tfrac{5}{3}\right) = 0$$
$$x^2 - \tfrac{10}{3}x + \tfrac{25}{9} = 0$$
$$9\left(x^2 - \tfrac{10}{3}x + \tfrac{25}{9}\right) = 9(0)$$
$$9x^2 - 30x + 25 = 0$$

7. Refer to the graph in the back of the textbook. The x-intercepts are at $x = 3$ and $x = -2.4$, so the equation in factored form is $y = (x-3)(x+2.4)$.

8. Refer to the graph in the back of the textbook. The x-intercepts are at $x = -1.3$ and $x = 2$, so the equation in factored form is $y = (-1)(x-2)(x+1.3)$.

9.
$$(2x - 5)^2 = 9$$
$$2x - 5 = \pm\sqrt{9}$$
$$2x - 5 = \pm 3$$
$$2x = 5 \pm 3$$
$$x = \tfrac{5 \pm 3}{2}$$
The solutions are $x = \tfrac{5+3}{2} = \tfrac{8}{2} = 4$ and $x = \tfrac{5-3}{2} = \tfrac{2}{2} = 1$.

10.
$$(7x - 1)^2 = 15$$
$$7x - 1 = \pm\sqrt{15}$$
$$7x = 1 \pm \sqrt{15}$$
$$x = \tfrac{1 \pm \sqrt{15}}{7}$$
The solutions are $x = \tfrac{1 + \sqrt{15}}{7}$ and $x = \tfrac{1 - \sqrt{15}}{7}$.

11.
$$x^2 - 4x - 6 = 0$$
$$x^2 - 4x = 6$$
Since one-half of -4 is -2, we add $(-2)^2 = 4$ to both sides.
$$x^2 - 4x + 4 = 6 + 4$$
$$(x-2)^2 = 10$$
$$x - 2 = \pm\sqrt{10}$$
$$x = 2 \pm \sqrt{10}$$
The solutions are $x = 2 + \sqrt{10}$ and $x = 2 - \sqrt{10}$.

12. $x^2 + 3x = 3$ Since one-half of 3 is $\frac{3}{2}$, we add $\left(\frac{3}{2}\right)^2 = \frac{9}{4}$ to both sides.
$$x^2 + 3x + \frac{9}{4} = 3 + \frac{9}{4}$$
$$\left(x + \frac{3}{2}\right)^2 = \frac{21}{4}$$
$$x + \frac{3}{2} = \pm\sqrt{\frac{21}{4}}$$
$$x = \frac{-3}{2} \pm \frac{\sqrt{21}}{2}$$
$$x = \frac{-3 \pm \sqrt{21}}{2}$$
The solutions are $x = \frac{-3 + \sqrt{21}}{2}$ and $x = \frac{-3 - \sqrt{21}}{2}$.

13. $$2x^2 + 3 = 6x$$
$$2x^2 - 6x = -3$$
$$\tfrac{1}{2}(2x^2 - 6x) = \tfrac{1}{2}(-3)$$
$$x^2 - 3x = -\tfrac{3}{2}$$
Since one-half of -3 is $-\frac{3}{2}$, we add $\left(-\frac{3}{2}\right)^2 = \frac{9}{4}$ to both sides.
$$x^2 - 3x + \frac{9}{4} = -\frac{3}{2} + \frac{9}{4}$$
$$\left(x - \frac{3}{2}\right)^2 = \frac{3}{4}$$
$$x - \frac{3}{2} = \pm\sqrt{\frac{3}{4}}$$
$$x = \frac{3}{2} \pm \frac{\sqrt{3}}{2}$$
$$x = \frac{3 \pm \sqrt{3}}{2}$$
The solutions are $x = \frac{3 + \sqrt{3}}{2}$ and $x = \frac{3 - \sqrt{3}}{2}$.

14. $$3x^2 = 2x + 3$$
$$3x^2 - 2x = 3$$
$$\tfrac{1}{3}(3x^2 - 2x) = \tfrac{1}{3}(3)$$
$$x^2 - \frac{2}{3}x = 1$$
Since one-half of $-\frac{2}{3}$ is $-\frac{1}{3}$, we add $\left(-\frac{1}{3}\right)^2 = \frac{1}{9}$ to both sides.
$$x^2 - \frac{2}{3}x + \frac{1}{9} = 1 + \frac{1}{9}$$
$$\left(x - \frac{1}{3}\right)^2 = \frac{10}{9}$$
$$x - \frac{1}{3} = \pm\sqrt{\frac{10}{9}}$$
$$x = \frac{1}{3} \pm \frac{\sqrt{10}}{3}$$
$$x = \frac{1 \pm \sqrt{10}}{3}$$
The solutions are $x = \frac{1 + \sqrt{10}}{3}$ and $x = \frac{1 - \sqrt{10}}{3}$.

15. $$\tfrac{1}{2}x^2 + 1 = \tfrac{3}{2}x$$
$$\tfrac{1}{2}x^2 - \tfrac{3}{2}x + 1 = 0$$
We have $a = \frac{1}{2}$, $b = -\frac{3}{2}$, and $c = 1$.
$$x = \frac{-\left(-\frac{3}{2}\right) \pm \sqrt{\left(-\frac{3}{2}\right)^2 - 4\left(\frac{1}{2}\right)(1)}}{2\left(\frac{1}{2}\right)}$$
$$= \frac{\frac{3}{2} \pm \sqrt{\frac{9}{4} - 2}}{1}$$
$$= \frac{3}{2} \pm \sqrt{\frac{1}{4}}$$
$$= \frac{3}{2} \pm \frac{1}{2}$$
$$= \frac{3 \pm 1}{2}$$

Chapter 4 Review

The solutions are $x = \dfrac{3+1}{2} = \dfrac{4}{2} = 2$
and $x = \dfrac{3-1}{2} = \dfrac{2}{2} = 1$.

16. In $x^2 - 3x + 1 = 0$, we have $a = 1$, $b = -3$, and $c = 1$.

$$x = \dfrac{-(-3) \pm \sqrt{(-3)^2 - 4(1)(1)}}{2(1)}$$
$$= \dfrac{3 \pm \sqrt{9-4}}{2}$$
$$= \dfrac{3 \pm \sqrt{5}}{2}$$

The solutions are $x = \dfrac{3+\sqrt{5}}{2} \approx 2.62$
and $x = \dfrac{3-\sqrt{5}}{2} \approx 0.38$.

17. In $x^2 - 4x + 2 = 0$, we have $a = 1$, $b = -4$, and $c = 2$.

$$x = \dfrac{-(-4) \pm \sqrt{(-4)^2 - 4(1)(2)}}{2(1)}$$
$$= \dfrac{4 \pm \sqrt{16-8}}{2}$$
$$= \dfrac{4 \pm \sqrt{8}}{2}$$
$$= \dfrac{4 \pm 2\sqrt{2}}{2}$$
$$= 2 \pm \sqrt{2}$$

The solutions are $x = 2 + \sqrt{2} \approx 3.41$
and $x = 2 - \sqrt{2} \approx 0.59$.

18. $2x^2 + 2x = 3$
$2x^2 + 2x - 3 = 0$
Here $a = 2$, $b = 2$, and $c = -3$.

$$x = \dfrac{-2 \pm \sqrt{(2)^2 - 4(2)(-3)}}{2(2)}$$
$$= \dfrac{-2 \pm \sqrt{4+24}}{4}$$
$$= \dfrac{-2 \pm \sqrt{28}}{4}$$
$$= \dfrac{-2 \pm 2\sqrt{7}}{4}$$
$$= \dfrac{-1 \pm \sqrt{7}}{2}$$

The solutions are
$x = \dfrac{-1+\sqrt{7}}{2} \approx 0.82$ and
$x = \dfrac{-1-\sqrt{7}}{2} \approx -1.82$.

19. $\dfrac{n(n-1)}{2} = 36$
$n(n-1) = 72$
$n^2 - n = 72$
$n^2 - n - 72 = 0$
$(n-9)(n+8) = 0$
The solutions are $n = 9$ and $n = -8$, so the organizers should invite 9 players.

20. $\dfrac{n(n+1)}{2} = 91$
$n(n+1) = 182$
$n^2 + n = 182$
$n^2 + n - 182 = 0$
$(n+14)(n-13) = 0$
The solutions are $n = -14$ and $n = 13$. The question asks how many positive integers add up to 91, so $n = -14$ makes no sense. Thus, the sum of the first 13 positive integers is 91.

Chapter 4 Review

21. $2464.20 = 2000(1+r)^2$
$1.2321 = (1+r)^2$
$\pm\sqrt{1.2321} = 1+r$
$-1 \pm 1.11 = r$

The solutions are $r = -1 + 1.11$
$= 0.11$ and $r = -1 - 1.11 = -2.11$. A negative interest rate makes no sense, so the solution is $r = 0.11$. The account had an interest rate of 11%.

22. $6474.74 = 5500(1+r)^2$
$1.1772255 = (1+r)^2$
$\pm\sqrt{1.1772255} = 1+r$
$-1 \pm 1.085 = r$

The solutions are $r = -1 + 1.085$
$= 0.085$ and $r = -1 - 1.085$
$= -2.085$. A negative interest rate makes no sense, so the answer is that $r = 0.085$. The interest rate on the loan was 8.5%.

23. Let x represent the number of members of the credit union this year. Then there were $x + 6$ members last year. The amount each member received last year was $\frac{12,600}{x+6}$ dollars. This year each of the x members received $\frac{12,600}{x+6} + 5$ dollars, so we have

$$x\left(\frac{12,600}{x+6} + 5\right) = 12,600$$
$$\frac{12,600x}{x+6} + 5x = 12,600$$
$$(x+6)\left(\frac{12,600x}{x+6} + 5x\right) = (x+6)12,600$$
$$12,600x + 5x(x+6) = 12,600x + 75,600$$
$$12,600x + 5x^2 + 30x = 12,600x + 75,600$$
$$5x^2 + 30x - 75,600 = 0$$
$$5\left(x^2 + 6x - 15,120\right) = 0$$
$$5(x - 120)(x + 126) = 0$$

The solutions are $x = 120$ and $x = -126$. Since the credit union cannot have a negative number of members, there are 120 members in the credit union this year.

Chapter 4 Review

24. Let x represent the width of the coops and l represent the length of the coops along the henhouse, as shown below.

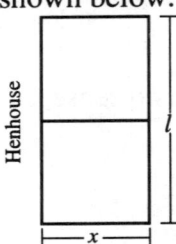

Then $l = 66 - 3x$. Since area $= lw$, and Irene wants the area of the coops to be 360 ft^2,
$$x(66 - 3x) = 360$$
$$66x - 3x^2 = 360$$
$$0 = 3x^2 - 66x + 360$$
$$0 = 3(x^2 - 22x + 120)$$
$$0 = 3(x - 12)(x - 10)$$
The solutions are $x = 12$ and $x = 10$. The dimensions of each coop are x feet by $\frac{1}{2}l$ feet.
When $x = 12$, $l = 66 - 3(12)$
$= 66 - 36 = 30$ and the coops are 12 feet by 15 feet.
When $x = 10$, $l = 66 - 3(10)$
$= 66 - 30 = 36$ and the coops are 10 feet by 18 feet.

25. Refer to the graph in the back of the textbook. In $y = \frac{1}{2}x^2$, $a = \frac{1}{2}$, $b = 0$, and $c = 0$. The vertex is where $x = \frac{-0}{2(\frac{1}{2})} = \frac{0}{1} = 0$. When $x = 0$, $y = \frac{1}{2}(0)^2 = \frac{1}{2}(0) = 0$, so the vertex is at (0, 0). This point is also the x-intercept and y-intercept.

26. Refer to the graph in the back of the textbook. In $y = x^2 - 4$, $a = 1$, $b = 0$, and $c = -4$. The vertex is where $x = \frac{-0}{2(1)} = \frac{0}{2} = 0$. When $x = 0$, $y = (0)^2 - 4 = -4$, so the vertex is at (0, -4). This is also the y-intercept. For the x-intercepts, solve:
$$0 = x^2 - 4$$
$$4 = x^2$$
$$\pm\sqrt{4} = x$$
$$\pm 2 = x$$
The x-intercepts are (2, 0) and (-2, 0).

27. Refer to the graph in the back of the textbook. In $y = x^2 - 9x$, $a = 1$, $b = -9$, and $c = 0$. The vertex is where $x = \frac{-(-9)}{2(1)} = \frac{9}{2}$. When $x = \frac{9}{2}$,
$$y = \left(\frac{9}{2}\right)^2 - 9\left(\frac{9}{2}\right) = \frac{81}{4} - \frac{81}{2} = -\frac{81}{4},$$
so the vertex is at $\left(\frac{9}{2}, -\frac{81}{4}\right)$. The y-intercept is at (0, 0). For the x-intercept, solve:
$$x^2 - 9x = 0$$
$$x(x - 9) = 0$$
The solutions are $x = 0$ and $x = 9$, so the x-intercepts are (0, 0) and (9, 0).

28. Refer to the graph in the back of the textbook. In $y = -2x^2 - 4x$, $a = -2$, $b = -4$, and $c = 0$. The vertex is where $x = \frac{-(-4)}{2(-2)} = \frac{4}{-4} = -1$. When $x = -1$, $y = -2(-1)^2 - 4(-1)$
$= -2(1) + 4 = -2 + 4 = 2$, so the vertex is at (-1, 2). The y-intercept

148

is at (0, 0). For the x-intercepts, solve:

$0 = -2x^2 - 4x$
$0 = -2x(x+2)$

The solutions are $x = 0$ and $x = -2$, so the x-intercepts are $(0, 0)$ and $(-2, 0)$.

29. Refer to the graph in the back of the textbook. In $y = x^2 - x - 12$, $a = 1$, $b = -1$, and $c = -12$. The vertex is where $x = \frac{-(-1)}{2(1)} = \frac{1}{2}$. When $x = \frac{1}{2}$,

$y = \left(\frac{1}{2}\right)^2 - \frac{1}{2} - 12 = \frac{1}{4} - \frac{25}{2} = \frac{-49}{4}$,

so the vertex is at $\left(\frac{1}{2}, \frac{-49}{4}\right)$. The y-intercept is at $(0, -12)$. For the x-intercepts, solve:

$0 = x^2 - x - 12$
$0 = (x-4)(x+3)$

The solutions are $x = 4$ and $x = -3$, so the x-intercepts are $(4, 0)$ and $(-3, 0)$.

30. Refer to the graph in the back of the textbook. In $y = -2x^2 + x - 4$, $a = -2$, $b = 1$, and $c = -4$. The vertex is where $x = \frac{-1}{2(-2)} = \frac{-1}{-4} = \frac{1}{4}$.

When $x = \frac{1}{4}$, $y = -2\left(\frac{1}{4}\right)^2 + \frac{1}{4} - 4$

$= -2\left(\frac{1}{16}\right) - \frac{15}{4} = -\frac{1}{8} - \frac{15}{4} = -\frac{31}{8}$, so

the vertex is at $\left(\frac{1}{4}, -\frac{31}{8}\right)$. The y-intercept is at $(0, -4)$. For the x-intercepts, use the quadratic formula.

$x = \frac{-1 \pm \sqrt{(1)^2 - 4(-2)(-4)}}{2(-2)}$

$= \frac{-1 \pm \sqrt{1-32}}{-4}$

$= \frac{-1 \pm \sqrt{-31}}{-4}$

Since there is a negative under the radical, there are no real solutions, hence no x-intercepts.

31. Refer to the graph in the back of the textbook. In $y = -x^2 + 2x + 4$, $a = -1$, $b = 2$, and $c = 4$. The vertex is where $x = \frac{-2}{2(-1)} = \frac{-2}{-2} = 1$. When $x = 1$,

$y = -(1)^2 + 2(1) + 4 = -1 + 2 + 4 = 5$

so the vertex is at $(1, 5)$. The y-intercept is at $(0, 4)$. For the x-intercepts, use the quadratic formula.

$x = \frac{-2 \pm \sqrt{(2)^2 - 4(-1)(4)}}{2(-1)}$

$= \frac{-2 \pm \sqrt{4+16}}{-2}$

$= \frac{-2 \pm \sqrt{20}}{-2}$

$= \frac{-2 \pm 2\sqrt{5}}{-2}$

$= 1 \pm \sqrt{5}$

The solutions are $x = 1 + \sqrt{5} \approx 3.24$ and $x = 1 - \sqrt{5} \approx -1.24$, so the x-intercepts are $(3.24, 0)$ and $(-1.24, 0)$.

Chapter 4 Review

32. Refer to the graph in the back of the textbook. In $y = x^2 - 3x + 4$, $a = 1$, $b = -3$, and $c = 4$. The vertex is where $x = \frac{-(-3)}{2(1)} = \frac{3}{2}$. When $x = \frac{3}{2}$,

$$y = \left(\frac{3}{2}\right)^2 - 3\left(\frac{3}{2}\right) + 4 = \frac{9}{4} - \frac{9}{2} + 4 = \frac{7}{4}$$

so the vertex is at $\left(\frac{3}{2}, \frac{7}{4}\right)$. The y-intercept is at $(0, 4)$. For the x-intercepts, use the quadratic formula

$$x = \frac{-(-3) \pm \sqrt{(-3)^2 - 4(1)(4)}}{2(1)}$$
$$= \frac{3 \pm \sqrt{9 - 16}}{2}$$
$$= \frac{3 \pm \sqrt{-7}}{2}$$

Since there is a negative under the radical, there are no real solutions, hence no x-intercepts.

33. To solve the inequality, first solve $(x - 3)(x + 2) = 0$. The solutions are $x = 3$ and $x = -2$. Refer to the graph in the back of the textbook. Since we are looking for where the y-coordinate is positive, the solution is where $x < -2$ or $x > 3$.

34. To solve the inequality, first solve

$$y^2 - y - 12 = 0$$
$$(y - 4)(y + 3) = 0$$

The solutions are $y = 4$ and $y = -3$. Refer to the graph in the back of the textbook. Since we are looking for where the graph is on or below the x-axis, the solution is $-3 \leq y \leq 4$.

35. An equivalent inequality is $2y^2 - y - 3 \leq 0$. To solve this inequality, first solve:

$$2y^2 - y - 3 = 0$$
$$(2y - 3)(y + 1) = 0$$

The solutions are

$$2y - 3 = 0 \quad \text{or} \quad y + 1 = 0$$
$$2y = 3 \qquad\qquad\qquad y = -1$$
$$y = \frac{3}{2}$$

Refer to the graph in the back of the textbook. Since we are looking for where the graph is on or below the x-axis, the solution is $-1 \leq y \leq \frac{3}{2}$.

36. An equivalent inequality is $3z^2 - 5z - 2 > 0$. To solve this inequality, first solve

$$3z^2 - 5z - 2 = 0$$
$$(3z + 1)(z - 2) = 0$$

The solutions are

$$3z + 1 = 0 \quad \text{or} \quad z - 2 = 0$$
$$3z = -1 \qquad\qquad\qquad z = 2$$
$$z = -\frac{1}{3}$$

Refer to the graph in the back of the textbook. Since we are looking for where the graph is above the z-axis, the solution is $z < -\frac{1}{3}$ or $z > 2$.

37. An equivalent inequality is $s^2 - 4 \leq 0$. To solve this inequality, first solve

$$s^2 - 4 = 0$$
$$(s + 2)(s - 2) = 0$$

The solutions are

$$s + 2 = 0 \quad \text{or} \quad s - 2 = 0$$
$$s = -2 \qquad\qquad\qquad s = 2$$

Chapter 4 Review

Refer to the graph in the back of the textbook. Since we are looking for where the graph is on or below the s-axis, the solution is $-2 \leq s \leq 2$.

38. An equivalent inequality is
$4t^2 - 12 > 0$. To solve this inequality, first solve
$$4t^2 - 12 = 0$$
$$4(t^2 - 3) = 0$$
$$4(t + \sqrt{3})(t - \sqrt{3}) = 0$$
The solutions are
$t - \sqrt{3} = 0 \quad$ or $\quad t + \sqrt{3} = 0$
$\quad t = \sqrt{3} \qquad\qquad t = -\sqrt{3}$
Refer to the graph in the back of the textbook. Since we are looking for where the graph is above the x-axis, the solution is $t < -\sqrt{3}$ or $t > \sqrt{3}$.

39. The revenue from selling $120 - \frac{1}{4}p$ sandwiches at p cents each is
$\left(120 - \frac{1}{4}p\right)p = 120p - \frac{1}{4}p^2$.
Therefore, solve
$120p - \frac{1}{4}p^2 > 14,000$ or
$-\frac{1}{4}p^2 + 120p - 14,000 > 0$. First solve
$$-\frac{1}{4}p^2 + 120p - 14,000 = 0$$
$$-\frac{1}{4}\left(p^2 - 480p + 56,000\right) = 0$$
$$-\frac{1}{4}(p - 200)(p - 280) = 0$$
The solutions are $p = 200$ and $p = 280$. Since the graph of the equality is a parabola which opens downward, and we are looking for the parts of the graph above the x-axis, the solution is $200 < p < 280$. The sandwiches can sell for between $2.00 and $2.80.

40. The revenue from selling $30 - \frac{1}{2}p$ screwdrivers at p dollars each is
$\left(30 - \frac{1}{2}p\right)p = 30p - \frac{1}{2}p^2$.
Therefore, solve $30p - \frac{1}{2}p^2 > 400$
or $-\frac{1}{2}p^2 + 30p - 400 > 0$. First solve
$$-\frac{1}{2}p^2 + 30p - 400 = 0$$
$$-\frac{1}{2}\left(p^2 - 60p + 800\right) = 0$$
$$-\frac{1}{2}(p - 40)(p - 20) = 0$$
The solutions are $p = 40$ and $p = 20$. Since the graph of the equality is a parabola which opens downward and we are looking for where the graph is above the x-axis, the solution is $20 < p < 40$. The price of the electric screwdrivers should be between $20 and $40.

41. Let x represent the number of trees removed. Thus the number of trees left in the orchard will be $60 - x$. After x trees are removed, the yield per tree will be $12 + \frac{1}{2}x$ bushels of apples, so the total harvest from the orchard will be
$$(60 - x)\left(12 + \frac{1}{2}x\right)$$
$$= 720 + 18x - \frac{1}{2}x^2$$
$$= \frac{-1}{2}x^2 + 18x + 720$$
The graph of this is a parabola which opens downward, so the maximum value occurs at the vertex. The vertex is where
$$x = \frac{-18}{2\left(\frac{-1}{2}\right)} = \frac{-18}{-1} = 18.\ \text{18 trees}$$

151

Chapter 4 Review

should be removed to maximize the apple harvest.

42. Let x represent the price increase in dollars. The number of radios sold per month will be $500 - 10x$ and the price of each radio will be $20 + x$, so the revenue from selling radios will be
$(20 + x)(500 - 10x)$
$= 10,000 + 300x - 10x^2$
$= -10x^2 + 300x + 10,000$
The graph of this is a parabola which opens downward, so the maximum value occurs at the vertex. The vertex is where $x = \dfrac{-300}{2(-10)} = \dfrac{-300}{-20} = 15$. Thus, to maximize the revenues, the company should raise the price by 15 dollars to $35 per radio.

43. Refer to the graph in the back of the textbook. Substitute the value $y = 3$ from the second equation into the first equation:
$3 + x^2 = 4$
$x^2 = 1$
$x = \pm\sqrt{1}$
$x = \pm 1$
so the solutions are $(1, 3)$ and $(-1, 3)$.

44. Refer to the graph in the back of the textbook. The given system is equivalent to
$y = 3 - x^2$
$y = 7 - 5x$
Equate the expressions for y:
$3 - x^2 = 7 - 5x$
$0 = x^2 - 5x + 4$
$0 = (x - 4)(x - 1)$
The solutions are when $x = 4$ and $x = 1$. When $x = 4$,
$y = 3 - (4)^2 = 3 - 16 = -13$.
When $x = 1$,
$y = 3 - (1)^2 = 3 - 1 = 2$. The solutions are $(4, -13)$ and $(1, 2)$.

45. Refer to the graph in the back of the textbook. Equate the expressions for y:
$x^2 - 5 = 4x$
$x^2 - 4x - 5 = 0$
$(x + 1)(x - 5) = 0$
The solutions are when $x = -1$ and $x = 5$. When $x = -1$, $y = 4(-1) = -4$ and when $x = 5$, $y = 4(5) = 20$. The solutions are $(-1, -4)$ and $(5, 20)$.

46. Refer to the graph in the back of the textbook. Equate the expressions for y:
$x^2 - 2x + 1 = 3 - x$
$x^2 - x - 2 = 0$
$(x - 2)(x + 1) = 0$
The solutions are when $x = 2$ and $x = -1$. When $x = 2$, $y = 3 - 2 = 1$ and when $x = -1$,
$y = 3 - (-1) = 3 + 1 = 4$. The solutions are $(2, 1)$ and $(-1, 4)$.

47. Refer to the graph in the back of the textbook. Equate the expressions for y:
$x^2 - 6x + 20 = 2x^2 - 2x - 25$
$0 = x^2 + 4x - 45$
$0 = (x + 9)(x - 5)$
The solutions are when $x = -9$ and $x = 5$. When $x = -9$,
$y = 2(-9)^2 - 2(-9) - 25$
$= 162 + 18 - 25$
$= 155$

Chapter 4 Review

When $x = 5$,
$$y = 2(5)^2 - 2(5) - 25$$
$$= 50 - 10 - 25$$
$$= 15$$
The solutions are $(-9, 155)$ and $(5, 15)$.

48. Refer to the graph in the back of the textbook. Equate the expressions for y:
$$x^2 - 5x - 28 = -x^2 + 4x + 28$$
$$2x^2 - 9x - 56 = 0$$
$$(2x + 7)(x - 8) = 0$$
The solutions are
$2x + 7 = 0 \quad$ or $\quad x - 8 = 0$
$\quad 2x = -7 \quad\quad\quad\quad x = 8$
$\quad x = -\frac{7}{2}$

When $x = -\frac{7}{2}$,
$$y = \left(-\frac{7}{2}\right)^2 - 5\left(-\frac{7}{2}\right) - 28$$
$$= \frac{49}{4} + \frac{35}{2} - 28$$
$$= \frac{7}{4}$$

When $x = 8$,
$$y = (8)^2 - 5(8) - 28$$
$$= 64 - 40 - 28$$
$$= -4$$

The solutions are $\left(-\frac{7}{2}, \frac{7}{4}\right)$ and $(8, -4)$.

49. From the points given, the equations are
$a - b + c = -4 \quad$ (1)
$c = -6 \quad$ (2)
$16a + 4b + c = 6 \quad$ (3)

From (2), $c = -6$. Put this value for c into (1):
$a - b - 6 = -4$
$a - b = 2 \quad$ (4)
and into (3):
$16a + 4b - 6 = 6$
$16a + 4b = 12 \quad$ (5)
Add 4 times (4) to (5):
$4a - 4b = 8 \quad$ (4a)
$16a + 4b = 12 \quad$ (5)
$20a = 20$
$a = 1$
Put this value into (4):
$1 - b = 2$
$-b = 1$
$b = -1$
The solution is $a = 1$, $b = -1$, and $c = -6$.

50. Using the points $(-4, 18)$, $(2, 0)$, and $(4, -14)$, the equations are
$16a - 4b + c = 18 \quad$ (1)
$4a + 2b + c = 0 \quad$ (2)
$16a + 4b + c = -14 \quad$ (3)
Subtract (3) from (1):
$16a - 4b + c = 18 \quad$ (1)
$-16a - 4b - c = 14 \quad$ (3a)
$-8b = 32$
$b = -4 \quad$ (4)
Subtract (2) from (1):
$16a - 4b + c = 18 \quad$ (1)
$-4a - 2b - c = 0 \quad$ (2a)
$12a - 6b = 18 \quad$ (5)
Put the value from (4) into (5):
$12a - 6(-4) = 18$
$12a + 24 = 18$
$12a = -6$
$a = -\frac{1}{2}$
Put the values for a and b into (2):

153

Chapter 4 Review

Thus, the parabola is
$y = -\frac{1}{2}x^2 - 4x + 10$. Since the fourth point (–8, 10) is on this parabola, we would get the same result if by using a different set of three points.

Chapter 5

Section 5.1

1. Function; the sales tax can be determined from the price by the formula $t = 0.04p$.

3. Not a function; the annual income may differ for people with the same number of years of education.

5. Function; the weight can be determined from the volume.

7. Independent variable: topic of the indexed item; dependent variable: page or pages on which the topic occurs; not a function: a topic can be found on more than one page.

9. Independent variable: topic of the indexed item; dependent variable: page or pages on which the topic occurs; not a function: a topic can be found on more than one page.

11. Independent variable: student's name; dependent variable: grade(s) on various tests; not a function: the same student may have different grades on different tests.

13. Independent variable: person stepping on scale; dependent variable: person's weight; function: each person has a unique weight (at least at any particular time of weighing).

15. Not a function: some values of x have more than one value of t. For example, when $x = -1$, t can be either 2 or 5.

17. Function: each value of x has a unique value of y.

19. Function: each value of r has a unique value of v.

21. Function: each value of p has a unique value of v.

23. Not a function: some values of T have more than one value of h. For example, on both Jan. 1 and Jan. 7.

25. Function: each value of I has a unique value of T.

27. a. $f(3) = 6 - 2(3) = 6 - 6 = 0$
    b. $f(-2) = 6 - 2(-2) = 6 + 4 = 10$
    c. $f(12.7) = 6 - 2(12.7) = 6 - 25.4 = -19.4$
    d. $f\left(\dfrac{2}{3}\right) = 6 - 2\left(\dfrac{2}{3}\right) = 6 - \dfrac{4}{3} = 4\dfrac{2}{3}$

29. a. $h(0) = 2(0)^2 - 3(0) + 1 = 0 - 0 + 1 = 1$
    b. $h(-1) = 2(-1)^2 - 3(-1) + 1 = 2 + 3 + 1 = 6$
    c. $h\left(\dfrac{1}{4}\right) = 2\left(\dfrac{1}{4}\right)^2 - 3\left(\dfrac{1}{4}\right) + 1$
       $= \dfrac{1}{8} - \dfrac{3}{4} + 1 = \dfrac{3}{8}$
    d. $h(-6.2) = 2(-6.2)^2 - 3(-6.2) + 1 = 76.88 + 18.6 + 1 = 96.48$

31. a. $H(4) = \dfrac{2(4) - 3}{(4) + 2} = \dfrac{5}{6}$
    b. $H(-3) = \dfrac{2(-3) - 3}{(-3) + 2} = \dfrac{-9}{-1} = 9$
    c. $H\left(\dfrac{4}{3}\right) = \dfrac{2\left(\frac{4}{3}\right) - 3}{\left(\frac{4}{3}\right) + 2} = \dfrac{-\frac{1}{3}}{\frac{10}{3}} = -\dfrac{1}{10}$

155

Chapter 5, Section 5.1

d. $H(4.5) = \dfrac{2(4.5) - 3}{(4.5) + 2} = \dfrac{6}{6.5}$
≈ 0.923

33. a. $E(16) = \sqrt{(16) - 4} = \sqrt{12} = 2\sqrt{3}$

b. $E(4) = \sqrt{(4) - 4} = \sqrt{0} = 0$

c. $E(7) = \sqrt{(7) - 4} = \sqrt{3}$

d. $E(4.2) = \sqrt{(4.2) - 4} = \sqrt{0.2}$
≈ 0.447

35. a. When $p = 25$, $v = 60.0$, so $g(25) = 60.0$.

b. When $p = 40$, $v = 37.5$, so $g(40) = 37.5$.

c. $x = 30$ (when $p = 30$, $v = 50.0$).

37. a. When
$I = 8750$ ($7010 < 8750 < 9169$),
$T = 15\%$, so $T(8750) = 15\%$.

b. When
$I = 6249$ ($4750 < 6249 < 7009$),
$T = 14\%$, so $T(6249) = 14\%$.

c. $x =$ any number between \$7010 and \$9169 inclusive (when $7010 \le I \le 9169$, $T = 15\%$).

39. a. Independent variable: t, the number of years; dependent variable: V, the value of the computer after t years.

b.

t	V
0 years	\$28,000
1 year	\$26,320
2 years	\$24,640
3 years	\$22,960
4 years	\$21,280

c. $V(10) = 28{,}000[1 - 0.06(10)]$
$= 28{,}000(1 - 0.6)$
$= 28{,}000(0.4) = \$11{,}200$

The value of the computer after 10 years will be \$11,200.

41. a. Independent variable: x, the number of clients; dependent variable: R, the revenue.

b.

x	R
0	800
10	3,800
20	16,800
30	39,800
40	72,800

c. $R(40) = 50(40)^2 - 200(40) + 800$
$= 50(1600) - 8000 + 800$
$= 80{,}000 - 7200 = 72{,}800$
When the company has 40 clients it will accrue 72,800 thousand dollars or \$72,800,000 in revenue.

43. a. Independent variable: p, the price of a compact car; dependent variable: N, the number of compact cars the dealership can sell.

b.

p	N
\$4000	3000
\$5000	2400
\$6000	2000
\$7000	≈ 1714
\$8000	1500

c. $N(6000) = \dfrac{12{,}000{,}000}{6000} = 2000$
If the dealership charges \$6000 for a compact car, it can sell 2000 of them.

45. a. Independent variable: d, the length of skid marks; dependent variable: v, the velocity of the car.

b.

d	v
150 ft	42.4 mph
200 ft	49.0 mph
250 ft	54.8 mph
300 ft	60.0 mph
350 ft	64.8 mph

c. $v(250) = \sqrt{12(250)} = \sqrt{3000}$
≈ 54.772
If the skid mark is 250 ft. long, the car was traveling 54.77 mph.

47. a. $G(3a) = 3(3a)^2 - 6(3a)$
$= 27a^2 - 18a$

b. $G(a+2) = 3(a+2)^2 - 6(a+2)$
$= (3a^2 + 12a + 12) + (-6a - 12)$
$= 3a^2 + 6a$

c. $G(a) + 2 = (3a^2 - 6a) + 2$
$= 3a^2 - 6a + 2$

d. $G(-a) = 3(-a)^2 - 6(-a)$
$= 3a^2 + 6a$

49. a. $g(2) = 8$
b. $g(8) = 8$
c. $g(a+1) = 8$
d. $g(-x) = 8$

51. a. $P(2x) = (2x)^3 - 1 = 8x^3 - 1$
b. $2P(x) = 2(x^3 - 1) = 2x^3 - 2$
c. $P(x^2) = (x^2)^3 - 1 = x^6 - 1$
d. $[P(x)]^2 = (x^3 - 1)^2 = x^6 - 2x^3 + 1$

53. a. $f(2) + f(3)$
$= [3(2) - 2] + [3(3) - 2]$
$= 4 + 7 = 11$

b. $f(2 + 3) = 3(2 + 3) - 2 = 15 - 2$
$= 13$

c. $f(a) + f(b)$
$= [3(a) - 2] + [3(b) - 2]$
$= (3a - 2) + (3b - 2)$
$= 3a + 3b - 4$

d. $f(a + b) = 3(a + b) - 2$
$= 3a + 3b - 2;$
$f(a + b) \neq f(a) + f(b)$

55. a. $f(2) + f(3)$
$= [(2)^2 + 3] + [(3)^2 + 3]$
$= 7 + 12 = 19$

b. $f(2 + 3) = (2 + 3)^2 + 3$
$= 25 + 3 = 28$

c. $f(a) + f(b)$
$= [(a)^2 + 3] + [(b)^2 + 3]$
$= (a^2 + 3) + (b^2 + 3)$
$= a^2 + b^2 + 3$

d. $f(a + b) = (a + b)^2 + 3$
$= (a^2 + 2ab + b^2) + 3$
$= a^2 + b^2 + 2ab + 3;$
$f(a + b) \neq f(a) + f(b)$

57. a. $f(2) + f(3) = \sqrt{(2) + 1} + \sqrt{(3) + 1}$
$= \sqrt{3} + \sqrt{4} = 2 + \sqrt{3}$

b. $f(2 + 3) = \sqrt{(2 + 3) + 1} = \sqrt{6}$

c. $f(a) + f(b) = \sqrt{(a) + 1} + \sqrt{(b) + 1}$
$= \sqrt{a + 1} + \sqrt{b + 1}$

d. $f(a + b) = \sqrt{(a + b) + 1}$
$= \sqrt{a + b + 1};$
$f(a + b) \neq f(a) + f(b)$

59. a. $f(2) + f(3) = \dfrac{-2}{(2)} + \dfrac{-2}{(3)}$
$= -1 - \dfrac{2}{3} = -\dfrac{5}{3}$

b. $f(2 + 3) = \dfrac{-2}{(2 + 3)} = -\dfrac{2}{5}$

Chapter 5, Section 5.1

c. $f(a) + f(b) = \dfrac{-2}{(a)} + \dfrac{-2}{(b)}$

$= \dfrac{-2}{a} + \dfrac{-2}{b} = \dfrac{-2(a+b)}{ab}$

d. $f(a+b) = \dfrac{-2}{(a+b)} = \dfrac{-2}{a+b}$;

$f(a+b) \neq f(a) + f(b)$

61. a. $f(0) = 2(0)^2 + 3(0) = 0 + 0 = 0$;
$g(0) = 5 - 6(0) = 5 - 0 = 5$

b. $f(x) = 0$
$2x^2 + 3x = 0$
$x(2x + 3) = 0$
$x = 0$ or $2x + 3 = 0$
$x = 0$ or $x = -\dfrac{3}{2}$

c. $g(x) = 0$
$5 - 6x = 0$
$x = \dfrac{5}{6}$

d. $f(x) = g(x)$
$2x^2 + 3x = 5 - 6x$
$2x^2 + 9x - 5 = 0$
$(x+5)(2x-1) = 0$
$x + 5 = 0$ or $2x - 1 = 0$
$x = -5$ or $x = \dfrac{1}{2}$

63. a. $f(0) = \sqrt{(0) + 2} = \sqrt{2}$;
$g(0) = 3(0) - 4 = 0 - 4 = -4$

b. $f(x) = 0$
$\sqrt{x+2} = 0$
$x + 2 = 0^2 = 0$
$x = -2$

c. $g(x) = 0$
$3x - 4 = 0$
$3x = 4$
$x = \dfrac{4}{3}$

d. $f(x) = g(x)$
$\sqrt{x+2} = 3x - 4$
$x + 2 = (3x - 4)^2$
$x + 2 = 9x^2 - 24x + 16$
$9x^2 - 25x + 14 = 0$
$(9x - 7)(x - 2) = 0$
$9x - 7 = 0$ or $x - 2 = 0$
$x = \dfrac{7}{9}$ or $x = 2$

65.

x	$f(x)$
1	$1 = 1 + 0 = 1 + 3(0)$ $= 1 + 3(1 - 1)$
2	$4 = 1 + 3 = 1 + 3(1)$ $= 1 + 3(2 - 1)$
3	$7 = 1 + 6 = 1 + 3(2)$ $= 1 + 3(3 - 1)$
4	$10 = 1 + 9 = 1 + 3(3)$ $= 1 + 3(4 - 1)$
5	$13 = 1 + 12 = 1 + 3(4)$ $= 1 + 3(5 - 1)$
x	$1 + 3(x - 1) = 1 + 3x - 3$ $= 3x - 2$

$f(x) = 3x - 2$

67.

t	$G(t)$
1	$2 = 1 + 1 = 1^2 + 1$
2	$5 = 4 + 1 = 2^2 + 1$
3	$10 = 9 + 1 = 3^2 + 1$
4	$17 = 16 + 1 = 4^2 + 1$
5	$26 = 25 + 1 = 5^2 + 1$
t	$t^2 + 1$

$G(t) = t^2 + 1$

Section 5.2

1. a. The points $(-3, -2)$, $(1, 0)$, and $(3, 5)$ lie on the graph, so $h(-3) = -2$, $h(1) = 0$, and $h(3) = 5$.

b. The point $(2, 3)$ lies on the graph, so $h(z) = 3$ when $z = 2$.

c. The points $(-2, 0)$, $(0, -2)$, and $(1, 0)$ lie on the graph, so the z-intercepts are $z = -2$ and $z = 1$, and the h-intercept is $h = -2$; $h(-2) = 0$, $h(0) = -2$, and $h(1) = 0$.

d. The highest point on the graph is $(3, 5)$ so the maximum value of $h(z)$ is 5.

e. The maximum value occurs when $z = 3$.

f. The first coordinates vary from -4 to 3, so the domain is $[-4, 3]$. The second coordinates vary from -5 to 5, so the range is $[-5, 5]$.

3. a. $(1, -1)$ is on the graph, so $R(1) = -1$; $(3, 2)$ is on the graph, so $R(3) = 2$.

b. $(3, 2)$ and $(\approx -1.5, 2)$ are on the graph, so $R(p) = 2$ when $p = 3$ and $p \approx -1.5$.

c. $(-2, 0)$, $(0, 4)$, $(2, 0)$, and $(4, 0)$ are on the graph, so the p-intercepts are -2, 2, and 4, and the R-intercept is 4. $R(-2) = 0$, $R(0) = 4$, $R(2) = 0$, and $R(4) = 0$.

d. The highest point is $(0, 4)$ and the lowest is $(5, -5)$, so $R(p)$ has a maximum of 4 and a minimum of -5.

e. The maximum occurs when $p = 0$ and the minimum occurs when $p = 5$.

f. The first coordinates vary from -3 to 5, so the domain is $[-3, 5]$. The second coordinates vary from -5 to 4 so the range is $[-5, 4]$.

5. a. $(0, 0)$, $\left(\dfrac{1}{6}, \dfrac{1}{2}\right)$, and $(-1, 0)$ are on the graph, so $S(0) = 0$, $S\left(\dfrac{1}{6}\right) = \dfrac{1}{2}$, and $S(-1) = 0$.

b. $\left(\dfrac{1}{3}, \approx \dfrac{5}{6}\right)$ is on the graph, so $S\left(\dfrac{1}{3}\right) \approx \dfrac{5}{6}$.

c. $\left(-\dfrac{5}{6}, -\dfrac{1}{2}\right)$, $\left(-\dfrac{1}{6}, -\dfrac{1}{2}\right)$, $\left(\dfrac{7}{6}, -\dfrac{1}{2}\right)$, and $\left(\dfrac{11}{6}, -\dfrac{1}{2}\right)$ are on the graph, so $S(x) = -\dfrac{1}{2}$ when $x = -\dfrac{5}{6}, -\dfrac{1}{6}, \dfrac{7}{6},$ and $\dfrac{11}{6}$.

d. The maximum value for $S(x)$ is 1. The minimum value is -1.

e. The maximum occurs when $x = -\dfrac{3}{2}$ and $x = \dfrac{1}{2}$. The minimum value occurs when $x = -\dfrac{1}{2}$ and $x = \dfrac{3}{2}$.

f. The domain is $[-2, 2]$. The range is $[-1, 1]$.

Chapter 5, Section 5.2

7. a. $(-3, 2)$, $(-2, 2)$, and $(2, 1)$ are on the graph, so $F(-3) = 2$, $F(-2) = 2$, and $F(2) = 1$.

b. $F(s) = -1$ for $-6 \leq s < -4$ and for $0 \leq s < 2$, i.e., s in $[-6, -4) \cup [0, 2)$.

c. The maximum is $F(s) = 2$. The minimum is $F(s) = -1$.

d. The maximum occurs when $-3 \leq s < -1$ or $3 \leq s < 5$. The minimum occurs when $-6 \leq s < -4$ or $0 \leq s < 2$.

e. The domain is $[-6, 5)$. The range is $s = -1, 2,$ or 2.

9.

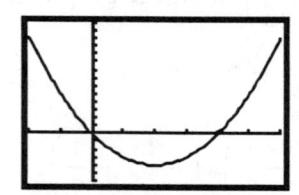

Xmin = -2, Xmax = 6,
Ymin = -6, and Ymax = 14
Domain: $[-2, 6]$
Range: $[-4, 12]$

11.

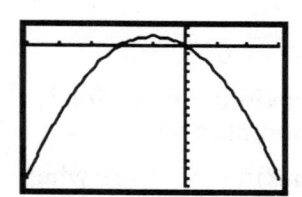

Xmin = -5, Xmax = 3,
Ymin = -16, and Ymax = 2
Domain: $[-5, 3]$
Range: $[-15, 1]$

13.

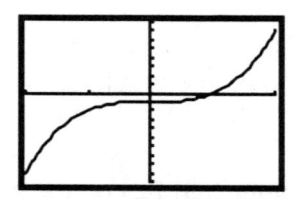

Xmin = -2, Xmax = 2,
Ymin = -10, and Ymax = 8
Domain: $[-2, 2]$
Range: $[-9, 7]$

15.

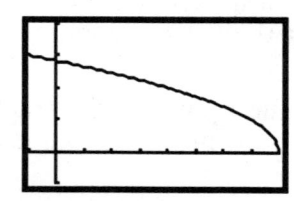

Xmin = -1, Xmax = 8,
Ymin = -1, and Ymax = 4
Domain: $[-1, 8]$
Range: $[0, 3]$

17.

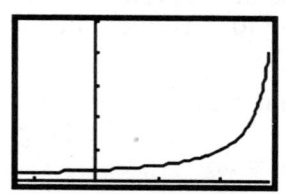

Xmin = -1.25, Xmax = 2.75,
Ymin = 0, and Ymax = 5
Domain: $[-1.25, 2.75]$
Range: $\left[\dfrac{4}{17}, 4\right]$

Chapter 5, Section 5.3

19.

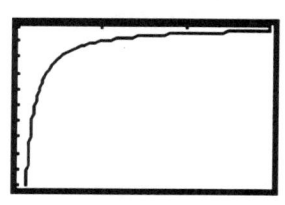

Xmin = 3, Xmax = 6,
Ymin = –10, and Ymax = 1
Domain: (3, 6]
Range: $\left(-\infty, -\dfrac{1}{3}\right)$

21. Function, since no vertical line intersects the graph in more than one point.

23. Not a function, since there are vertical lines which intersect the graph in more than one point.

25. Not a function, since there are vertical lines which intersect the graph in more than one point.

27. Function, since no vertical line intersects the graph in more than one point.

29. Not a function, since there is (exactly one) vertical line which intersects the graph in more than one point.

Section 5.3

1. Refer to the graph in the back of the textbook.

3. Refer to the graph in the back of the textbook.

5. Refer to the graph in the back of the textbook.

7. Refer to the graph in the back of the textbook.

9. Refer to the graph in the back of the textbook.

11. Refer to the graph in the back of the textbook.

13. 8; $8^3 = 512$

15. –0.4; $(-0.4)^3 = -0.064$

17. ≈2.080; $2.080^3 = 8.998912$

19. ≈ –0.126; $(-0.126^3) = -0.002000376$

21. a. $-|-9| = -(+9) = -9$
    b. $-(-9) = 9$

23. a. $|-8| - |12| = 8 - 12 = -4$
    b. $|-8 - 12| = |-20| = 20$

25. $4 - 9|2 - 8| = 4 - 9|-6| = 4 - 9 \cdot 6$
    $= 4 - 54 = -50$

27. $|-4 - 5| |1 - 3(-5)| = |-9| |1 + 15|$
    $= 9|16| = 9 \cdot 16 = 144$

29. $||-5| - |-6|| = |5 - 6| = |-1| = 1$

31. a.

x	f(x)
–3	–27
–2	–8
–1	–1
0	0
1	1
2	8
3	27

Chapter 5, Section 5.3

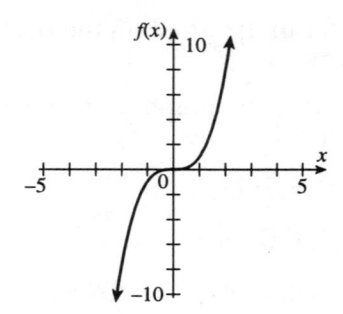

    b. Domain: $(-\infty, \infty)$
       Range: $[-\infty, \infty)$

    c.

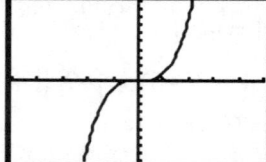

Xmin = –5, Xmax = 5,
Ymin = –10, and Ymax = 10

33. a.

x	f(x)
0	0
1	1
4	2
9	3

    b. Domain: $[0, \infty)$
       Range: $[0, \infty)$

    c.

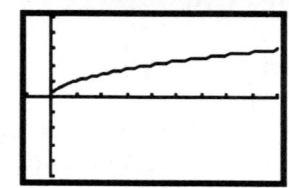

Xmin = –1, Xmax = 9,
Ymin = –5, and Ymax = 5

35. a.

x	f(x)
–10	$-\dfrac{1}{10}$
–2	$-\dfrac{1}{2}$
–1	–1
$-\dfrac{1}{2}$	–2
$-\dfrac{1}{10}$	–10
0	Undefined
$\dfrac{1}{10}$	10
$\dfrac{1}{2}$	2
1	1
2	$\dfrac{1}{2}$
10	$\dfrac{1}{10}$

162

Chapter 5, Section 5.3

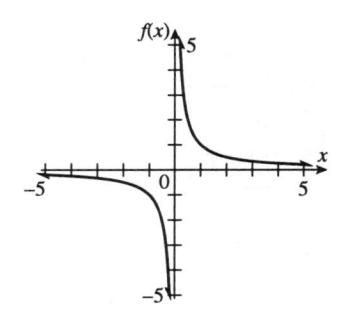

b. Domain: $(-\infty, 0) \cup (0, \infty)$
Range: $(-\infty, 0) \cup (0, \infty)$

c.

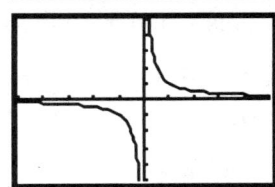

Xmin = –5, Xmax = 5,
Ymin = –5, and Ymax = 5

37. a. 3.2, 6.3
  b. ±2.6, ±3.5
  c. 3.2, –2.5
  d. ±3.9, ±1.9
  e. 10.2, 5.3
  f. $2 \leq x \leq 3.2, -3.2 \leq x \leq -2$

39. a. 0.3, –1.3
  b. –0.4, 0.9
  c. 0.2, –5
  d. $0.2 \leq h \leq 3.3$

41. Refer to the graph in the back of the textbook.
43. Refer to the graph in the back of the textbook.
45. Refer to the graph in the back of the textbook.
47. Refer to the graph in the back of the textbook.
49. Refer to the graph in the back of the textbook.
51. Refer to the graph in the back of the textbook.
53. $y = \sqrt{x}$
55. $y = |x|$
57. $y = x^3$
59. $y = \sqrt[3]{x}$
61. $y = \dfrac{1}{x}$
63. $y = \dfrac{1}{x^2}$
65.

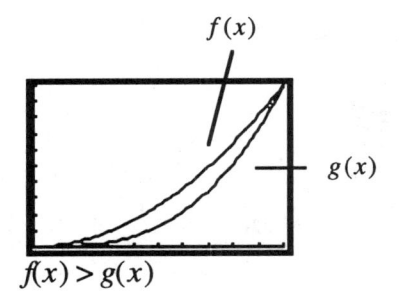

$f(x) > g(x)$

163

Chapter 5, Section 5.4

67.

$g(x) > f(x)$

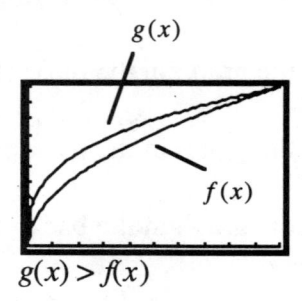

69.

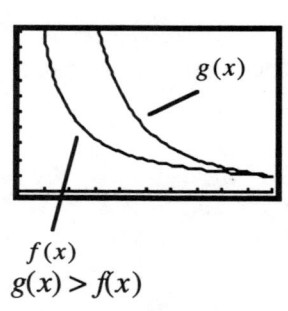

$g(x) > f(x)$

71. Increasing on entire domain:
$y = x$, $y = x^3$, $y = \sqrt{x}$, $y = \sqrt[3]{x}$
Decreasing on entire domain:
$y = \dfrac{1}{x}$

73. All *except* $y = \dfrac{1}{x}$ and $y = \dfrac{1}{x^2}$ can be graphed in one piece.

Section 5.4

1. a. $y = kx$; but $y = 1.5$ when $x = 5$, so $1.5 = k(5)$, so $k = \dfrac{1.5}{5} = 0.3$. Thus, $y = 0.3x$.

b.

x	y
2	0.6
5	1.5
8	2.4
12	3.6
15	4.5

c.

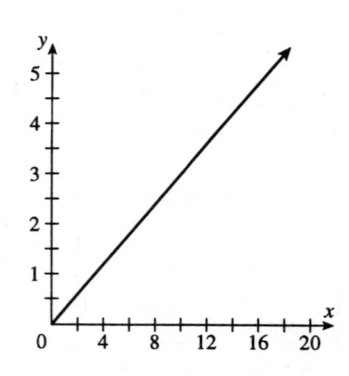

d. *y* doubles when *x* doubles.

3. a. $y = kx^2$; but $y = 24$ when $x = 6$, so $24 = k(6)^2 = 36k$, so $k = \dfrac{24}{36} = \dfrac{2}{3}$.
Thus, $y = \dfrac{2}{3}x^2$.

b.

x	y
3	6
6	24
9	54
12	96
15	150

164

c.

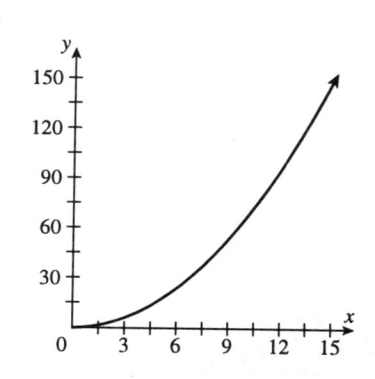

d. y quadruples when x doubles.

5. a. $y = \dfrac{k}{x}$; but $y = 6$ when $x = 20$, so $6 = \dfrac{k}{20}$, so $k = 120$. Thus, $y = \dfrac{120}{x}$.

b.

x	y
4	30
8	15
20	6
30	4
40	3

c.

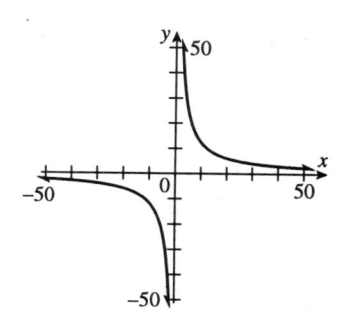

d. y becomes half as big when x doubles.

7. (b) goes up continuously, starting from 0.

9. Only (c) has y getting larger without limit as x approaches 0, and y approaching 0 as x gets larger without limit, so only (c) could represent inverse variation.

11.

x	y	$\dfrac{y}{x}$	$\dfrac{y}{x^2}$
2	2	1	0.5
3	4.5	1.5	0.5
5	12.5	2.5	0.5
8	32	4	0.5

(b) y varies directly with x^2, with $k = 0.5$ ($y = 0.5x^2$).

13.

x	y	$\dfrac{y}{x}$	$\dfrac{y}{x^2}$	$\dfrac{y}{x^3}$	etc.
1.5	3	2	1.33	0.89	etc.
2.4	7.2	3	1.25	0.52	etc.

(c) y divided by powers of x does not yield constants, so y does not vary directly as a power of x.

15.

x	y	xy	x^2y
0.5	288	144	72
2	18	36	72
3	8	24	72
6	2	12	72

(b) y varies inversely as x^2, with $k = 72$ $\left(y = \dfrac{72}{x^2}\right)$.

Chapter 5, Section 5.4

17.

x	y	x + y
1	4	5
1.3	3.7	5
3	2	5
4	1	5

(c) $y = 5 - x$, rather than $y = \dfrac{k}{x^n}$, so y does not vary inversely with a power of x.

19. a.

Length	Width	Perimeter	Area	$\dfrac{\text{Perimeter}}{\text{Length}}$	$\dfrac{\text{Area}}{\text{Length}}$
10	8	36	80	3.6	8
12	8	40	96	≈ 3.33	8
15	8	46	120	-	8
20	8	56	160	-	8

b. The perimeter does not vary directly with the length.

c. $P = 2L + 16$

d. The area does vary directly with the length.

e. $A = 8L$

21. a. $m = kw$; but $m = 24.75$ when $w = 150$, so $k = \dfrac{24.75}{150} = 0.165$. Thus $m = 0.165w$.

b. $m = 0.165(120) = 19.8$ lb.

c. $w = \dfrac{50}{0.165} \approx 303$ lb.

d.

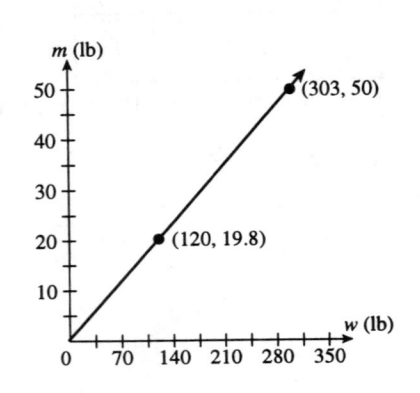

23. a. $L = kT^2$; but $L = 3.25$ ft. when $T = 2$ sec, so $3.25 = k(2)^2$
$k = 0.8125$. Thus, $L = 0.8125T^2$.

b. $L = 0.8125(17)^2 \approx 234.8$ ft.

c. $\dfrac{9}{12} = 0.8125T^2$, so

$T^2 = \dfrac{\frac{9}{12}}{0.8125} \approx 0.923$

$T \approx \sqrt{0.923} \approx 0.96$ sec.

d.

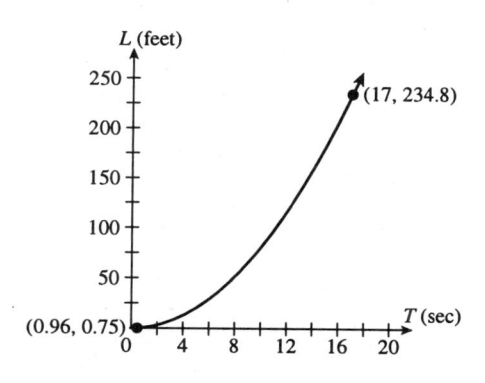

Graph: L (feet) vs T (sec), curve from (0.96, 0.75) to (17, 234.8)

25. a. $F = \dfrac{k}{d}$; but $F = 22$ milligauss when $d = 4$ in., so $22 = \dfrac{k}{4}$

$k = 22(4) = 88$. Thus, $F = \dfrac{88}{d}$.

b. $L = \dfrac{88}{10} = 8.8$ milligauss.

c. $L = 4.3 = \dfrac{88}{d}$, so $d = \dfrac{88}{4.3} \approx 20.5$ in. The computer user should sit at least 20.5 inches from the screen.

d.

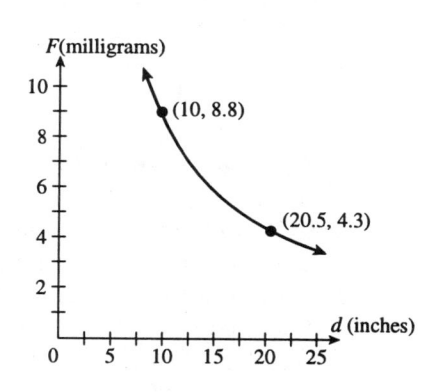

Graph: F (milligrams) vs d (inches), curve through (10, 8.8) and (20.5, 4.3)

27. a. $P = kw^3$; but $P = 7300$ kW when $w = 32$ mph, so $7300 = k(32)^3$

$k = \dfrac{7300}{32^3} \approx 0.223$. Thus, $P \approx 0.223 w^3$.

b. $P \approx 0.223(15)^3 \approx 752$ kW.

c. $P = 10,000 \approx 0.223 w^3$, so

$w^3 = \dfrac{10,000}{0.223} \approx 44,888.67$

$w \approx \sqrt[3]{44,887.67} \approx 35.5$ mph.

d.

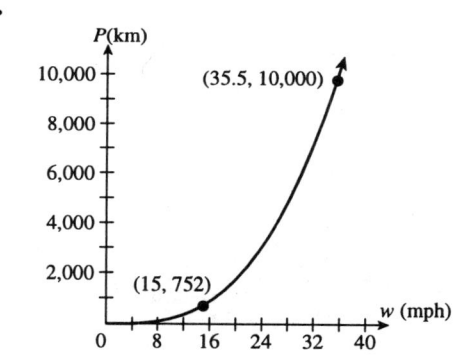

Graph: P(km) vs w (mph), curve through (15, 752) and (35.5, 10,000)

29.

v	d	$\dfrac{d}{v^2}$
10	0.5	0.005
20	2	0.005
40	8	0.005

a. $d = 0.005 v^2$

b. $d = 0.005(100)^2 = 50$ m.

31.

p	m	pm
1	8	8
2	4	8
4	2	8
8	1	8

Chapter 5, Section 5.5

a. $m = \dfrac{8}{p}$

b. $m = \dfrac{8}{10} = 0.8$ tons.

33.

d	t	dt
0.5	12	6
1	6	6
2	3	6
3	2	6

a. $t = \dfrac{6}{d}$

b. $t = \dfrac{6}{6} = 1°$ C.

35.

d	w	$\dfrac{w}{d^2}$
0.5	150	600
1.0	600	600
1.5	1350	600
2.0	2400	600

a. $w = 600 d^2$

b. $w = 600(1.2)^2 = 864$ newtons.

37. The illumination is only $\dfrac{1}{4}$ as strong.

39. Since the speed is now 90% = 0.9 times what it was, the resistance will be $(0.9)^2 = 0.81 = 81\%$ of what it was, so the resistance will decrease by 19%.

41. If $y_1 = kx_1$ and $y_2 = kx_2$, where $x_2 = cx_1$, then
$y_2 = k(cx_1) = c(kx_1) = cy_1$.

43. If $y = kx^2$ then $\dfrac{y}{x^2} = \dfrac{kx^2}{x^2} = k$.

45. If $x = k_1 y$ and $y = k_2 z$, then $x = k_1(k_2 z) = (k_1 k_2)z = k_3 z$, so x varies directly with z.

Section 5.5

1. (b); your pulse rate rises during the class and falls as you rest after the class.

3. (a); your income rises at a steady rate as the number of hours you work increases.

5. Refer to the graph in the back of the textbook. (See the explanation of Example 7 in the text.)

7. Refer to the graph in the back of the textbook. Your distance from the English classroom increases as you leave class, decreases to zero as you return for the book, and then increases as you walk to math class.

9. Refer to the graph in the back of the textbook. The distance increases as she drives to the gym, stays constant while she is at the gym, and increases as she drives to her friend's home, then decreases back to 0 as she returns home.

11. (NOTE: In all of the following, the difference between successive x's is always 1, so the successive slopes are the same as the successive Δy's.)
1-c-D; the function values increase at an increasing rate. $m = 12, 14, 18, 20, 25, 30, 36$. The graph gets steeper and steeper.
2-d-A; the function values continue to increase, but at a decreasing rate. $m = 25, 18, 17, 14, 13, 12, 10$. The graph starts out steep but gets less and less steep.

Chapter 5, Section 5.5

3-b-B; the function values increase at the same rate. $m = 60, 60, 60, ..., 60$. Graph B has a constant slope.

4-a-C; the function values increase at a decreasing rate and level off toward the end. $m = 36, 22, 13, 7, 5, 3, 1$. The graph starts out steep but levels off.

13. The distance from x to the origin is six units. $|x| = 6$

15. $|p - (-3)| = 5$, so $|p + 3| = 5$.

17. The distance from t to 6 is less than or equal to 3, so $|t - 6| \leq 3$.

19. The distance from b to -1 is greater than or equal to 0.5, so $|b - (-1)| \geq 0.5$, so $|b + 1| \geq 0.5$.

21. a. The distance between x and -12 is $|x - (-12)| = |x + 12| = |x + 4|$. The distance between x and -4 is $|x - (-4)| = |x + 4|$. The distance between x and $+24$ is $|x - 24|$.

b. $|x + 12| + |x + 4| + |x - 24|$

c.

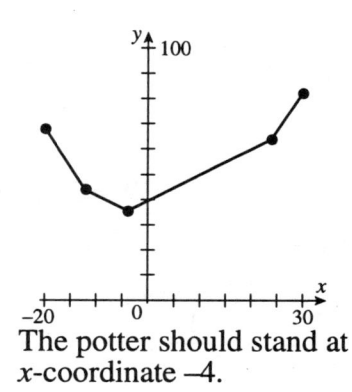

The potter should stand at x-coordinate -4.

23. Let the road be the x-axis and the point where the river crosses the road be the point 0. Then Richard works at $+10$, Marian works at -6, and the health club is at $+2$. Let x be the coordinate of their apartment (obviously the apartment should be between the two work places, that is, $-6 \leq x \leq 10$). Since Marian's distance to work and back is $2|x + 6|$, which is $2(x + 6)$ for $x \geq -6$, and Richard's is $2|x - 10|$, which is $2(10 - x)$ for $x \leq 10$, their total driving distance to and from work is $2(x + 6) + 2(10 - x) = 32$ miles regardless of where they live. Each of them drives $2|x - 2|$ miles to the health club and back, for a total of $4|x - 2|$ miles, which would be a minimum of 0 for $x = 2$, i.e., they should live as close to the health club as possible.

25.

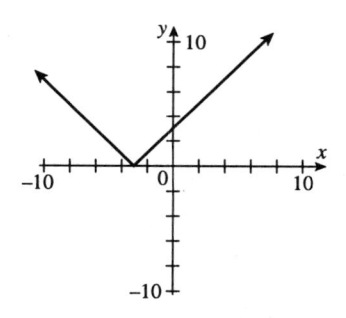

a. $x = -5$ or -1

b. $-7 \leq x \leq 1$

c. $x < -8$ or $x > 2$

27.

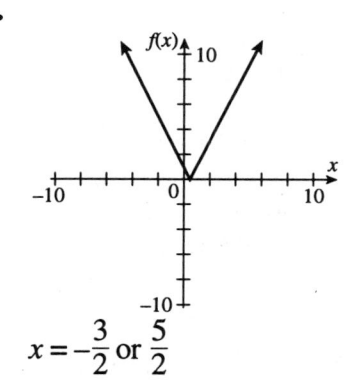

$x = -\dfrac{3}{2}$ or $\dfrac{5}{2}$

Chapter 5, Section 5.5

29.

$-\dfrac{9}{2} < x < -\dfrac{3}{2}$

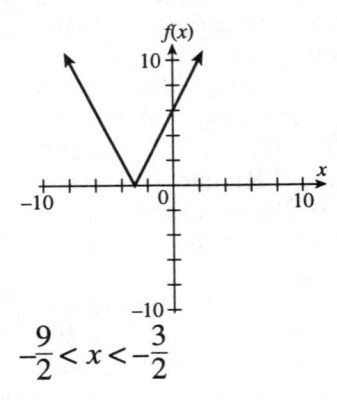

31.

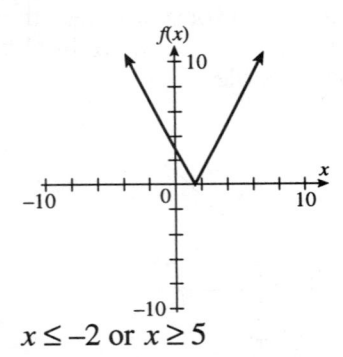

$x \leq -2$ or $x \geq 5$

33. a. $s = 50\sqrt{d}$

   b. $s = 50\sqrt{0.36} = 30$ cm/sec.

35. a. $d = 1.225\sqrt{h}$

   b. $d = 1.225\sqrt{20{,}000} \approx 173.24$ mi.

37. a. $L = 2a$

   b. $H = \sqrt{16 - a^2}$

   c. $A = LH = 2a\sqrt{16 - a^2}$

   d. If $a \leq 0$ the rectangle has no length and if $a \geq 4$ it has no height, so $0 < a < 4$.

   e.

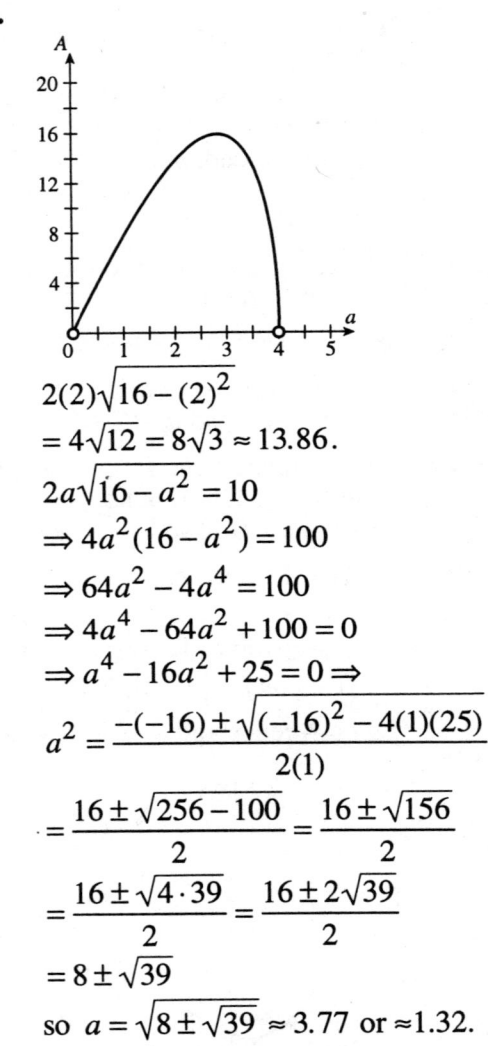

   $2(2)\sqrt{16 - (2)^2}$
   $= 4\sqrt{12} = 8\sqrt{3} \approx 13.86$.
   $2a\sqrt{16 - a^2} = 10$
   $\Rightarrow 4a^2(16 - a^2) = 100$
   $\Rightarrow 64a^2 - 4a^4 = 100$
   $\Rightarrow 4a^4 - 64a^2 + 100 = 0$
   $\Rightarrow a^4 - 16a^2 + 25 = 0 \Rightarrow$

   $a^2 = \dfrac{-(-16) \pm \sqrt{(-16)^2 - 4(1)(25)}}{2(1)}$

   $= \dfrac{16 \pm \sqrt{256 - 100}}{2} = \dfrac{16 \pm \sqrt{156}}{2}$

   $= \dfrac{16 \pm \sqrt{4 \cdot 39}}{2} = \dfrac{16 \pm 2\sqrt{39}}{2}$

   $= 8 \pm \sqrt{39}$

   so $a = \sqrt{8 \pm \sqrt{39}} \approx 3.77$ or ≈ 1.32.

   f. $a \approx 2.83$ gives a maximum of ≈ 16

39. a. The slope from $(2, 1)$ to $(a, 0)$ is
   $\dfrac{1 - 0}{2 - a} = \dfrac{1}{2 - a}$.

b. The slope from $(0, b)$ to $(2, 1)$ is
$$\frac{1-b}{2-0} = \frac{1-b}{2}, \text{ so}$$
$$\frac{1-b}{2} = \frac{1}{2-a} \Rightarrow$$
$$1-b = \frac{2}{2-a} \Rightarrow 1 - \frac{2}{2-a} = b \Rightarrow$$
$$b = \frac{(2-a)-2}{2-a} = \frac{-a}{2-a} = \frac{a}{a-2}.$$

c. $A = \frac{1}{2}ab = \frac{1}{2}a\left(\frac{a}{a-2}\right) = \frac{a^2}{2(a-2)}$

d. If $a \leq 2$, the line will not intersect the y-axis in such a way as to make a triangle. As b gets lower, a gets bigger without limit, so the domain is $(2, \infty)$.

e.

[Graph showing area A vs a, with axes labeled from 0 to 25]

If $a = 3$, the area is 4.5:
$\left(\frac{(3)^2}{2(3-2)} = \frac{9}{2}\right)$; 2.4 or 12.

f. The smallest value (4 sq. units) occurs when $a = 4$. The triangle is undefined for $a = 2$. As a increases from 2 to 4, the area decreases. As a increases past 4, the area increases (with no limit).

41. a. $C = 2\pi r$, so $r = \frac{C}{2\pi}$.

b. $A = \pi r^2 = \pi\left(\frac{C}{2\pi}\right)^2 = \frac{C^2 \pi}{4\pi^2} = \frac{C^2}{4\pi}$

c. Varies directly as the square; Domain: $(0, \infty)$

d. $\frac{(100)^2}{4\pi} \approx 795.8$ sq. yd.

43. Refer to the graph in the back of the textbook.

Chapter 5 Review

1. Function; x determines y.

2. Not a function; when $p = -2$, q can be -4 or 3.

3. Not a function; two students with IQ test scores of 98 had different SAT scores. (However, IQ test score is a function of SAT score, and each is a function of student's ID letter.)

4. Function; number of answers on math quiz determines quiz grade.

5. $N(10) = 2000 + 500(10)$
$= 2000 + 5000 = 7000$.
Ten days after the well is opened, 7000 barrels of oil are pumped.

6. $H(16) = \frac{24}{(16) - 8} = \frac{24}{8} = 3$
If the top speed is 16 mph, it takes 3 hours to travel upstream.

Chapter 5 Review

7. $F(0) = \sqrt{1 + 4(0)^2} = \sqrt{1 + 4 \cdot 0}$
$= \sqrt{1 + 0} = \sqrt{1} = 1$
$F(-3) = \sqrt{1 + 4(-3)^2} = \sqrt{1 + 4 \cdot 9}$
$= \sqrt{1 + 36} = \sqrt{37} \approx 6.08$

8. $H(2a) = (2a)^2 + 2(2a) = 4a^2 + 4a$
$H(a+1) = (a+1)^2 + 2(a+1)$
$= (a^2 + 2a + 1) + (2a + 2)$
$= a^2 + 4a + 3$

9. $f(2) + f(3) = [2 - 3(2)] + [2 - 3(3)]$
$= (2 - 6) + (2 - 9)$
$= (-4) + (-7) = -11$
$f(2 + 3) = 2 - 3(2 + 3) = 2 - 3(5)$
$= 2 - 15 = -13$

10. $f(a) + f(b)$
$= [2(a)^2 - 4] + [2(b)^2 - 4]$
$= (2a^2 - 4) + (2b^2 - 4)$
$= 2a^2 + 2b^2 - 8$
$f(a + b) = 2(a + b)^2 - 4$
$= 2(a^2 + 2ab + b^2) - 4$
$= 2a^2 + 4ab + 2b^2 - 4$

11. a. $P(0) = (0)^2 - 6(0) + 5$
$= 0 - 0 + 5 = 5$

b. $x^2 - 6x + 5 = 0$
$(x - 1)(x - 5) = 0$
$x - 1 = 0$ or $x - 5 = 0$
$x = 1$ or $x = 5$

12. a. $R(0) = \sqrt{4 - 0^2} = \sqrt{4} = 2$

b. $\sqrt{4 - x^2} = 0$
$4 - x^2 = 0^2$
$4 = x^2$
$x = \pm\sqrt{4} = \pm 2$

13. a. Since $(-2, 3)$ is on the graph, $f(-2) = 3$. Since $(2, 5)$ is on the graph, $f(2) = 5$.

b. Since $(1, 4)$ and $(3, 4)$ are on the graph, $f(1) = 4$ and $f(3) = 4$.

c. Since $(-3, 0)$ and $(4, 0)$ are on the graph, the t-intercepts are $t = -3$ and $t = 4$. Since $(0, 2)$ is on the graph, the $f(t)$-intercept is $f(t) = 2$.

d. Since $(2, 5)$ is the highest point on the graph, the maximum value of f is $f(t) = 5$ when $t = 2$.

14. a. Since $(-3, -2)$ is on the graph, $P(-3) = -2$. Since $(3, 3)$ is on the graph, $P(3) = 3$.

b. Since $(-5, 2)$, $\left(-\frac{1}{2}, 2\right)$, and $(4, 2)$ are on the graph, $P(z) = 2$ for $z = -5, -\frac{1}{2},$ and 4.

c. Since $(-4, 0)$, $(-1, 0)$, and $(5, 0)$ are on the graph, the z-intercepts are $z = -4$, $z = -1$, and $z = 5$. Since $(0, 3)$ is on the graph, the $P(z)$-intercept is $P(z) = 3$.

d. Since $(-2, -3)$ is the lowest point on the graph, the minimum value of P is $P(z) = -3$ when $z = -2$.

Chapter 5 Review

15. Refer to the graph in the back of the textbook. Domain: [–2, 4].
The highest point is $\left(\frac{3}{2}, \frac{9}{4}\right)$;
$f(-2) = -(-2)^2 + 3(-2)$
$= -4 - 6 = -10$ and
$f(4) = -(4)^2 + 3(4) = -16 + 12 = -4$,
so the lowest point is $(-2, -10)$. The range is $\left[-10, \frac{9}{4}\right]$.

16. Refer to the graph in the back of the textbook. Domain: [2, 6];
$f(2) = \sqrt{(2) - 2} = \sqrt{0} = 0$ and
$f(6) = \sqrt{(6) - 2} = \sqrt{4} = 2$, so the lowest point is $(2, 0)$ and the highest is $(6, 2)$. The range is $[0, 2]$.

17. Refer to the graph in the back of the textbook. Domain: $(-2, 4]$; $F(x)$ grows without limit as x approaches -2 from the right, so there is no highest point. $F(4) = \dfrac{1}{(4) + 2} = \dfrac{1}{6}$,
so the lowest point is $\left(4, \dfrac{1}{6}\right)$.
The range is $\left[\dfrac{1}{6}, \infty\right)$.

18. Refer to the graph in the back of the textbook. Domain: $[-4, 2)$;
$H(-4) = \dfrac{1}{2 - (-4)} = \dfrac{1}{6}$, so the
lowest point is $\left(-4, \dfrac{1}{6}\right)$. $H(x)$ grows without limit as x approaches 2 from the left, so there is no highest point. The range is $\left[\dfrac{1}{6}, \infty\right)$.

19. Function; each vertical line crosses the graph only once.

20. Not a function; there are vertical lines which cross the graph twice.

21. Not a function; a vertical line at the right end of the graph crosses it infinitely many times.

22. Function; each vertical line crosses the graph only once.

23. Refer to the graph in the back of the textbook.

24. Refer to the graph in the back of the textbook.

25. Refer to the graph in the back of the textbook.

26. Refer to the graph in the back of the textbook.

27. Refer to the graph in the back of the textbook.

28. Refer to the graph in the back of the textbook.

29. Refer to the graph in the back of the textbook.

30. Refer to the graph in the back of the textbook.

31. Refer to the graph in the back of the textbook.

32. Refer to the graph in the back of the textbook.

33. Refer to the graph in the back of the textbook.
    a. $x \approx 0.5$ $(x = 0.512)$
    b. $x \approx 3.4$ $(x = 3.375)$
    c. $x > \approx 4.9$ $(x > 4.913)$
    d. $x \leq \approx 2.0$ $(x \leq 2.000376)$

173

Chapter 5 Review

34. Refer to the graph in the back of the textbook.
 a. $x = 0.4$
 b. $x = 3.2$
 c. $0 < x \le \dfrac{9}{2}$
 d. $x > \dfrac{1}{5}$ or $x < 0$

35. Refer to the graph in the back of the textbook
 a. $x \approx \pm 5.8 \left(x = \pm \dfrac{10\sqrt{3}}{3} \approx \pm 5.774\right)$
 b. $x = \pm 0.4$
 c. $-2.5 < x < 0$ or $0 < x < 2.5$
 d. $x \le -0.5$ or $x \ge 0.5$

36. Refer to the graph in the back of the textbook
 a. $x \approx 0.5$ (probably $0.707 \approx \dfrac{\sqrt{2}}{2}$, so $x = 0.5$)
 b. $x \approx 2.9$ ($x = 2.89$)
 c. $0 < x < 2.25$
 d. $x \ge \approx 1.7$ ($x \ge 1.69$)

37.

x	y	$\dfrac{y}{x^2}$
2	4.8	1.2
5	30	1.2
8	76.8	1.2
11	145.2	1.2

$y = 1.2x^2$

38.

x	y	$\dfrac{y}{x}$
1.4	75.6	54
2.3	124.2	54
5.9	318.6	54
8.3	448.2	54

$y = 54x$

39.

x	y	xy
0.5	40	20
2	10	20
4	5	20
8	2.5	20

$y = \dfrac{20}{x}$

40.

x	y	x^2y
1.5	320	720
2.5	115.2	720
4	45	720
6	20	720

$y = \dfrac{720}{x^2}$

41. a. $s = kt^2$; but $s = 28$ cm when $t = 4$ sec, so $28 = k(4)^2 = 16k$, so
$k = \dfrac{28}{16} = 1.75$
$s = 1.75t^2$.
 b. $s = 1.75(6)^2 = 1.75(36) = 63$ cm

Chapter 5 Review

42. a. $V = \dfrac{kT}{P}$; but $V = 40$ when $T = 300$ and $P = 30$, so
$(40) = \dfrac{k(300)}{(30)}$, $k = 40 \cdot \dfrac{30}{300} = 4$.
$V = \dfrac{4T}{P}$.

b. $V = \dfrac{4(320)}{(40)} = 32$

43. Refer to the graph in the back of the textbook. $d = \dfrac{k}{p}$; but $d = 600$ bottles when $p = \$8$, so $(600) = \dfrac{k}{(8)}$, so $k = 600 \cdot 8 = 4800$, so $d = \dfrac{4800}{p}$;
$d(10) = \dfrac{4800}{(10)} = 480$ bottles.

44.

d	I
3	100
6	$\dfrac{100}{(2)^2} = 25$
9	$\dfrac{100}{(3)^2} \approx 11.1$

$I = \dfrac{k}{d^2}$; but $I = 100$ lumens when $d = 3$ ft, so $(100) = \dfrac{k}{(3)^2}$.
$k = 100(3)^2 = 900$, so $I = \dfrac{900}{d^2}$;
$I(8) = \dfrac{900}{(8)^2} = 14.0625$ lumens.

45. a. $w(r) = \dfrac{k}{r^2}$

b. Refer to the graph in the back of the textbook.

c. For w to be $\dfrac{1}{3}$ as large, r^2 must be 3 times as large, so r must be $\sqrt{3}$ times as large, or $3960\sqrt{3} \approx 6858.92$ mi.

46. a. $T = k\sqrt{L}$

b. Refer to the graph in the back of the textbook.

c. The period will be
$\sqrt{\dfrac{4}{5}} = \dfrac{2\sqrt{5}}{5} \approx 0.894$ times as long.

47. Refer to the graph in the back of the textbook.

48. Refer to the graph in the back of the textbook.

49. a. Refer to the graph in the back of the textbook.

b.

x	g(x)	xg(x)
2	12	24
3	8	24
4	6	24
6	4	24
8	3	24
12	2	24

$g(x) = \dfrac{24}{x}$

50. a. Refer to the graph in the back of the textbook.

Chapter 5 Review

b.

x	$F(x)$	$\dfrac{F(x)}{x^3}$
−2	8	−1
−1	1	−1
0	0	Undefined
1	−1	−1
2	−8	−1
3	−27	−1

$F(x) = -x^3$

51. $|x| = 4$

52. $|y + 5| = 3$

53. $|p - 7| < 4$

54. $|q + 4| \geq 0.3$

55. Refer to the graph in the back of the textbook. $-\dfrac{2}{3} < x < 2$

56. Refer to the graph in the back of the textbook. $-0.4 \leq x \leq 0.1$

57. Refer to the graph in the back of the textbook. $y \leq -0.9$ or $y \geq 0.1$

58. Refer to the graph in the back of the textbook. $z < -\dfrac{5}{18}$ or $z > -\dfrac{1}{18}$.

59. a. $a = \dfrac{s\sqrt{3}}{2}$

  b. $A(s) = \dfrac{1}{2} as = \dfrac{1}{2} \cdot \dfrac{s\sqrt{3}}{2} \cdot s = \dfrac{s^2 \sqrt{3}}{4}$

  c. Refer to the graph in the back of the textbook.

  d. $A(4) = \dfrac{(4)^2 \sqrt{3}}{4} = 4\sqrt{3}$
  ≈ 6.928 sq. cm

  e. $\dfrac{s^2 \sqrt{3}}{4} = 2.7$, so
  $s = \sqrt{\dfrac{4(2.7)}{\sqrt{3}}} \approx 2.5$ ft.

60. a. Length $= a$; height $= b = 10 - \dfrac{1}{2}a$

  b. $A(a) =$ length $\cdot$ height $= ab$
  $= a\left(10 - \dfrac{1}{2}a\right) = 10a - \dfrac{1}{2}a^2$

  c. The lines intersect the axes at (0, 10) and (20, 0) so the domain is (0, 20).

  d. Refer to the graph in the back of the textbook.

  e. $P(a) = 2(a + b)$
  $= 2\left[a + \left(10 - \dfrac{1}{2}a\right)\right] = a + 20$

Chapter 6

Section 6.1

1. $b^4 \cdot b^5 = b^{4+5} = b^9$ By the first law of exponents

3. $6^5 \cdot 6^3 = 6^{5+3} = 6^8$ By the first law of exponents

5. $(q^3)(q)(q^5) = q^{3+1+5} = q^9$ By the first law of exponents

7. $(4y)(-6y) = (-24)(y \cdot y)$ Rearrange factors
   $= (-24)(y^{1+1})$ Apply first law
   $= -24y^2$ Add exponents

9. $(2wz^3)(-8z) = (-16)(w)(z^3 \cdot z)$ Rearrange factors
   $= (-16)(w)(z^{3+1})$ Apply first law
   $= -16wz^4$ Add exponents

11. $-4x(3xy)(xy^3) = -12(x \cdot x \cdot x)(y \cdot y^3)$ Rearrange factors
    $= -12(x^{1+1+1})(y^{1+3})$ Apply first law
    $= -12x^3 y^4$ Add exponents

13. $-7ab^2(-3ab^3) = 21(a \cdot a)(b^2 \cdot b^3)$ Rearrange factors
    $= 21(a^{1+1})(b^{2+3})$ Apply first law
    $= 21a^2 b^5$ Add exponents

15. $(-2x^3)(x^2 y)(-4y^2) = 8(x^3 \cdot x^2)(y \cdot y^2)$ Rearrange factors
    $= 8(x^{3+2})(y^{1+2})$ Apply first law
    $= 8x^5 y^3$ Add exponents

17. $y^2 z(-3x^2 z^3)(-y^4 z) = 3(x^2)(y^2 \cdot y^4)(z \cdot z^3 \cdot z)$ Rearrange factors
    $= 3(x^2)(y^{2+4})(z^{1+3+1})$ Apply first law
    $= 3x^2 y^6 z^5$ Add exponents

19. $\dfrac{w^6}{w^3} = w^{6-3}$ Apply second law ($6 > 3$)
    $= w^3$ Subtract exponents

Chapter 6, Section 6.1

21. $\dfrac{2^9}{2^4} = 2^{9-4}$ Apply second law $(9 > 4)$
   $\phantom{\dfrac{2^9}{2^4}} = 2^5$ Subtract exponents

23. $\dfrac{z^6}{z^9} = \dfrac{1}{z^{9-6}}$ Apply second law $(6 < 9)$
   $\phantom{\dfrac{z^6}{z^9}} = \dfrac{1}{z^3}$ Subtract exponents

25. $\dfrac{2a^3 b}{8a^4 b^5} = \left(\dfrac{1}{4}\right)\left(\dfrac{a^3}{a^4}\right)\left(\dfrac{b}{b^5}\right)$ Rearrange factors
   $\phantom{\dfrac{2a^3 b}{8a^4 b^5}} = \left(\dfrac{1}{4}\right)\left(\dfrac{1}{a^{4-3}}\right)\left(\dfrac{1}{b^{5-1}}\right)$ Apply second law $(3 < 4 \text{ and } 1 < 5)$
   $\phantom{\dfrac{2a^3 b}{8a^4 b^5}} = \dfrac{1}{4ab^4}$ Subtract exponents

27. $\dfrac{-12qw^4}{8qw^2} = \left(-\dfrac{3}{2}\right)\left(\dfrac{q}{q}\right)\left(\dfrac{w^4}{w^2}\right)$ Rearrange factors
   $\phantom{\dfrac{-12qw^4}{8qw^2}} = \left(-\dfrac{3}{2}\right)\left(\dfrac{w^{4-2}}{1}\right)$ Cancel q's and apply second law $(4 > 2)$
   $\phantom{\dfrac{-12qw^4}{8qw^2}} = \dfrac{-3w^2}{2}$ Subtract exponents

29. $\dfrac{14x^4 yz^6}{10x^6 y^2 z^2} = \left(\dfrac{7}{5}\right)\left(\dfrac{x^4}{x^6}\right)\left(\dfrac{y}{y^2}\right)\left(\dfrac{z^6}{z^2}\right)$ Rearrange factors
   $\phantom{\dfrac{14x^4 yz^6}{10x^6 y^2 z^2}} = \left(\dfrac{7}{5}\right)\left(\dfrac{1}{x^{6-4}}\right)\left(\dfrac{1}{y^{2-1}}\right)\left(\dfrac{z^{6-2}}{1}\right)$ Apply second law
   $\phantom{\dfrac{14x^4 yz^6}{10x^6 y^2 z^2}} = \dfrac{7z^4}{5x^2 y}$ Subtract exponents

31. $\dfrac{-15b^3 c^2}{-3b^3 c^4} = \left(\dfrac{5}{1}\right)\left(\dfrac{b^3}{b^3}\right)\left(\dfrac{c^2}{c^4}\right)$ Rearrange factors
   $\phantom{\dfrac{-15b^3 c^2}{-3b^3 c^4}} = \left(\dfrac{5}{1}\right)\left(\dfrac{1}{c^{4-2}}\right)$ Cancel b's and apply second law
   $\phantom{\dfrac{-15b^3 c^2}{-3b^3 c^4}} = \dfrac{5}{c^2}$ Subtract exponents

Chapter 6, Section 6.1

33. $(d^3)^5 = d^{3\cdot 5}$ Apply third law
    $ = d^{15}$ Multiply exponents

35. $(5^4)^3 = 5^{4\cdot 3}$ Apply third law
    $ = 5^{12}$ Multiply exponents

37. $(t^8)^4 = t^{8\cdot 4}$ Apply third law
    $ = t^{32}$ Multiply exponents

39. $(6x)^3 = 216x^3$ Apply fourth law

41. $(2t^3)^5 = (2^5)(t^3)^5$ Apply fourth law
    $ = 2^5 t^{3\cdot 5}$ Apply third law
    $ = 32t^{15}$ Multiply exponents

43. $(-4a^2 b^4)^4 = (-4)^4 (a^2)^4 (b^4)^4$ Apply fourth law
    $ = 4^4 a^{2\cdot 4} b^{4\cdot 4}$ Apply third law
    $ = 256 a^8 b^{16}$ Multiply exponents

45. $b^3 (b^2)^5 = b^3 (b^{2\cdot 5})$ Apply third law
    $ = b^3 (b^{10})$ Multiply exponents
    $ = b^{10+3}$ Apply first law
    $ = b^{13}$ Add exponents

47. $(p^2 q)^3 (pq^3) = ((p^2)^3 q^3)(pq^3)$ Apply fourth law
    $ = (p^{2\cdot 3} q^3)(pq^3)$ Apply third law
    $ = (p^6 \cdot p)(q^3 \cdot q^3)$ Multiply exponents and rearrange
    $ = p^{6+1} q^{3+3}$ Apply first law
    $ = p^7 q^6$ Add exponents

Chapter 6, Section 6.1

49. $(2x^3y)^2(xy^3)^4 = \left(2^2(x^3)^2 y^2\right)\left(x^4(y^3)^4\right)$ Apply fourth law
$ = \left(2^2(x^{3\cdot 2})y^2\right)\left(x^4(y^{3\cdot 4})\right)$ Apply third law
$ = (2^2 x^6 y^2)(x^4 y^{12})$ Multiply exponents
$ = 2^2 (x^6 \cdot x^4)(y^2 \cdot y^{12})$ Rearrange factors
$ = 2^2 x^{6+4} y^{2+12}$ Apply first law
$ = 4x^{10} y^{14}$ Add exponents

51. $(yz^2)^3 (zw^3)(-2yzw) = \left(y^3 (z^2)^3\right)(zw^3)(-2yzw)$ Apply fourth law
$ = (y^3 z^{2\cdot 3})(zw^3)(-2yzw)$ Apply third law
$ = (y^3 z^6)(zw^3)(-2yzw)$ Multiply exponents
$ = -2(w^3 \cdot w)(y^3 \cdot y)(z^6 \cdot z \cdot z)$ Rearrange factors
$ = -2(w^{3+1})(y^{3+1})(z^{6+1+1})$ Apply first law
$ = -2w^4 y^4 z^8$ Add exponents

53. $-a^2(-a)^2 = -a^2(-1)^2 (a)^2$ Apply fourth law
$ = -a^{2+2}$ Apply first law
$ = -a^4$ Add exponents

55. $-(-xy^2)^3 (-x^3)^3 = -\left((-1)^3 (x)^3 (y^2)^3\right)\left((-1)^3 (x^3)^3\right)$ Apply fourth law
$ = -\left(-x^3 (y^{2\cdot 3})\right)\left(-(x^{3\cdot 3})\right)$ Apply third law
$ = -(x^3 \cdot x^9) y^6$ Multiply exponents and rearrange
$ = -x^{3+9} y^6$ Apply first law
$ = -x^{12} y^6$ Add exponents

57. $\left(\dfrac{w}{6}\right)^6 = \left(\dfrac{w^6}{2^6}\right)$ Apply fifth law
$\phantom{\left(\dfrac{w}{6}\right)^6} = \left(\dfrac{w^6}{64}\right)$ Evaluate 2^6

Chapter 6, Section 6.1

59. $\left(\dfrac{-4}{p^5}\right)^3 = \dfrac{(-4)^3}{(p^5)^3}$ Apply fifth law

$\phantom{\left(\dfrac{-4}{p^5}\right)^3} = \dfrac{-64}{p^{5 \cdot 3}}$ Apply third law and evaluate $(-4)^3$

$\phantom{\left(\dfrac{-4}{p^5}\right)^3} = \dfrac{-64}{p^{15}}$ Multiply exponents

61. $\left(\dfrac{h^2}{m^3}\right)^4 = \dfrac{(h^2)^4}{(m^3)^4}$ Apply fifth law

$\phantom{\left(\dfrac{h^2}{m^3}\right)^4} = \dfrac{h^{2 \cdot 4}}{m^{3 \cdot 4}}$ Apply third law

$\phantom{\left(\dfrac{h^2}{m^3}\right)^4} = \dfrac{h^8}{m^{12}}$ Multiply exponents

63. $\left(\dfrac{-2x}{3y^2}\right)^3 = \left(\dfrac{(-2)^3 x^3}{3^3 (y^2)^3}\right)$ Apply fifth law

$\phantom{\left(\dfrac{-2x}{3y^2}\right)^3} = \dfrac{-8x^3}{27 y^{2 \cdot 3}}$ Apply third law

$\phantom{\left(\dfrac{-2x}{3y^2}\right)^3} = \dfrac{-8x^3}{27 y^6}$ Multiply exponents

65. $\dfrac{(4x)^3}{(-2x^2)^2} = \dfrac{(4)^3 (x)^3}{(-2)^2 (x^2)^2}$ Apply fifth law

$\phantom{\dfrac{(4x)^3}{(-2x^2)^2}} = \dfrac{64 x^3}{4 x^{2 \cdot 2}}$ Apply third law

$\phantom{\dfrac{(4x)^3}{(-2x^2)^2}} = \dfrac{16 x^3}{x^4}$ Multiply exponents

$\phantom{\dfrac{(4x)^3}{(-2x^2)^2}} = \dfrac{16}{x^{4-3}}$ Apply second law

$\phantom{\dfrac{(4x)^3}{(-2x^2)^2}} = \dfrac{16}{x}$ Subtract exponents

Chapter 6, Section 6.1

67.

$$\frac{(xy)^2(-x^2y)^3}{(x^2y^2)^2} = \frac{(x^2y^2)\left((-1)^3(x^2)^3y^3\right)}{(x^2)^2(y^2)^2} \quad \text{Apply fifth law}$$

$$= \frac{(x^2y^2)\left(-(x^{2\cdot3})y^3\right)}{x^{2\cdot2}y^{2\cdot2}} \quad \text{Apply third law}$$

$$= \frac{(x^2y^2)(-x^6y^3)}{x^4y^4} \quad \text{Multiply exponents}$$

$$= \frac{-(x^2 \cdot x^6)(y^2 \cdot y^3)}{x^4y^4} \quad \text{Rearrange factors}$$

$$= \frac{-(x^{2+6})(y^{2+3})}{x^4y^4} \quad \text{Apply first law}$$

$$= -\left(\frac{x^8}{x^4}\right)\left(\frac{y^5}{y^4}\right) \quad \text{Add exponents and rearrange}$$

$$= -x^{8-4}y^{5-4} \quad \text{Apply second law}$$

$$= -x^4y \quad \text{Subtract exponents}$$

69.

$$\left(\frac{-2x}{y^2}\right)^3\left(\frac{y^2}{3x}\right)^2 = \frac{(-2)^3x^3(y^2)^2}{(y^2)^3(3)^2x^2} \quad \text{Apply fifth law}$$

$$= \frac{-8x^3y^{2\cdot2}}{9y^{2\cdot3}x^2} \quad \text{Apply third law}$$

$$= \left(\frac{-8}{9}\right)\left(\frac{x^3}{x^2}\right)\left(\frac{y^4}{y^6}\right) \quad \text{Multiply exponents and rearrange}$$

$$= \frac{-8x^{3-2}}{9y^{6-4}} \quad \text{Apply second law}$$

$$= \frac{-8x}{9y^2} \quad \text{Subtract exponents}$$

71. a. $w + w = 2w$

b. $w(w) = w^{1+1}$ First law

$ = w^2$ Add exponents

182

Chapter 6, Section 6.1

73. a. $4z^2 - 6z^2 = -2z^2$

  b. $4z^2(-6z^2) = -24z^{2+2}$ First law
                $= -24z^4$ Add exponents

75. a. $p^2 + 3p^3$ Cannot be simplified

  b. $4p^2(3p^3) = 12p^{2+3}$ First law
              $= 12p^5$ Add exponents

77. a. $3^9 \cdot 3^8 = 3^{9+8}$ First law
              $= 3^{17}$ Add exponents

  b. $3^9 + 3^8$ Cannot be simplified

79. a. $f(a^2) = (a^2)^3$ By definition of f
            $= a^6$ By third law

  b. $a^3 \cdot f(a^3) = a^3 \cdot (a^3)^3$ By part a.
                 $= a^3 \cdot a^9$ By third law
                 $= a^{12}$ By first law

  c. $f(ab) = (ab)^3$ By definition of f
            $= a^3 b^3$ By fourth law

  d. $f(a+b) = (a+b)^3$ By definition of f, cannot be simplified further

81. a. $F(2a) = 3(2a)^5$ By definition of F
             $= 3(2^5)a^5$ By fourth law
             $= 96a^5$

  b. $2F(a) = 2\left(3(a)^5\right)$ By definition of F
            $= 6a^5$

  c. $F(a^2) = 3(a^2)^5$ By definition of F
            $= 3a^{10}$ By third law

Chapter 6, Section 6.2

    d. $[F(a)]^2 = [3a^5]^2$ By definition of F
                 $= 3^2(a^5)^2$ By fourth law
                 $= 9a^{10}$ By third law

Section 6.2

1. $2^{-1} = \frac{1}{2}$ Rewrite without negative exponents

3. $(-5)^{-2} = \frac{1}{(-5)^2}$ Rewrite without negative exponents
             $= \frac{1}{25}$

5. $\left(\frac{1}{3}\right)^{-3} = \left(\frac{3}{1}\right)^3$ Rewrite without negative exponents
           $= \frac{27}{1}$ Fifth law
           $= 27$

7. $\frac{1}{(-2)^{-4}} = (-2)^4$ Rewrite without negative exponents
           $= 16$ Evaluate $(-2)^4$

9. $\frac{5}{4^{-3}} = 5 \cdot 4^3$ Rewrite without negative exponents
          $= 5 \cdot 64$ Evaluate 4^3
          $= 320$

11. $(2q)^{-5} = \frac{1}{(2q)^5}$ Rewrite without negative exponents
            $= \frac{1}{2^5 q^5}$ Fourth law
            $= \frac{1}{32q^5}$ Evaluate 2^5

13. $-4x^{-2} = \frac{-4}{x^2}$ Rewrite without negative exponents

15. $y^{-2} + y^{-3} = \frac{1}{y^2} + \frac{1}{y^3}$ Rewrite without negative exponents

Chapter 6, Section 6.2

17. $(m-n)^{-2} = \dfrac{1}{(m-n)^2}$ Rewrite without negative exponents

19. $\dfrac{-5y^{-2}}{x^{-5}} = -5(y^{-2})\left(\dfrac{1}{x^{-5}}\right)$ Rearrange terms

$\phantom{\dfrac{-5y^{-2}}{x^{-5}}} = -5\left(\dfrac{1}{y^2}\right)(x^5)$ Rewrite without negative exponents

$\phantom{\dfrac{-5y^{-2}}{x^{-5}}} = \dfrac{-5x^5}{y^2}$ Combine factors

21. $\dfrac{1}{5^{13}} = 5^{-13}$ Rewrite with negative exponents

23. $\dfrac{7}{3^{24}} = 7 \cdot 3^{-24}$ Rewrite with negative exponents

25. $\dfrac{3}{r^4} = 3r^{-4}$ Rewrite with negative exponents

27. $\dfrac{2}{5w^3} = 2 \cdot 5^{-1} w^{-3}$ Rewrite with negative exponents

29. a. $2^3 = 8$

   b. $(-2)^3 = (-1)^3 2^3 = -8$

   c. $2^{-3} = \dfrac{1}{2^3} = \dfrac{1}{8}$

   d. $(-2)^{-3} = \dfrac{1}{(-2)^3} = -\dfrac{1}{8}$

31. a. $\left(\dfrac{1}{2}\right)^3 = \dfrac{1}{8}$

   b. $\left(-\dfrac{1}{2}\right)^3 = (-1)^3\left(\dfrac{1}{2}\right)^3 = -\dfrac{1}{8}$

   c. $\left(\dfrac{1}{2}\right)^{-3} = \left(\dfrac{2}{1}\right)^3 = 8$

   d. $\left(-\dfrac{1}{2}\right)^{-3} = \left(-\dfrac{2}{1}\right)^3 = (-1)^3 2^3 = -8$

185

Chapter 6, Section 6.2

33. **a.** Refer to the answers in the back of the textbook. To evaluate $f(1.2)$, enter the following keystrokes: 1.2 ^ ((-) 2) ENTER
Compute the other values in the same manner.

b. Notice that the values of $f(x)$ decrease as the values of x increase. The reason for this is $x^{-2} = \frac{1}{x^2}$. So, as larger values are plugged in for x, $\frac{1}{x^2}$ gets smaller.

35. **a.** Refer to the answers in the back of the textbook. To evaluate $f(2.5)$, enter the following keystrokes: 2.5 ^ ((-) 2) ENTER
Compute the other values in the same manner.

b. Notice that the values of $f(x)$ increase toward infinity as the values of x decrease toward 0. The reason for this is $x^{-2} = \frac{1}{x^2}$. So, if x is very small, then $\frac{1}{x^2}$ is very large.

37. See the back of the textbook for graphs of a., b., c., and d.

e. Graphs of b., c., and d. are the same. We expect this since $x^{-2} = \frac{1}{x^2} = \left(\frac{1}{x}\right)^2$.

39. We use the graph from Problem 37b with the window setting given.

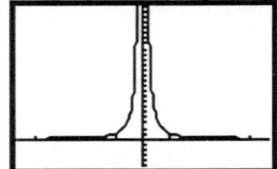

Use this graph to approximate the values of x^{-2} for the value of x given. One way to do this is by using the TRACE and ZOOM keys. Approximate answers are given in the back of the text.

41. $a^{-3} \cdot a^8 = a^{-3+8}$ First law
$\phantom{a^{-3} \cdot a^8} = a^5$ Add exponents

43. $5^{-4} \cdot 5^{-3} = 5^{-4+(-3)}$ First law
$\phantom{5^{-4} \cdot 5^{-3}} = 5^{-7}$ Add exponents
$\phantom{5^{-4} \cdot 5^{-3}} = \frac{1}{5^7}$ Rewrite without negative exponents

45. $(4x^{-5})(5x^2) = 20x^{-5+2}$ Rearrange factors and first law
$\phantom{(4x^{-5})(5x^2)} = 20x^{-3}$ Add exponents
$\phantom{(4x^{-5})(5x^2)} = \frac{20}{x^3}$ Rewrite without negative exponents

47. $\dfrac{p^{-7}}{p^{-4}} = p^{-7-(-4)}$ Second law

$\quad\quad = p^{-3}$ Subtract exponents

$\quad\quad = \dfrac{1}{p^3}$ Rewrite without negative exponents

49. $\dfrac{3u^{-3}}{9u^9} = \dfrac{1}{3}u^{-3-(9)}$ Second law

$\quad\quad = \dfrac{1}{3}u^{-12}$ Subtract exponents

$\quad\quad = \dfrac{1}{3u^{12}}$ Rewrite without negative exponents

51. $\dfrac{5^6 t^0}{5^{-2} t^{-1}} = (5^{6-(-2)})(t^{0-(-1)})$ Second law

$\quad\quad = 5^8 t$ Subtract exponents

53. $(7^{-2})^5 = 7^{-2 \cdot 5}$ Third law

$\quad\quad = 7^{-10}$ Multiply exponents

$\quad\quad = \dfrac{1}{7^{10}}$ Rewrite without negative exponents

55. $(3x^{-2}y^3)^{-2} = 3^{-2}(x^{-2})^{-2}(y^3)^{-2}$ Apply fourth law

$\quad\quad = 3^{-2} x^{-2(-2)} y^{3(-2)}$ Apply third law

$\quad\quad = 3^{-2} x^4 y^{-6}$ Multiply exponents

$\quad\quad = \dfrac{x^4}{9y^6}$ Rewrite without negative exponents

57. $\left(\dfrac{6a^{-3}}{b^2}\right)^{-2} = \dfrac{6^{-2}(a^{-3})^{-2}}{(b^2)^{-2}}$ Fourth and fifth laws

$\quad\quad = \dfrac{6^{-2} a^6}{b^{-4}}$ Multiply exponents

$\quad\quad = \dfrac{a^6 b^4}{36}$ Rewrite without negative exponents

Chapter 6, Section 6.2

59.
$$\frac{5h^{-3}(h^4)^{-2}}{6h^{-5}} = \frac{5h^{-3}(h^{4(-2)})}{6h^{-5}} \quad \text{Third law}$$
$$= \frac{5h^{-3} \cdot h^{-8}}{6h^{-5}} \quad \text{Multiply exponents}$$
$$= \frac{5h^{-3+(-8)}}{6h^{-5}} \quad \text{First law}$$
$$= \frac{5h^{-11}}{6h^{-5}} \quad \text{Add exponents}$$
$$= \frac{5}{6}h^{-11-(-5)} \quad \text{Second law}$$
$$= \frac{5}{6}h^{-6} \quad \text{Subtract exponents}$$
$$= \frac{5}{6h^6} \quad \text{Rewrite without negative exponents}$$

61. $x^{-1}(x^2 - 3x + 2) = (x^{-1})x^2 - (x^{-1})3x + (x^{-1})2$ Distributive law
$$= x^{-1+2} - 3x^{-1+1} + 2x^{-1} \quad \text{First law}$$
$$= x - 3x^0 + 2x^{-1} \quad \text{Add exponents}$$
$$= x - 3 + 2x^{-1} \quad x^0 = 1$$

63. $-3t^{-2}(t^2 - 2 - 4t^{-2}) = (-3t^{-2})t^2 - (-3t^{-2})2 - (-3t^{-2})4t^{-2}$ Distributive law
$$= -3t^{-2+2} + 6t^{-2} + 12t^{-2+(-2)} \quad \text{First law}$$
$$= -3t^0 + 6t^{-2} + 12t^{-4} \quad \text{Add exponents}$$
$$= -3 + 6t^{-2} + 12t^{-4} \quad t^0 = 1$$

65. $2u^{-3}(-2u^3 - u^2 + 3u) = -(2u^{-3})2u^3 - (2u^{-3})u^2 + (2u^{-3})3u$ Distributive law
$$= -4u^{-3+3} - 2u^{-3+2} + 6u^{-3+1} \quad \text{First law}$$
$$= -4u^0 - 2u^{-1} + 6u^{-2} \quad \text{Add exponents}$$
$$= -4 - 2u^{-1} + 6u^{-2} \quad u^0 = 1$$

67. To factor out $4x^{-2}$ from $4x^2 + 16x^{-2}$, we divide each term of $4x^2 + 16x^{-2}$ by $4x^{-2}$.
$$\frac{4x^2}{4x^{-2}} + \frac{16x^{-2}}{4x^{-2}} = x^{2-(-2)} + 4x^{-2-(-2)} \quad \text{Second law}$$
$$= x^4 + 4x^0 \quad \text{Subtract exponents}$$
$$= x^4 + 4 \quad x^0 = 1$$

Chapter 6, Section 6.2

69. To factor out p^{-1} from $6p^3 - 3 - 12p^{-1}$, we divide each term of $6p^3 - 3 - 12p^{-1}$ by p^{-1}.

$$\frac{6p^3}{p^{-1}} - \frac{3}{p^{-1}} - \frac{12p^{-1}}{p^{-1}}$$

$= 6p^{3-(-1)} - \frac{3}{p^{-1}} - 12p^{-1-(-1)}$ Second law

$= 6p^4 - 3p - 12p^0$ Subtract exponents

$= 6p^4 - 3p - 12$ $p^0 = 1$

71. To factor out $\frac{a^{-3}}{2}$ from $\frac{3}{2}a^{-3} - 3a + a^3$, we divide each term of $\frac{3}{2}a^{-3} - 3a + a^3$ by $\frac{a^{-3}}{2}$. This is the same as multiplying each term of $\frac{3}{2}a^{-3} - 3a + a^3$ by the reciprocal of $\frac{a^{-3}}{2}$ which is $\frac{2}{a^{-3}} = 2a^3$.

$(2a^3)\frac{3}{2}a^{-3} - (2a^3)3a + (2a^3)a^3 = 3a^{3+(-3)} - 6a^{3+1} + 2a^{3+3}$ First law

$\qquad = 3a^0 - 6a^4 + 2a^6$ Add exponents

$\qquad = 3 - 6a^4 + 2a^6$ $a^0 = 1$

73. $x^{-2} + y^{-2} = \frac{1}{x^2} + \frac{1}{y^2}$ Rewrite with positive exponents

75. $2w^{-1} - (2w)^{-2} = 2w^{-1} - 2^{-2}w^{-2}$ Fourth law

$\qquad = \frac{2}{w} - \frac{1}{4w^2}$ Rewrite with positive exponents

77. $a^{-1}b - ab^{-1} = \frac{b}{a} - \frac{a}{b}$ Rewrite with positive exponents

79. $(x^{-1} + y^{-1})^{-1} = \frac{1}{x^{-1} + y^{-1}}$ Rewrite with positive exponents

$\qquad = \frac{1}{\frac{1}{x} + \frac{1}{y}}$ Rewrite with positive exponents

189

Chapter 6, Section 6.2

81. $\dfrac{x+x^{-2}}{x} = \dfrac{x}{x} + \dfrac{x^{-2}}{x}$ Divide each term by x

$= 1 + x^{-2-1}$ Second law

$= 1 + x^{-3}$ Subtract exponents

$= 1 + \dfrac{1}{x^3}$ Rewrite with positive exponents

83. $\dfrac{a^{-1}+b^{-1}}{(ab)^{-1}} = \dfrac{a^{-1}+b^{-1}}{a^{-1}b^{-1}}$ Fourth law

$= \dfrac{a^{-1}}{a^{-1}b^{-1}} + \dfrac{b^{-1}}{a^{-1}b^{-1}}$ Divide each term by $a^{-1}b^{-1}$

$= \dfrac{1}{b^{-1}} + \dfrac{1}{a^{-1}}$ Cancel terms on top and bottom

$= b + a$ Rewrite with positive exponents

85. To write 285 in scientific notation we first locate the decimal point so that there is only one digit to the left: 2.85. The power of ten depends on the number of places we moved the decimal point: 2, and the power is positive since 285 is greater than 1. So the answer is: 2.85×10^2.

87. To write 8,372,000 in scientific notation we first locate the decimal point so that there is only one digit to the left: 8.372. The power of ten depends on the number of places we moved the decimal point: 6, and the power is positive since 8,372,000 is greater than 1. So the answer is: 8.372×10^6.

89. To write 0.024 in scientific notation we first locate the decimal point so that there is only one digit to the left: 2.4. The power of ten depends on the number of places we moved the decimal point: 2, and the power is negative since 0.024 is less than 1. So the answer is: 2.4×10^{-2}.

91. To write 0.000523 in scientific notation we first locate the decimal point so that there is only one digit to the left: 5.23. The power of ten depends on the number of places we moved the decimal point: 4, and the power is negative since 0.000523 is less than 1. So the answer is: 5.23×10^{-4}.

93. To convert 2.4×10^2 to standard notation we move the decimal point 2 places to the right: 240.

95. To convert 6.87×10^{15} to standard notation we move the decimal point 15 places to the right: 6,870,000,000,000,000.

97. To convert 5.0×10^{-3} to standard notation we move the decimal point 3 places to the left: 0.005.

99. To convert 2.02×10^{-4} to standard notation we move the decimal point 4 places to the left: 0.000202.

101. We enter the calculation as:
2.4 [2nd] [EE] [(-)] 8 [×] 6.5 [2nd] [EE] 32 [÷] 5.2 [2nd] [EE] 18 [ENTER]
The answer is 3,000,000.

103. We enter the calculation as:
[(] 7.5 [2nd] [EE] [(-)] 13 [×] 3.6 [2nd] [EE] [(-)] 9 [)] [÷] [(] 1.5 [2nd] [EE] [(-)] 15 [×] 1.6 [2nd] [EE] [(-)] 11 [)] [ENTER]
The answer is 112,500.

105. a. To write 4,351,200,000,000 in scientific notation we first locate the decimal point so that there is only one digit to the left: 4.3512. The power of ten depends on the number of places we moved the decimal point: 12, and the power is positive since 4,351,200,000,000 is greater than 1. Thus in scientific notation:
$4,351,200,000,000 = 4.3512 \times 10^{12}$.

b. We first write 257,908,000 in scientific notation. Locate the decimal point so that there is only one digit to the left of it: 2.57908, then count the number of places we moved the decimal point: 8. Finally, 257,908,000 is greater than 1 so the power of 10 is positive. $257,908,000 = 2.57908 \times 10^8$.
To find the per capita debt, we divide the total debt by the population:
$$\frac{4.3512 \times 10^{12}}{2.57908 \times 10^8} = 16871.13$$
So the per capita debt in 1993 was $16,871.13 per person.

107. a. There are 60 seconds in a minute and 60 minutes in an hour and 24 hours in a day and 365 days in a year so in one year there is $60 \cdot 60 \cdot 24 \cdot 365 = 31,536,000$ seconds. Light travels at 186,000 miles per second so in one year light travels $31,536,000 \cdot 186,000 = 5.8657 \times 10^{12}$ miles.

b. Proxima Centauri is 4.3 light years away, and from a., 1 light year is equal to 5.8657×10^{12} miles, therefore Proxima Centauri is
$4.3 \cdot (5.8657 \times 10^{12}) = 2.5223 \times 10^{13}$ miles away from Earth. Given that Pioneer 10 travels at 32,114 miles per hour, it will take this vehicle
$$\frac{2.5223 \times 10^{13}}{32,114} = 7.8542 \times 10^8 \text{ hours to reach this star.}$$

Chapter 6, Section 6.3

109. a. $400,338,326,350,000,000,000 = 4.0034 \times 10^{20}$

b. The area covered would be $15(4.0034 \times 10^{20}) = 6.0051 \times 10^{21}$ square inches. A square foot is 144 square inches, so 6.0051×10^{21} square inches $= \dfrac{6.0051 \times 10^{21}}{144} = 4.1702 \times 10^{19}$ square feet.

c. One square mile is 2.7878×10^{7} square feet, so 196,937,400 square miles equals $196,937,400(2.7878 \times 10^{7}) = 5.4903 \times 10^{15}$ square feet. Therefore, the answer from b. divided by the total surface area of the Earth is: $\dfrac{4.1702 \times 10^{19}}{5.4903 \times 10^{15}} = 7595$. So the Reichsbank marks would cover the Earth about over 7,595 times.

Section 6.3

The keying sequence is given for problems 1 through 23.

1. [2nd] [√] 121 [ENTER]
yields 11

3. [(-)] 27 [^] [(] 1 [÷] 3 [)] [ENTER]
yields −3

5. $\sqrt[4]{-625}$ is not a real number.

7. [(-)] [(] 32 [^] [(] 1 [÷] 5 [)] [)] [ENTER]
yields −2

9. 16 [^] [(] 1 [÷] 4 [)] [ENTER]
yields 2

11. 729 [^] [(] 1 [÷] 3 [)] [ENTER]
yields 9

13. 9 [^] [(] 1 [÷] 2 [)] [ENTER]
yields 3

15. $(-81)^{1/4}$ is not a real number.

17. [(-)] 64 [^] [(] 1 [÷] 6 [)] [ENTER]
yields −2

19. [(] [(-)] 32 [)] [^] [(] 1 [÷] 5 [)] [ENTER]
yields −2

Chapter 6, Section 6.3

21. ((-) 8) ^ (((-) 1 ÷ 3) ENTER
 yields −0.5

23. 64 ^ ((-) 0.5) ENTER
 yields 0.125

25. $3^{1/2} = \sqrt{3}$

27. $4x^{1/3} = 4\sqrt[3]{x}$

29. $(4x)^{0.2} = (4x)^{1/5} = \sqrt[5]{4x}$

31. $(8)^{-1/4} = \dfrac{1}{8^{1/4}} = \dfrac{1}{\sqrt[4]{8}}$

33. $3(xy)^{-1/3} = \dfrac{3}{(xy)^{1/3}} = \dfrac{3}{\sqrt[3]{xy}}$

35. $(x-2)^{1/4} = \sqrt[4]{x-2}$

37. $\sqrt{7} = 7^{1/2}$

39. $\sqrt[3]{2x} = (2x)^{1/3}$

41. $2\sqrt[5]{z} = 2(z)^{1/5}$

43. $\dfrac{-3}{\sqrt[4]{6}} = \dfrac{-3}{6^{1/4}} = -3(6^{-1/4})$

45. $\sqrt[4]{x-3y} = (x-3y)^{1/4}$

47. $\dfrac{-1}{\sqrt[3]{1+3b}} = \dfrac{-1}{(1+3b)^{1/3}}$
 $= -(1+3b)^{-1/3}$

49. See back of textbook for answers.

51. See back of textbook for answers.

53. See back of textbook for answers.

55. See back of textbook for answers. The keying sequence is given for Problems 57 through 67.

57. 2 ^ (1 ÷ 2) ENTER
 yields 1.414

59. 75 ^ (1 ÷ 3) ENTER
 yields 4.217

61. ((-) 43) ^ ((1 ÷ 5) ENTER
 yields −2.122

63. 1.6 ^ (1 ÷ 4) ENTER
 yields 1.125

65. 365 ^ ((-) 1 ÷ 3) ENTER
 yields 0.140

67. .006 ^ ((-) .2) ENTER
 yields 2.782

69. Evaluate $w = 4\left(\dfrac{Th}{m}\right)^{1/2}$ at $T = 293$, $h = 15$, and $m = 4$.

 $w = 4\left(\dfrac{293(15)}{4}\right)^{1/2} = 4(1098.75)^{1/2} = 132.6$

 Therefore, the width of the sonic boom is 132.6 kilometers.

71. a. The number of members in 1960 is given by $M(t)$ where t is the number of years since 1950. So, $t = 1960 - 1950 = 10$, and

 $M(10) = 72 + 100(10)^{1/3} = 72 + 215.443 = 287.443 \approx 287$

 The number of members in 1970 is given by $M(t)$ where t is the number of years since 1950. So, $t = 1970 - 1950 = 20$, and

 $M(20) = 72 + 100(20)^{1/3} = 72 + 271.442 = 343.442 \approx 343$

Chapter 6, Section 6.3

   b. In order to determine in which year the membership was 400, we solve for t in the equation $M(t) = 400$.

$72 + 100t^{1/3} = 400$ Substitute $M(t) = 72 + 100t^{1/3}$

$100t^{1/3} = 328$ Subtract 72 from both sides

$t^{1/3} = 3.28$ Divide both sides by 100

$(t^{1/3})^3 = (3.28)^3$ Cube both sides

$t = 35.288$ $(t^{1/3})^3 = t$ by the third law of exponents

≈ 35

So, the membership was 400 approximately 35 years after 1950 or about 1985. In order to determine in which year the membership was 500, we solve for t in the equation $M(t) = 500$.

$72 + 100t^{1/3} = 500$ Substitute $M(t) = 72 + 100t^{1/3}$

$100t^{1/3} = 428$ Subtract 72 from both sides

$t^{1/3} = 4.28$ Divide both sides by 100

$(t^{1/3})^3 = (4.28)^3$ Cube both sides

$t = 78.403$ $(t^{1/3})^3 = t$ by the third law of exponents

≈ 78

So, the membership was 500 approximately 78 years after 1950 or about 2028.

   c. Graph $Y_1 = 72 + 100X^{\wedge}(1/3)$

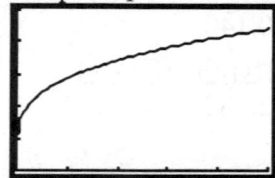

Xmin = 0, Xmax = 50, Ymin = 0, Ymax = 500
From the graph we notice that since 1950 the membership has grown each year but by less and less.

73. a.

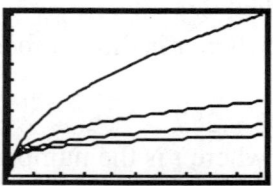

All of the graphs are increasing but by less and less as x gets larger. When taking a larger root, the graph is lower and flatter.

Chapter 6, Section 6.3

b. Use the [TRACE] and arrow keys to move the cursor until $x = 100$.
$100^{1/2} = 10 \qquad 100^{1/3} = 4.642 \qquad 100^{1/4} = 3.162 \qquad 100^{1/5} = 2.512$

c. The keying sequence for $n = 10$ is: 100 [^] [(] 1 [÷] 10 [)] [ENTER]
The sequence for other values of n is similar.
$100^{1/10} = 1.585 \qquad 100^{1/100} = 1.047 \qquad 100^{1/1000} = 1.005$
As the values for n get larger the values of $100^{1/n}$ get very close to 1.

75.

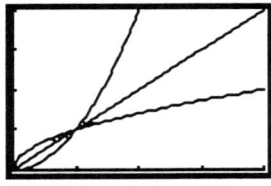

The graph of $\sqrt{x}$ is the same as the graph of x^2 reflected through the line $y = x$.

77.

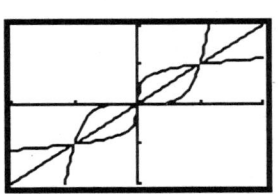

The graph of $\sqrt[5]{x}$ is the same as the graph of x^5 reflected through the line $y = x$.

79. $\sqrt{\sqrt{x}} = ((x)^{1/2})^{1/2}$ Rewrite radicals with exponents
$\phantom{\sqrt{\sqrt{x}}} = (x)^{1/2(1/2)}$ Third law of exponents
$\phantom{\sqrt{\sqrt{x}}} = x^{1/4}$ Multiply exponents
$\phantom{\sqrt{\sqrt{x}}} = \sqrt[4]{x}$ Rewrite exponent with radical

81. $\left(\sqrt[3]{125}\right)^3 = 125$ Cubing and cube root are inverse operations

83. $\left(\sqrt[4]{2}\right)^4 = 2$ Raising to the 4th power and taking 4th root are inverse operations

85. $(3\sqrt{7})^2 = 3^2(\sqrt{7})^2$ Fourth law of exponents
$\phantom{(3\sqrt{7})^2} = 9(7)$ Squaring and square root are inverse operations
$\phantom{(3\sqrt{7})^2} = 63$ Multiply factors

Chapter 6, Section 6.3

87. $(-x^2\sqrt[3]{2x})^3 = (-1)^3(x^2)^3(\sqrt[3]{2x})^3$ Fourth law of exponents
$\phantom{(-x^2\sqrt[3]{2x})^3} = -x^{2\cdot 3}(2x)$ Third law of exponents and cube, cube root are inverse operations
$\phantom{(-x^2\sqrt[3]{2x})^3} = -x^6(2x)$ Multiply factors
$\phantom{(-x^2\sqrt[3]{2x})^3} = -2x^{6+1}$ First law of exponents
$\phantom{(-x^2\sqrt[3]{2x})^3} = -2x^7$ Add exponents

89. $2\sqrt[3]{x} - 5 = -17$
$2\sqrt[3]{x} = -12$ Add 5 to both sides
$\sqrt[3]{x} = -6$ Divide by 2
$(\sqrt[3]{x})^3 = (-6)^3$ Cube both sides
$x = -216$ Evaluate cubes

91. $\sqrt[4]{x-1} = 2$
$(\sqrt[4]{x-1})^4 = 2^4$ Raise each side to the fourth power
$x - 1 = 16$ Evaluate fourth powers
$x = 17$ Add 1 to both sides

93. $4(x+2)^{1/5} = 12$
$(x+2)^{1/5} = 3$ Divide both sides by 4
$((x+2)^{1/5})^5 = 3^5$ Raise each side to the fifth power
$x + 2 = 243$ Evaluate fifth powers
$x = 241$ Subtract 2 from both sides

95. $(2x-3)^{-1/4} = \dfrac{1}{2}$
$((2x-3)^{-1/4})^{-4} = \left(\dfrac{1}{2}\right)^{-4}$ Raise each side to the -4 power
$2x - 3 = 16$ Evaluate exponents
$2x = 19$ Add 3 to both sides
$x = \dfrac{19}{2}$ Divide both sides by 2

Chapter 6, Section 6.3

97. $\sqrt[3]{x^2 - 3} = 3$

$\left(\sqrt[3]{x^2 - 3}\right)^3 = 3^3$ Cube both sides

$x^2 - 3 = 27$ Evaluate cubes

$x^2 = 30$ Add 3 to both sides

$x = \pm\sqrt{30} \approx \pm 5.477$ Take square roots of both sides

99. $\sqrt[3]{2x^2 - 15x} = 5$

$\left(\sqrt[3]{2x^2 - 15x}\right)^3 = 5^3$ Cube both sides

$2x^2 - 15x = 125$ Evaluate cubes

$2x^2 - 15x - 125 = 0$ Subtract 125 from both sides

$(2x - 25)(x + 5) = 0$ Factor the quadratic

$x = \dfrac{25}{2}, -5$ Solve for x in each factor

101. a. $M(1) = 2(1) - (1)^{1/3} = 2 - 1 = 1$

    b. $M(8) = 2(8) - (8)^{1/3} = 16 - 2 = 14$

    c. $M(0.001) = 2(0.001) - (0.001)^{1/3} = 0.002 - 0.1 = -0.098$

    d. $M(24) = 2(24) - (24)^{1/3} = 48 - 2.885 = 45.115$

103. We solve for a in the equation $816.814 = \pi a^3$.

    $260 = a^3$ Divide both sides by π

    $6.383 = a$ Take cube roots of both sides

    So the height is 6.383 meters.

Chapter 6, Section 6.3

105. We solve for T in the equation $(5.7 \times 10^{-5})T^4 = \dfrac{3.9 \times 10^{33}}{4\pi(9.96 \times 10^{10})^2}$.

$(5.7)(10^{-5})T^4 = \dfrac{3.9(10^{33})}{4\pi(9.96)^2(10^{10})^2}$ Rearrange and apply fourth law

$(5.7)(10^{-5})T^4 = \dfrac{3.9(10^{33})}{4\pi(99.202)(10^{20})}$ Apply third law

$T^4 = \dfrac{3.9}{4\pi(99.202)(5.7)}\left(\dfrac{10^{33}}{10^{20}(10^{-5})}\right)$ Divide by $(5.7)(10^{-5})$ and rearrange

$T^4 = (5.489 \times 10^{-4})(10^{18})$ Simplify each factor; use first and second laws of exponents

$T^4 = (5.489 \times 10^{14})$ Combine factors; use first law of exponents

$T = 4840.312$ Take fourth roots of both sides

The temperature of the sun is approximately $4840°$ K.

107. $r = \sqrt[3]{\dfrac{3V}{4\pi}}$

$r^3 = \dfrac{3V}{4\pi}$ Cube both sides

$\dfrac{4\pi r^3}{3} = V$ Multiply both sides by $\dfrac{4\pi}{3}$

109. $R = \sqrt[4]{\dfrac{8Lvf}{\pi p}}$

$R^4 = \dfrac{8Lvf}{\pi p}$ Raise both sides to the fourth power

$pR^4 = \dfrac{8Lvf}{\pi}$ Multiply both sides by p

$p = \dfrac{8Lvf}{\pi R^4}$ Divide both sides by R^4

Section 6.4

1. $81^{3/4} = \left(81^{1/4}\right)^3 = 3^3 = 27$

3. $-8^{2/3} = -\left(8^{1/3}\right)^2 = -2^2 = -4$

5. $16^{-3/2} = \left(16^{1/2}\right)^{-3} = 4^{-3} = \dfrac{1}{64}$

7. $-125^{-4/3} = -\left(125^{1/3}\right)^{-4} = -5^{-4} = -\dfrac{1}{625}$

9. $625^{0.75} = 625^{3/4} = \left(625^{1/4}\right)^3 = 5^3 = 125$

11. $32^{-1.6} = 32^{-8/5} = \left(32^{1/5}\right)^{-8} = 2^{-8} = \dfrac{1}{256}$

13. $x^{4/5} = \left(x^4\right)^{1/5} = \sqrt[5]{x^4}$

15. $3x^{2/5} = 3\left(x^2\right)^{1/5} = 3\sqrt[5]{x^2}$

17. $b^{-5/6} = \left(b^{-5}\right)^{1/6} = \left(\dfrac{1}{b^5}\right)^{1/6} = \dfrac{1}{\sqrt[6]{b^5}}$

19. $(pq)^{-2/3} = \left((pq)^{-2}\right)^{1/3} = \left(\dfrac{1}{(pq)^2}\right)^{1/3} = \dfrac{1}{\sqrt[3]{(pq)^2}}$

21. $4z^{-2/3} = 4\left(z^{-2}\right)^{1/3} = 4\left(\dfrac{1}{z^2}\right)^{1/3} = \dfrac{4}{\sqrt[3]{z^2}}$

23. $-2x^{1/4}y^{3/4} = -2\left(xy^3\right)^{1/4} = -2\sqrt[4]{xy^3}$

25. $\sqrt[3]{x^2} = \left(x^2\right)^{1/3} = x^{2/3}$

Chapter 6, Section 6.4

27. $\sqrt[3]{(ab)^2} = ((ab)^2)^{1/3} = (ab)^{2/3}$

29. $2\sqrt[5]{ab^3} = 2(ab^3)^{1/5} = 2a^{1/5}(b^3)^{1/5} = 2a^{1/5}b^{3/5}$

31. $\dfrac{8}{\sqrt[4]{x^3}} = \dfrac{8}{(x^3)^{1/4}} = \dfrac{8}{x^{3/4}} = 8x^{-3/4}$

33. $\dfrac{-4m}{\sqrt[6]{p^7}} = \dfrac{-4m}{(p^7)^{1/6}} = \dfrac{-4m}{p^{7/6}} = -4mp^{-7/6}$

35. $\dfrac{R}{3\sqrt{TK^5}} = \dfrac{R}{3(TK^5)^{1/2}} = \dfrac{R}{3T^{1/2}K^{5/2}} = \dfrac{RT^{-1/2}K^{-5/2}}{3}$

37. $\sqrt[5]{32^3} = (32^3)^{1/5} = (32^{1/5})^3 = 2^3 = 8$

39. $-\sqrt[3]{27^4} = -(27^4)^{1/3} = -(27^{1/3})^4 = -3^4 = -81$

41. $\sqrt[4]{16y^{12}} = (16y^{12})^{1/4} = (16)^{1/4}(y^{12})^{1/4} = 2y^{12/4} = 2y^3$

43. $-\sqrt{a^8 b^{16}} = -(a^8 b^{16})^{1/2} = -(a^8)^{1/2}(b^{16})^{1/2} = -a^{8/2}b^{16/2} = -a^4 b^8$

45. $\sqrt[3]{8x^9 y^{27}} = (8x^9 y^{27})^{1/3} = 8^{1/3}(x^9)^{1/3}(y^{27})^{1/3} = 2x^{9/3}y^{27/3} = 2x^3 y^9$

47. $-\sqrt[4]{81a^8 b^{12}} = -(81a^8 b^{12})^{1/4} = -81^{1/4}(a^8)^{1/4}(b^{12})^{1/4} = -3a^{8/4}b^{12/4} = -3a^2 b^3$

The keying sequence is given for Problems 49 - 55.

49. 12 ▲ (5 ÷ 6) ENTER
    yields 7.931

51. 6 ▲ (4 ÷ 3) ENTER
    yields 10.903

53. 37 ▲ (− 2 ÷ 3) ENTER
    yields .090

55. 4.7 ▲ 2.3 ENTER
    yields 35.142

Chapter 6, Section 6.4

57. $4a^{6/5}a^{4/5} = 4a^{6/5+4/5}$ First law of exponents
$\phantom{4a^{6/5}a^{4/5}} = 4a^{10/5}$ Add exponents
$\phantom{4a^{6/5}a^{4/5}} = 4a^2$ Simplify

59. $\left(-2m^{2/3}\right)^4 = (-2)^4\left(m^{2/3}\right)^4$ Fourth law of exponents
$\phantom{\left(-2m^{2/3}\right)^4} = 16m^{(2/3)4}$ Third law of exponents
$\phantom{\left(-2m^{2/3}\right)^4} = 16m^{8/3}$ Simplify

61. $\dfrac{8w^{9/4}}{2w^{3/4}} = 4w^{9/4-3/4}$ Second law of exponents
$\phantom{\dfrac{8w^{9/4}}{2w^{3/4}}} = 4w^{6/4}$ Subtract exponents
$\phantom{\dfrac{8w^{9/4}}{2w^{3/4}}} = 4w^{3/2}$ Simplify

63. $\left(-3u^{5/3}\right)\left(5u^{-2/3}\right) = -15u^{5/3+(-2/3)}$ First law of exponents
$\phantom{\left(-3u^{5/3}\right)\left(5u^{-2/3}\right)} = -15u^{3/3}$ Add exponents
$\phantom{\left(-3u^{5/3}\right)\left(5u^{-2/3}\right)} = -15u$ Simplify

65. $\dfrac{k^{3/4}}{2k} = \dfrac{1}{2k^{1-(3/4)}}$ Second law of exponents
$\phantom{\dfrac{k^{3/4}}{2k}} = \dfrac{1}{2k^{1/4}}$ Subtract exponents

67. $c^{-2/3}\left(\dfrac{2}{3}c^2\right) = \dfrac{2}{3}c^{(-2/3)+2}$ First law of exponents
$\phantom{c^{-2/3}\left(\dfrac{2}{3}c^2\right)} = \dfrac{2}{3}c^{4/3}$ Add exponents

69. $2x^{1/2}\left(x - x^{1/2}\right) = \left(2x^{1/2}\right)x - \left(2x^{1/2}\right)x^{1/2}$ Distributive law
$\phantom{2x^{1/2}\left(x - x^{1/2}\right)} = 2x^{1/2+1} - 2x^{1/2+1/2}$ First law of exponents
$\phantom{2x^{1/2}\left(x - x^{1/2}\right)} = 2x^{3/2} - 2x$ Add exponents

Chapter 6, Section 6.4

71. $\frac{1}{2}y^{-1/3}\left(y^{2/3} + 3y^{-5/6}\right) = \left(\frac{1}{2}y^{-1/3}\right)y^{2/3} + \left(\frac{1}{2}y^{-1/3}\right)3y^{-5/6}$ Distributive law

$= \frac{1}{2}y^{-1/3+2/3} + \frac{3}{2}y^{-1/3+(-5/6)}$ First law of exponents

$= \frac{1}{2}y^{1/3} + \frac{3}{2}y^{-7/6}$ Add exponents

73. $\left(2x^{1/4} + 1\right)\left(x^{1/4} - 1\right) = \left(2x^{1/4}\right)x^{1/4} - \left(2x^{1/4}\right) + \left(x^{1/4}\right) - 1$ Distributive law

$= 2x^{1/4+1/4} - 2x^{1/4} + x^{1/4} - 1$ First law of exponents

$= 2x^{1/2} - x^{1/4} - 1$ Add exponents and Combine like terms

75. $\left(a^{3/4} - 2\right)^2 = \left(a^{3/4}\right)a^{3/4} - 2a^{3/4} - 2a^{3/4} + 4$ Distributive law

$= a^{3/4+3/4} - 2a^{3/4} - 2a^{3/4} + 4$ First law of exponents

$= a^{3/2} - 2a^{3/4} - 2a^{3/4} + 4$ Add exponents

$= a^{3/2} - 4a^{3/4} + 4$ Combine like terms

77. Factor x out of $x^{3/2} + x$:

$\frac{x^{3/2}}{x} + \frac{x}{x} = x^{3/2-1} + 1 = x^{1/2} + 1$ Apply second law of exponents

So $x^{3/2} + x = x\left(x^{1/2} + 1\right)$

79. Factor $y^{-1/4}$ out of $y^{3/4} - y^{-1/4}$:

$\frac{y^{3/4}}{y^{-1/4}} - \frac{y^{-1/4}}{y^{-1/4}} = y^{3/4-(-1/4)} - 1 = y - 1$ Apply second law of exponents

So $y^{3/4} - y^{-1/4} = y^{-1/4}(y - 1)$

81. Factor $a^{-1/3}$ out of $a^{1/3} + 3 - a^{-1/3}$:

$\frac{a^{1/3}}{a^{-1/3}} + \frac{3}{a^{-1/3}} - \frac{a^{-1/3}}{a^{-1/3}} = a^{1/3-(-1/3)} + 3a^{1/3} - 1$ Apply second law of exponents

$= a^{2/3} + 3a^{1/3} - 1$

So $a^{1/3} + 3 - a^{-1/3} = a^{-1/3}\left(a^{2/3} + 3a^{1/3} - 1\right)$

Chapter 6, Section 6.4

83. a. $Q(16) = 4(16)^{5/2} = 4\left(16^{1/2}\right)^5 = 4(4)^5 = 4(1024) = 4096$

b. $Q\left(\dfrac{1}{4}\right) = 4\left(\dfrac{1}{4}\right)^{5/2} = 4\left(\dfrac{1}{4^{1/2}}\right)^5 = 4\left(\dfrac{1}{2}\right)^5 = 4\left(\dfrac{1}{32}\right) = \dfrac{1}{8}$

c. $Q(3) = 4(3)^{5/2}$, to compute this value we enter the following keying sequence:
4 [×] 3 [^] [(] 5 [÷] 2 [)] [ENTER]
yields 62.354

d. $Q(100) = 4(100)^{5/2} = 4\left(100^{1/2}\right)^5 = 4(10)^5 = 4(100,000) = 400,000$

85. $x^{2/3} - 1 = 15$
$x^{2/3} = 16$ Add 1 to both sides
$\left(x^{2/3}\right)^{3/2} = 16^{3/2}$ Raise both sides to the 3/2
$x = \left(16^{1/2}\right)^3 = (4)^3 = 64$ Evaluate $16^{3/2}$

87. $x^{-2/5} = 9$
$\left(x^{-2/5}\right)^{-5/2} = 9^{-5/2}$ Raise both sides to the $-5/2$
$x = \left(9^{1/2}\right)^{-5} = (3)^{-5} = \dfrac{1}{243}$ $\left(x^{-2/5}\right)^{-5/2} = x$; Evaluate $9^{-5/2}$

89. $2\left(5.2 - x^{5/3}\right) = 1.4$
$5.2 - x^{5/3} = 0.7$ Divide both sides by 2
$-x^{5/3} = -4.5$ Subtract 5.2 from both sides
$x^{5/3} = 4.5$ Multiply both sides by -1
$\left(x^{5/3}\right)^{3/5} = 4.5^{3/5}$ Raise both sides to the 3/5
$x = 4.5^{3/5}$ $\left(x^{5/3}\right)^{3/5} = x$

To evaluate $4.5^{3/5}$ enter the following keying sequence:
4.5 [^] [(] 3 [÷] 5 [)] [ENTER]
yields 2.466. So $x = 2.466$.

Chapter 6, Section 6.4

91. In order to find x so that $f(x) = 27$, we solve for x in the equation $(3x-4)^{3/2} = 27$.

$(3x-4)^{3/2} = 27$

$\left((3x-4)^{3/2}\right)^{2/3} = 27^{2/3}$ Raise both sides to the $2/3$

$3x - 4 = 9$ Evaluate exponents

$3x = 13$ Add 4 to both sides

$x = \dfrac{13}{3}$ Divide both sides by 3

93. In order to find x so that $S(x) = 20$, we solve for x in the equation $12x^{-5/4} = 20$.

$12x^{-5/4} = 20$

$x^{-5/4} = \dfrac{20}{12} = \dfrac{5}{3}$

$\left(x^{-5/4}\right)^{-4/5} = \left(\dfrac{5}{3}\right)^{-4/5}$

$x = \left(\dfrac{5}{3}\right)^{-4/5}$

To evaluate $\left(\dfrac{5}{3}\right)^{-4/5}$ enter the following keying sequence into the graphing calculator:

(5 ÷ 3) ^ ((-) 4 ÷ 5) ENTER

yields 0.665, so $x = 0.665$.

95. a. The number of people infected after 5 days is $I(5) = 50(5)^{3/5} = 131$, after 10 days $I(10) = 50(10)^{3/5} = 199$, and after 15 days $I(15) = 50(15)^{3/5} = 254$.

  b. Solve for t in the equation $I(t) = 300$.

  $I(t) = 50(t)^{3/5} = 300$

  $(t)^{3/5} = \dfrac{300}{50} = 6$ Divide both sides by 50

  $\left((t)^{3/5}\right)^{5/3} = 6^{5/3}$ Raise both sides to the $5/3$

  $t = 19.8 \approx 20$ Evaluate exponents

  After approximately 20 days 300 people are ill.

c.

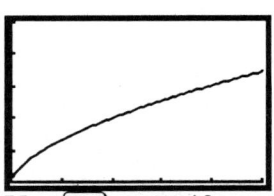

Use [TRACE] to verify answers to a. and b. .

97. We want to compute p where $p = K^{1/2}a^{3/2}$, $K = 1.243 \times 10^{-24}$ and $a = 1.417 \times 10^8$.

$p = \left(1.243 \times 10^{-24}\right)^{1/2}\left(1.417 \times 10^8\right)^{3/2}$

$p = \left(1.243^{1/2}\right)\left(10^{-24(1/2)}\right)\left(1.417^{3/2}\right)\left(10^{8(3/2)}\right)$ Apply third and fourth laws of exponents

$p = \left(1.243^{1/2}\right)\left(1.417^{3/2}\right)\left(10^{-12+12}\right)$ Rearrange and apply first law of exponents

$p = \left(1.243^{1/2}\right)\left(1.417^{3/2}\right)$ $10^0 = 1$

$p = 1.88$ Evaluate

Mars revolves around the sun once every 1.88 years.

99. The accountant would budget $Cr^{0.6}$ dollars with $C = 5{,}000$ and $r = 1.8$.
$5000(1.8)^{0.6} = 7{,}114.32$. So the accountant should budget \$7,114.32.

101. a. When $P = 20{,}000$, $R = 0.015(20000)^{1.2} = 2174.34$. When $P = 40{,}000$, $R = 0.015(40000)^{1.2} = 4995.32$. When $P = 60{,}000$, $R = 0.015(60000)^{1.2} = 8125.92$.

  b. Solve for P in the equation $5000 = 0.015P^{1.2}$.

$P^{1.2} = \dfrac{5000}{0.015}$ Divide both sides by 0.015

$\left(P^{1.2}\right)^{1/1.2} = \left(\dfrac{5000}{0.015}\right)^{1/1.2}$ Raise both sides to the $1/1.2$

$P = 40{,}031$ Evaluate exponents

Chapter 6, Section 6.5

c. Using a graphing calculator, graph $Y_1 = 0.015X\wedge(1.2)$

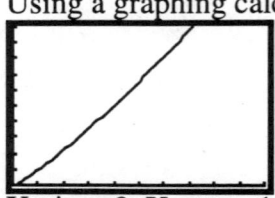

Xmin = 0, Xmax = 100,000, Ymin = 0, and Ymax = 10,000
Use [TRACE] to verify answers to a. and b.

Section 6.5

1. $\sqrt{18} = \sqrt{9}\sqrt{2}$ Factor out perfect squares
   $= 3\sqrt{2}$ Simplify

3. $\sqrt[3]{24} = \sqrt[3]{8}\sqrt[3]{3}$ Factor out perfect cubes
   $= 2\sqrt[3]{3}$ Simplify

5. $-\sqrt[4]{64} = -\sqrt[4]{16}\sqrt[4]{4}$ Factor out perfect fourth roots
   $= -2\sqrt[4]{4}$ Simplify

7. $\sqrt{60,000} = \sqrt{10,000}\sqrt{6}$ Factor out perfect squares
   $= 100\sqrt{6}$ Simplify

9. $\sqrt[3]{900,000} = \sqrt[3]{1,000}\sqrt[3]{900}$ Factor out perfect cubes
   $= 10\sqrt[3]{900}$ Simplify

11. $\sqrt[3]{\dfrac{-40}{27}} = \sqrt[3]{\dfrac{-8}{27}}\sqrt[3]{5}$ Factor out perfect cubes
    $= -\dfrac{2}{3}\sqrt[3]{5}$ Simplify

13. $\sqrt[3]{x^{10}} = \sqrt[3]{x^9}\sqrt[3]{x}$ Factor out perfect cubes
    $= x^3\sqrt[3]{x}$ Simplify

15. $\sqrt{27z^3} = \sqrt{9z^2}\sqrt{3z}$ Factor out perfect squares
    $= 3z\sqrt{3z}$ Simplify

17. $\sqrt[4]{48a^9b^{12}} = \sqrt[4]{16a^8b^{12}}\sqrt[4]{3a}$ Factor out perfect fourth roots
    $= 2a^2b^3\sqrt[4]{3a}$ Simplify

Chapter 6, Section 6.5

19. $-\sqrt[5]{96p^7q^9} = -\sqrt[5]{32p^5q^5}\sqrt[5]{3p^2q^4}$ Factor out perfect fifth roots
    $= -2pq\sqrt[5]{3p^2q^4}$ Simplify

21. $-\sqrt{18s}\sqrt{2s^3} = -\sqrt{36s^4}$ Apply property (1)
    $= -6s^2$ Simplify

23. $\sqrt[3]{7h^2}\sqrt[3]{-49h} = \sqrt[3]{-343h^3}$ Apply property (1)
    $= -7h$ Simplify

25. $\sqrt{16-4x^2} = \sqrt{4}\sqrt{4-x^2}$ Factor out perfect squares
    $= 2\sqrt{4-x^2}$ Simplify

27. $\sqrt[3]{8A^3 + A^6} = \sqrt[3]{A^3}\sqrt[3]{8+A^3}$ Factor out perfect cubes
    $= A\sqrt[3]{8+A^3}$ Simplify

29. $\sqrt{\dfrac{125p^{13}}{a^4}} = \sqrt{\dfrac{25p^{12}}{a^4}}\sqrt{5p}$ Factor out perfect squares
    $= \dfrac{5p^6}{a^2}\sqrt{5p}$ Simplify

31. $\sqrt[3]{\dfrac{56v^2}{w^6}} = \sqrt[3]{\dfrac{8}{w^6}}\sqrt[3]{7v^2}$ Factor out perfect cubes
    $= \dfrac{2}{w^2}\sqrt[3]{7v^2}$ Simplify

33. $\dfrac{\sqrt{a^5b^3}}{\sqrt{ab}} = \sqrt{a^4b^2}$ Apply property (2)
    $= a^2b$ Simplify

35. $\dfrac{\sqrt{98x^2y^3}}{\sqrt{xy}} = \sqrt{98xy^2}$ Apply property (2)
    $= \sqrt{49y^2}\sqrt{2x}$ Factor out perfect squares
    $= 7y\sqrt{2x}$ Simplify

Chapter 6, Section 6.5

37. $\dfrac{\sqrt[3]{8b^7}}{\sqrt[3]{a^6 b^2}} = \sqrt[3]{\dfrac{8b^5}{a^6}}$ Apply property (2)

$= \sqrt[3]{\dfrac{8b^3}{a^6}} \sqrt[3]{b^2}$ Factor out perfect cubes

$= \dfrac{2b}{a^2} \sqrt[3]{b^2}$ Simplify

39. $\dfrac{\sqrt[5]{a}\sqrt[5]{b^2}}{\sqrt[5]{ab}} = \dfrac{\sqrt[5]{ab^2}}{\sqrt[5]{ab}}$ Apply property (1)

$= \sqrt[5]{b}$ Apply property (2)

41. $\sqrt{4x^2} = 2|x|$

43. $\sqrt{(x-5)^2} = |x-5|$

45. $\sqrt{x^2 - 6x + 9} = \sqrt{(x-3)^2} = |x-3|$

47. a.

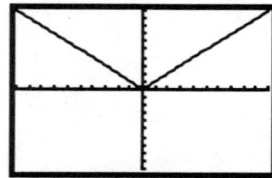

Xmin = –10, Xmax = 10, Ymin = –10, and Ymax = 10

This matches the graph of $y = |x|$ since $\sqrt{x^2} = |x|$ when x is not assumed to be a positive number.

b.

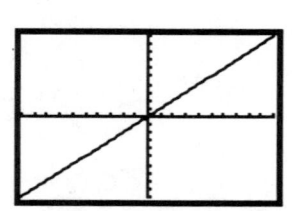

This matches the graph of $y = x$ since $\sqrt[3]{x^3} = x$ regardless of whether or not x is assumed to be a positive number.

49. $3\sqrt{7} + 2\sqrt{7} = 5\sqrt{7}$ Combine like terms

51. $4\sqrt{3} - \sqrt{27} = 4\sqrt{3} - \sqrt{9}\sqrt{3}$ Factor out perfect squares
$\phantom{4\sqrt{3} - \sqrt{27}} = 4\sqrt{3} - 3\sqrt{3}$ Simplify
$\phantom{4\sqrt{3} - \sqrt{27}} = \sqrt{3}$ Combine like terms

53. $\sqrt{50x} + \sqrt{32x} = \sqrt{25}\sqrt{2x} + \sqrt{16}\sqrt{2x}$ Factor out perfect squares
$\phantom{\sqrt{50x} + \sqrt{32x}} = 5\sqrt{2x} + 4\sqrt{2x}$ Simplify
$\phantom{\sqrt{50x} + \sqrt{32x}} = 9\sqrt{2x}$ Combine like terms

55. $3\sqrt[3]{16} - \sqrt[3]{2} - 2\sqrt[3]{54} = 3\sqrt[3]{8}\sqrt[3]{2} - \sqrt[3]{2} - 2\sqrt[3]{27}\sqrt[3]{2}$ Factor out perfect cubes
$\phantom{3\sqrt[3]{16} - \sqrt[3]{2} - 2\sqrt[3]{54}} = 6\sqrt[3]{2} - \sqrt[3]{2} - 6\sqrt[3]{2}$ Simplify
$\phantom{3\sqrt[3]{16} - \sqrt[3]{2} - 2\sqrt[3]{54}} = -\sqrt[3]{2}$ Combine like terms

57. $4\sqrt[3]{40} + 6\sqrt{80} - 5\sqrt{45} - \sqrt[3]{135}$
$= 4\sqrt[3]{8}\sqrt[3]{5} + 6\sqrt{16}\sqrt{5} - 5\sqrt{9}\sqrt{5} - \sqrt[3]{27}\sqrt[3]{5}$ Factor out perfect squares and cubes
$= 8\sqrt[3]{5} + 24\sqrt{5} - 15\sqrt{5} - 3\sqrt[3]{5}$ Simplify
$= 5\sqrt[3]{5} + 9\sqrt{5}$ Combine like terms

59. $3\sqrt{4xy^2} - 4\sqrt{9xy^2} + 2\sqrt{4x^2y}$
$= 3\sqrt{4y^2}\sqrt{x} - 4\sqrt{9y^2}\sqrt{x} + 2\sqrt{4x^2}\sqrt{y}$ Factor out perfect squares
$= 6y\sqrt{x} - 12y\sqrt{x} + 4x\sqrt{y}$ Simplify
$= -6y\sqrt{x} + 4x\sqrt{y}$ Combine like terms

61. $2(3 - \sqrt{5}) = (2)3 - (2)\sqrt{5} = 6 - 2\sqrt{5}$ Distribute

63. $\sqrt{2}(\sqrt{6} + \sqrt{10}) = (\sqrt{2})\sqrt{6} + (\sqrt{2})\sqrt{10}$ Distribute
$\phantom{\sqrt{2}(\sqrt{6} + \sqrt{10})} = \sqrt{12} + \sqrt{20}$ Apply property (1)
$\phantom{\sqrt{2}(\sqrt{6} + \sqrt{10})} = \sqrt{4}\sqrt{3} + \sqrt{4}\sqrt{5}$ Factor out perfect squares
$\phantom{\sqrt{2}(\sqrt{6} + \sqrt{10})} = 2\sqrt{3} + 2\sqrt{5}$ Simplify

65. $\sqrt[3]{2}(\sqrt[3]{20} - 2\sqrt[3]{12}) = (\sqrt[3]{2})\sqrt[3]{20} - (\sqrt[3]{2})2\sqrt[3]{12}$ Distribute
$\phantom{\sqrt[3]{2}(\sqrt[3]{20} - 2\sqrt[3]{12})} = \sqrt[3]{40} - 2\sqrt[3]{24}$ Apply property (1)
$\phantom{\sqrt[3]{2}(\sqrt[3]{20} - 2\sqrt[3]{12})} = \sqrt[3]{8}\sqrt[3]{5} - 2\sqrt[3]{8}\sqrt[3]{3}$ Factor out perfect cubes
$\phantom{\sqrt[3]{2}(\sqrt[3]{20} - 2\sqrt[3]{12})} = 2\sqrt[3]{5} - 4\sqrt[3]{3}$ Simplify

Chapter 6, Section 6.5

67. $2\sqrt{x}(\sqrt{24x} + \sqrt{12}) = (2\sqrt{x})\sqrt{24x} + (2\sqrt{x})\sqrt{12}$ Distribute
$\qquad = 2\sqrt{24x^2} + 2\sqrt{12x}$ Apply property (1)
$\qquad = 2\sqrt{4x^2}\sqrt{6} + 2\sqrt{4}\sqrt{3x}$ Factor out perfect squares
$\qquad = 4x\sqrt{6} + 4\sqrt{3x}$ Simplify

69. $(\sqrt{x} - 3)(\sqrt{x} + 3) = \sqrt{x}\sqrt{x} + 3\sqrt{x} - 3\sqrt{x} - 9$ Distribute
$\qquad = \sqrt{x^2} + 3\sqrt{x} - 3\sqrt{x} - 9$ Apply property (1)
$\qquad = x + 3\sqrt{x} - 3\sqrt{x} - 9$ Simplify
$\qquad = x - 9$ Combine like terms

71. $(\sqrt{2} - \sqrt{3})(\sqrt{2} + 2\sqrt{3}) = \sqrt{2}\sqrt{2} + 2\sqrt{2}\sqrt{3} - \sqrt{2}\sqrt{3} - 2\sqrt{3}\sqrt{3}$ Distribute
$\qquad = \sqrt{4} + 2\sqrt{6} - \sqrt{6} - 2\sqrt{9}$ Apply property (1)
$\qquad = 2 + 2\sqrt{6} - \sqrt{6} - 6$ Simplify
$\qquad = \sqrt{6} - 4$ Combine like terms

73. $(\sqrt{5} - \sqrt{2})^2 = (\sqrt{5} - \sqrt{2})(\sqrt{5} - \sqrt{2})$
$\qquad = \sqrt{5}\sqrt{5} - \sqrt{2}\sqrt{5} - \sqrt{2}\sqrt{5} + \sqrt{2}\sqrt{2}$ Distribute
$\qquad = \sqrt{25} - \sqrt{10} - \sqrt{10} + \sqrt{4}$ Apply property (1)
$\qquad = 5 - \sqrt{10} - \sqrt{10} + 2$ Simplify
$\qquad = 7 - 2\sqrt{10}$ Combine like terms

75. $(3\sqrt{x} + \sqrt{2y})(2\sqrt{x} - 3\sqrt{2y})$
$\qquad = 6\sqrt{x}\sqrt{x} - 9\sqrt{x}\sqrt{2y} + 2\sqrt{x}\sqrt{2y} - 3\sqrt{2y}\sqrt{2y}$ Distribute
$\qquad = 6\sqrt{x^2} - 9\sqrt{2xy} + 2\sqrt{2xy} - 3\sqrt{4y^2}$ Apply property (1)
$\qquad = 6x - 9\sqrt{2xy} + 2\sqrt{2xy} - 6y$ Simplify
$\qquad = 6x - 7\sqrt{2xy} - 6y$ Combine like terms

77. $(\sqrt{a} - 2\sqrt{b})^2 = (\sqrt{a} - 2\sqrt{b})(\sqrt{a} - 2\sqrt{b})$
$\qquad = \sqrt{a}\sqrt{a} - 2\sqrt{a}\sqrt{b} - 2\sqrt{a}\sqrt{b} + 4\sqrt{b}\sqrt{b}$ Distribute
$\qquad = \sqrt{a^2} - 2\sqrt{ab} - 2\sqrt{ab} + 4\sqrt{b^2}$ Apply property (1)
$\qquad = a - 2\sqrt{ab} - 2\sqrt{ab} + 4b$ Simplify
$\qquad = a - 4\sqrt{ab} + 4b$ Combine like terms

Chapter 6, Section 6.5

79. $\dfrac{2}{2} + \dfrac{2\sqrt{3}}{2} = 1 + \sqrt{3}$ Divide both terms by 2 and simplify

$2 + 2\sqrt{3} = 2(1 + \sqrt{3})$

81. $\dfrac{2\sqrt{27}}{6} + \dfrac{6}{6} = \dfrac{\sqrt{27}}{3} + 1$ Divide both terms by 6 and simplify

$\phantom{\dfrac{2\sqrt{27}}{6} + \dfrac{6}{6}} = \dfrac{\sqrt{9}\sqrt{3}}{3} + 1$ Factor out perfect squares

$\phantom{\dfrac{2\sqrt{27}}{6} + \dfrac{6}{6}} = \sqrt{3} + 1$ Simplify

$2\sqrt{27} + 6 = 6(\sqrt{3} + 1)$

83. $\dfrac{4}{4} + \dfrac{\sqrt{16y}}{4} = 1 + \dfrac{\sqrt{16y}}{4}$ Divide both terms by 4 and simplify

$\phantom{\dfrac{4}{4} + \dfrac{\sqrt{16y}}{4}} = 1 + \dfrac{\sqrt{16}\sqrt{y}}{4}$ Factor out perfect squares

$\phantom{\dfrac{4}{4} + \dfrac{\sqrt{16y}}{4}} = 1 + \sqrt{y}$ Simplify

$4 + \sqrt{16y} = 4(1 + \sqrt{y})$

85. $\dfrac{\sqrt{2}}{\sqrt{2}} - \dfrac{\sqrt{6}}{\sqrt{2}} = \dfrac{\sqrt{2}}{\sqrt{2}} - \sqrt{\dfrac{6}{2}}$ Divide both terms by $\sqrt{2}$. Apply property (2)

$\phantom{\dfrac{\sqrt{2}}{\sqrt{2}} - \dfrac{\sqrt{6}}{\sqrt{2}}} = 1 - \sqrt{3}$ Simplify

$\sqrt{2} - \sqrt{6} = \sqrt{2}(1 - \sqrt{3})$

87. $\dfrac{2y\sqrt{x}}{\sqrt{x}} + \dfrac{3\sqrt{xy}}{\sqrt{x}} = \dfrac{2y\sqrt{x}}{\sqrt{x}} + 3\sqrt{\dfrac{xy}{x}}$ Divide both terms by $\sqrt{x}$. Apply property (2)

$\phantom{\dfrac{2y\sqrt{x}}{\sqrt{x}} + \dfrac{3\sqrt{xy}}{\sqrt{x}}} = 2y + 3\sqrt{y}$ Simplify

$2y\sqrt{x} + 3\sqrt{xy} = \sqrt{x}(2y + 3\sqrt{y})$

Chapter 6, Section 6.5

89. $\dfrac{4x}{2\sqrt{x}} - \dfrac{\sqrt{12x}}{2\sqrt{x}}$ Divide both terms by $2\sqrt{x}$

$= \dfrac{4x}{2\sqrt{x}} - \dfrac{\sqrt{4}\sqrt{3x}}{2\sqrt{x}}$ Factor out perfect squares

$= \dfrac{2x}{\sqrt{x}} - \dfrac{\sqrt{3x}}{\sqrt{x}}$ Simplify and apply the second law of exponents

$= 2x^{1-1/2} - \sqrt{\dfrac{3x}{x}}$ Apply property (2)

$= 2\sqrt{x} - \sqrt{3}$ Simplify

$4x - \sqrt{12x} = 2\sqrt{x}\left(2\sqrt{x} - \sqrt{3}\right)$

91. $\dfrac{6}{\sqrt{3}} = \dfrac{6\sqrt{3}}{\sqrt{3}\sqrt{3}}$ Multiply numerator and denominator by $\sqrt{3}$

$= \dfrac{6\sqrt{3}}{3}$ Simplify

$= 2\sqrt{3}$ Simplify

93. $\dfrac{-\sqrt{3}}{\sqrt{7}} = \dfrac{-\sqrt{3}\sqrt{7}}{\sqrt{7}\sqrt{7}}$ Multiply numerator and denominator by $\sqrt{7}$

$= \dfrac{-\sqrt{21}}{7}$ Apply property (1) and simplify

95. $\sqrt{\dfrac{7x}{18}} = \dfrac{\sqrt{7x}}{\sqrt{18}}$ Apply property (2)

$= \dfrac{\sqrt{7x}}{\sqrt{9}\sqrt{2}}$ Factor out perfect squares

$= \dfrac{\sqrt{7x}\sqrt{2}}{3\sqrt{2}\sqrt{2}}$ Multiply numerator and denominator by $\sqrt{2}$

$= \dfrac{\sqrt{14x}}{6}$ Simplify

97. $\sqrt{\dfrac{2a}{b}} = \dfrac{\sqrt{2a}}{\sqrt{b}}$ Apply property (2)

$= \dfrac{\sqrt{2a}\sqrt{b}}{\sqrt{b}\sqrt{b}}$ Multiply numerator and denominator by $\sqrt{b}$

$= \dfrac{\sqrt{2ab}}{b}$ Apply property (1) and simplify

Chapter 6, Section 6.5

99. $\dfrac{2\sqrt{3}}{\sqrt{2k}} = \dfrac{2\sqrt{3}\sqrt{2k}}{\sqrt{2k}\sqrt{2k}}$ Multiply numerator and denominator by $\sqrt{2k}$

$= \dfrac{2\sqrt{6k}}{2k}$ Apply property (1) and simplify

$= \dfrac{\sqrt{6k}}{k}$ Simplify

101. $\dfrac{-9x^2\sqrt{5x^3}}{2\sqrt{6x}} = \dfrac{-9x^2\sqrt{5x^3}\sqrt{6x}}{2\sqrt{6x}\sqrt{6x}}$ Multiply numerator and denominator by $\sqrt{6x}$

$= \dfrac{-9x^2\sqrt{30x^4}}{12x}$ Apply property (1) and simplify

$= \dfrac{-9x^2\sqrt{x^4}\sqrt{30}}{12x}$ Factor out perfect squares

$= \dfrac{-9x^2 x^2 \sqrt{30}}{12x}$ Evaluate perfect squares

$= \dfrac{-9x^{2+2-1}\sqrt{30}}{12}$ Apply first and second laws of exponents

$= \dfrac{-3x^3\sqrt{30}}{4}$ Simplify

103. $\dfrac{4}{(1+\sqrt{3})} = \dfrac{4(1-\sqrt{3})}{(1+\sqrt{3})(1-\sqrt{3})}$ Multiply numerator and denominator by conjugate

$= \dfrac{4(1-\sqrt{3})}{1^2 - (\sqrt{3})^2}$ Simplify denominator

$= \dfrac{4(1-\sqrt{3})}{1-3}$ Simplify

$= \dfrac{4(1-\sqrt{3})}{-2}$

$= -2(1-\sqrt{3})$

Chapter 6, Section 6.5

105. $\dfrac{x}{x-\sqrt{3}} = \dfrac{x(x+\sqrt{3})}{(x-\sqrt{3})(x+\sqrt{3})}$ Multiply numerator and denominator by conjugate

$= \dfrac{x(x+\sqrt{3})}{x^2-(\sqrt{3})^2}$ Simplify

$= \dfrac{x(x+\sqrt{3})}{x^2-3}$

107. $\dfrac{\sqrt{6}-3}{2-\sqrt{6}}$

$= \dfrac{(\sqrt{6}-3)(2+\sqrt{6})}{(2-\sqrt{6})(2+\sqrt{6})}$ Multiply numerator and denominator by conjugate

$= \dfrac{2\sqrt{6}+(\sqrt{6})^2-6-3\sqrt{6}}{2^2-(\sqrt{6})^2}$ Simplify

$= \dfrac{2\sqrt{6}+6-6-3\sqrt{6}}{4-6}$

$= \dfrac{\sqrt{6}}{2}$

109. $\dfrac{\sqrt{5}}{5\sqrt{3}+3\sqrt{5}}$

$= \dfrac{\sqrt{5}(5\sqrt{3}-3\sqrt{5})}{(5\sqrt{3}+3\sqrt{5})(5\sqrt{3}-3\sqrt{5})}$ Multiply numerator and denominator by conjugate

$= \dfrac{5\sqrt{3}\sqrt{5}-3\sqrt{5}\sqrt{5}}{(5\sqrt{3})^2-(3\sqrt{5})^2}$ Simplify

$= \dfrac{5\sqrt{15}-15}{75-45}$

$= \dfrac{5\sqrt{15}-15}{30}$

$= \dfrac{\sqrt{15}-3}{6}$

Chapter 6, Section 6.5

111. $\dfrac{1}{\sqrt[3]{2z}} = \dfrac{\sqrt[3]{2x} \cdot \sqrt[3]{2x}}{\sqrt[3]{2x} \cdot \sqrt[3]{2x} \cdot \sqrt[3]{2x}}$

$= \dfrac{\sqrt[3]{(2x)^2}}{2x}$ Apply property (1)

$= \dfrac{\sqrt[3]{4x^2}}{2x}$ Simplify

113. $\dfrac{1}{\sqrt[3]{x^2}} = \dfrac{\sqrt[3]{x^2} \cdot \sqrt[3]{x^2}}{\sqrt[3]{x^2} \cdot \sqrt[3]{x^2} \cdot \sqrt[3]{x^2}}$ Multiply numerator and denominator by $\sqrt[3]{x^2}\sqrt[3]{x^2}$

$= \dfrac{\sqrt[3]{x^2 \cdot x^2}}{x^2}$ Apply property (1) and simplify

$= \dfrac{\sqrt[3]{x^4}}{x^2}$

$= \dfrac{\sqrt[3]{x^3}\sqrt[3]{x}}{x^2}$ Factor out perfect cubes

$= \dfrac{x\sqrt[3]{x}}{x^2}$ Apply second law of exponents

$= \dfrac{\sqrt[3]{x}}{x}$ Simplify

115. $\sqrt[3]{\dfrac{2}{3y}} = \dfrac{\sqrt[3]{2}}{\sqrt[3]{3y}}$ Apply property (2)

$= \dfrac{\sqrt[3]{2} \cdot \sqrt[3]{3y} \cdot \sqrt[3]{3y}}{\sqrt[3]{3y} \cdot \sqrt[3]{3y} \cdot \sqrt[3]{3y}}$ Multiply numerator and denominator by $\sqrt[3]{3y}\sqrt[3]{3y}$

$= \dfrac{\sqrt[3]{2 \cdot 3y \cdot 3y}}{3y}$ Simplify

$= \dfrac{\sqrt[3]{18y^2}}{3y}$

Chapter 6, Section 6.6

117. $\sqrt[4]{\dfrac{x}{8y^3}} = \dfrac{\sqrt[4]{x}}{\sqrt[4]{8y^3}}$ Apply property (2)

$= \dfrac{\sqrt[4]{x} \cdot \sqrt[4]{2y}}{\sqrt[4]{8y^3} \cdot \sqrt[4]{2y}}$ Multiply numerator and denominator by $\sqrt[4]{2y}$

$= \dfrac{\sqrt[4]{2xy}}{\sqrt[4]{16y^4}}$ Apply property (1)

$= \dfrac{\sqrt[4]{2xy}}{\sqrt[4]{2^4 y^4}}$ Rewrite denominator

$= \dfrac{\sqrt[4]{2xy}}{2y}$ Simplify

119. $\dfrac{9x^3}{\sqrt[4]{27x}} = \dfrac{9x^3 \cdot \sqrt[4]{3x^3}}{\sqrt[4]{27x} \cdot \sqrt[4]{3x^3}}$ Multiply numerator and denominator by $\sqrt[4]{3x^3}$

$= \dfrac{9x^3 \cdot \sqrt[4]{3x^3}}{\sqrt[4]{81x^4}}$ Apply property (1)

$= \dfrac{9x^3 \cdot \sqrt[4]{3x^3}}{\sqrt[4]{3^4 x^4}}$ Rewrite denominator

$= \dfrac{9x^3 \cdot \sqrt[4]{3x^3}}{3x}$ Simplify

$= 3x^2 \cdot \sqrt[4]{3x^3}$

Section 6.6

1. $3z + 4 = \sqrt{3z + 10}$

$(3z+4)^2 = \left(\sqrt{3z+10}\right)^2$ Square both sides

$9z^2 + 24z + 16 = 3z + 10$

$9z^2 + 21z + 6 = 0$ Subtract $3z + 10$ from both sides

$3(3z+1)(z+2) = 0$ Factor

$z = -\dfrac{1}{3}$ or $z = -2$ Set each factor equal to 0

216

Chapter 6, Section 6.6

Check:

$3\left(-\dfrac{1}{3}\right)+4 = \sqrt{3\left(-\dfrac{1}{3}\right)+10}$?

$-1+4 = \sqrt{-1+10}$?

$3 = 3$ Yes; $-\dfrac{1}{3}$ is a solution

$3(-2)+4 = \sqrt{3(-2)+10}$?

$-2 = \sqrt{-6+10}$?

$-2 = 2$ No; -2 is not a solution

Thus, the solution to the original equation is $-\dfrac{1}{3}$.

3. $2x+1 = \sqrt{10x+5}$

   $(2x+1)^2 = \left(\sqrt{10x+5}\right)^2$ Square both sides

   $4x^2 + 4x + 1 = 10x + 5$

   $4x^2 - 6x - 4 = 0$ Subtract $10x+5$ from both sides

   $2(x-2)(2x+1) = 0$ Factor

   $x = 2$ or $x = -\dfrac{1}{2}$ Set each factor equal to 0

Check:

$2(2)+1 = \sqrt{10(2)+5}$?

$5 = 5$ Yes; 2 is a solution

$2\left(-\dfrac{1}{2}\right)+1 = \sqrt{10\left(-\dfrac{1}{2}\right)+5}$?

$0 = 0$ Yes; $-\dfrac{1}{2}$ is a solution

Thus, the solutions to the original equation are 2 and $-\dfrac{1}{2}$.

Chapter 6, Section 6.6

5. $\sqrt{y+4} = y-8$
   $(\sqrt{y+4})^2 = (y-8)^2$ Square both sides
   $y+4 = y^2 - 16y + 64$
   $0 = y^2 - 17y + 60$ Subtract $y+4$ from both sides
   $0 = (y-5)(y-12)$ Factor
   $y = 5$ or $y = 12$ Set each factor equal to 0
   Check:
   $\sqrt{5+4} = 5-8$?
   $3 = -3$ No; 5 is not a solution
   $\sqrt{12+4} = 12-8$?
   $4 = 4$ Yes; 12 is a solution
   Thus, the solution to the original equation is 12.

7. $\sqrt{2y-1} = \sqrt{3y-6}$
   $(\sqrt{2y-1})^2 = (\sqrt{3y-6})^2$ Square both sides
   $2y - 1 = 3y - 6$
   $5 = y$ Subtract $2y-6$ from both sides
   Check:
   $\sqrt{2(5)-1} = \sqrt{3(5)-6}$?
   $3 = 3$ Yes; 5 is a solution
   Thus, the solution to the original equation is 5.

9. $\sqrt{x-3}\sqrt{x} = 2$
   $(\sqrt{x-3}\sqrt{x})^2 = 2^2$ Square both sides
   $(\sqrt{x-3})^2(\sqrt{x})^2 = 4$ Apply property (1)
   $(x-3)(x) = 4$ Evaluate squares
   $x^2 - 3x = 4$ Multiply factors
   $x^2 - 3x - 4 = 0$ Subtract 4 from both sides
   $(x+1)(x-4) = 0$ Factor
   $x = -1$ or $x = 4$ Set each factor equal to 0

Chapter 6, Section 6.6

Check:
$\sqrt{-1} - 3\sqrt{-1} = 2$?
undefined   No; −1 is not a solution
$\sqrt{4} - 3\sqrt{4} = 2$?
$2 = 2$ Yes; 4 is a solution
Thus, the solution to the original equation is 4.

11. $\sqrt{y+4} = \sqrt{y+20} - 2$

   $\left(\sqrt{y+4}\right)^2 = \left(\sqrt{y+20} - 2\right)^2$ Square both sides

   $y + 4 = y + 20 - 4\sqrt{y+20} + 4$

   $4\sqrt{y+20} = 20$ Simplify

   $\sqrt{y+20} = 5$ Divide both sides by 4

   $\left(\sqrt{y+20}\right)^2 = 5^2$ Square both sides

   $y + 20 = 25$

   $y = 5$ Subtract 20 from both sides

   Check:
   $\sqrt{5+4} = \sqrt{5+20} - 2$?

   $3 = 3$ Yes; 5 is a solution
   Thus, the solution to the original equation is 5.

13. $\sqrt{x} + \sqrt{2} = \sqrt{x+2}$

   $\left(\sqrt{x} + \sqrt{2}\right)^2 = \left(\sqrt{x+2}\right)^2$ Square both sides

   $x + 2\sqrt{2x} + 2 = x + 2$

   $2\sqrt{2x} = 0$ Simplify

   $\left(2\sqrt{2x}\right)^2 = 0^2$ Square both sides

   $8x = 0$

   $x = 0$ Divide both sides by 8

   Check:
   $\sqrt{0} + \sqrt{2} = \sqrt{0+2}$?

   $\sqrt{2} = \sqrt{2}$ Yes; 0 is a solution
   Thus, the solution to the original equation is 0.

219

Chapter 6, Section 6.6

15.
$$\sqrt{5+x} + \sqrt{x} = 5$$
$$\sqrt{5+x} = 5 - \sqrt{x} \qquad \text{Subtract } \sqrt{x} \text{ from both sides}$$
$$\left(\sqrt{5+x}\right)^2 = \left(5 - \sqrt{x}\right)^2 \qquad \text{Square both sides}$$
$$5 + x = 25 - 10\sqrt{x} + x$$
$$10\sqrt{x} = 20 \qquad \text{Simplify}$$
$$\sqrt{x} = 2 \qquad \text{Divide both sides by 10}$$
$$\left(\sqrt{x}\right)^2 = 2^2 \qquad \text{Square both sides}$$
$$x = 4$$
Check:
$$\sqrt{5+4} + \sqrt{4} = 5 \ ?$$
$$5 = 5 \qquad \text{Yes; 4 is a solution}$$
Thus, the solution to the original equation is 4.

17. a. Francine is traveling at 1 mile per minute on the highway. Therefore, t minutes after turning onto the highway she has traveled t miles.

b. Let $d(t)$ be Francine's distance from the college t minutes after turning onto the highway. By the Pythagorean theorem, $d(t) = \sqrt{8^2 + t^2} = \sqrt{64 + t^2}$.

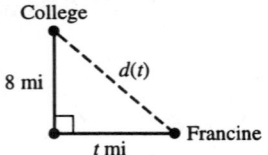

c.

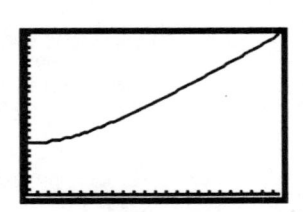

d. The range of KGVC is 15 miles so we solve the equation $\sqrt{64 + t^2} = 15$ for t.
$$\left(\sqrt{64 + t^2}\right)^2 = 15^2 \qquad \text{Square both sides}$$
$$64 + t^2 = 225$$
$$t^2 = 161 \qquad \text{Subtract 64 from both sides}$$
$$t = \sqrt{161} = 12.7 \qquad \text{Take the square root of both sides}$$
Francine will be out of range in 12.7 minutes.

Chapter 6, Section 6.6

19. a. The UFO descends at a rate of 10 feet per second from an altitude of 700 feet. Therefore, after t seconds, the altitude of the UFO is $700 - 10t$ feet.

b. You are running at 15 feet per second. Therefore, after t seconds, you have run $15t$ feet.

c.

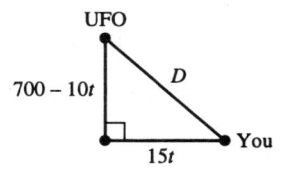

```
        UFO
         •
         |\
         | \  D
700 – 10t|  \
         |   \
         |____\• You
           15t
```

d. By the Pythagorean theorem, $D(t) = \sqrt{(700-10t)^2 + (15t)^2}$.

e. $D(10) = \sqrt{(700-10(10))^2 + (15(10))^2} = \sqrt{360,000 + 22,500} = 618.5$ feet.

f.

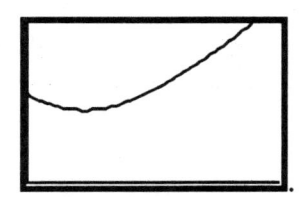

$D(45) = 719.8$.

g. From the graph of $D(t)$, the distance is smallest when t is approximately 22 seconds. We find this by using the [TRACE] key to move the cursor to the lowest point of the graph.

h. To find when the UFO reaches the ground we solve for t in the equation
$700 - 10t = 0$
$700 = 10t$ Add $10t$ to both sides
$70 = t$ Divide both sides by 10
Therefore, a more realistic Xmax would be 70.

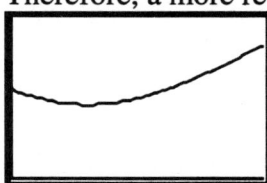

21. a. The total distance along the road to the station is 6 miles. Since Delbert can walk along the road at a rate of 4 miles per hour, it will take him $\dfrac{6}{4} = 1.5$ hours to walk back to the station.

Chapter 6, Section 6.6

b. By the Pythagorean theorem, the station is $\sqrt{5^2 + 1^2} = \sqrt{26} = 5.1$ miles from Delbert through the fields. Since Delbert can walk at a rate of 3 miles per hour through the fields, it would take him $\frac{5.3}{3} = 1.7$ hours to walk to the station this way.

c. **i.** The distance he walks along the road is $5 - x$ miles. By the Pythagorean theorem, the distance he walks through the field is $\sqrt{x^2 + 1^2}$ or just $\sqrt{x^2 + 1}$ miles provided $x > 0$.

ii. The road part of the walk will take $\frac{5-x}{4}$ hours, since Delbert walks 4 miles per hour along the road. The field part of the walk will take $\frac{\sqrt{x^2 + 1}}{3}$ hours, since he walks 3 miles per hour through the fields.

iii. The sum of the two times in ii. is $\frac{5-x}{4} + \frac{\sqrt{x^2+1}}{3}$ hours.

d.

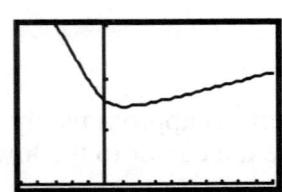

No point corresponds to the answer from a. the point (5, 1.7) corresponds to the answer from b.

e. The appropriate domain is $0 < x \leq 5$. We can leave the range at $0 < y \leq 3$.

f.

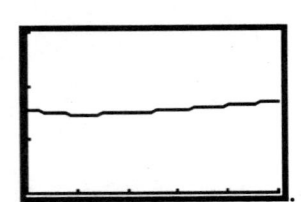

If $x = 1.15$ miles then it takes him about 1.47 hours.

Chapter 6, Section 6.6

23. Distance = $\sqrt{(4-1)^2+(5-1)^2} = \sqrt{25} = 5$

Midpoint = $\left(\dfrac{1+4}{2}, \dfrac{1+5}{2}\right) = \left(\dfrac{5}{2}, 3\right)$

25. Distance = $\sqrt{(-2-2)^2+(-1-(-3))^2} = \sqrt{20} = 2\sqrt{5}$

Midpoint = $\left(\dfrac{2+(-2)}{2}, \dfrac{(-3)+(-1)}{2}\right) = (0,-2)$

27. Distance = $\sqrt{(-2-3)^2+(5-5)^2} = \sqrt{25} = 5$

Midpoint = $\left(\dfrac{3+(-2)}{2}, \dfrac{5+5}{2}\right) = \left(\dfrac{1}{2}, 5\right)$

29. To find the perimeter of this triangle we add together the three distances between vertices.

$d_1 = \sqrt{(3-10)^2+(1-1)^2} = \sqrt{49} = 7$

$d_2 = \sqrt{(5-3)^2+(9-1)^2} = \sqrt{68} = 2\sqrt{17}$

$d_3 = \sqrt{(10-5)^2+(1-9)^2} = \sqrt{89}$

perimeter = $7 + 2\sqrt{17} + \sqrt{89}$

31.

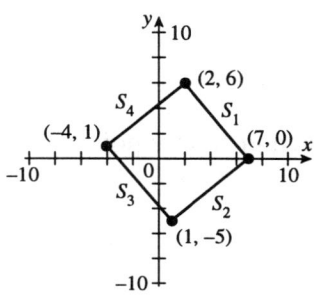

Let S_1, S_2, S_3, and S_4 be the lengths of the four sides of the rectangle as shown. To ensure that this rectangle is a square we must check that $S_1 = S_2 = S_3 = S_4$.

$S_1 = \sqrt{(7-2)^2+(0-6)^2} = \sqrt{61}$

$S_2 = \sqrt{(1-7)^2+(-5-0)^2} = \sqrt{61}$

$S_3 = \sqrt{(-4-1)^2+(1-(-5))^2} = \sqrt{61}$

$S_4 = \sqrt{(2-(-4))^2+(6-1)^2} = \sqrt{61}$

223

Chapter 6, Section 6.6

33. Refer to the graph in the back of the textbook.

35. $4x^2 + 4y^2 = 16$
$x^2 + y^2 = 4$ Divide both sides by 4
Refer to the graph in the back of the textbook.

37. Refer to the graph in the back of the textbook.

39. Refer to the graph in the back of the textbook.

41. To convert to standard form we complete the square for both variables.
$x^2 + 2x + (1) + y^2 - 4y + (4) - 6 = (1) + (4)$ Add 5 to both sides
$(x+1)^2 + (y-2)^2 - 6 = 5$ Factor
$(x+1)^2 + (y-2)^2 = 11$ Add 6 to both sides
Refer to the graph in the back of the textbook.

43. To convert to standard form we complete the square for both variables.
$x^2 + 8x + (16) + y^2 = 4 + (16)$ Add 16 to both sides
$(x+4)^2 + (y-0)^2 = 20$ Factor
Refer to the graph in the back of the textbook.

45. $(x-(-2))^2 + (y-5)^2 = (2\sqrt{3})^2$
$(x+2)^2 + (y-5)^2 = 12$ Simplify
$x^2 + 4x + 4 + y^2 - 10y + 25 = 12$ Compute squares
$x^2 + y^2 + 4x - 10y + 17 = 0$ Simplify

47. To find the radius we compute the distance from the center to any point on the circle.

$$\text{radius} = \sqrt{\left(4 - \frac{3}{2}\right)^2 + (-3 - (-4))^2}$$
$$= \sqrt{\frac{25}{4} + 1}$$
$$= \frac{\sqrt{29}}{2}$$

Thus, the equation of the circle is:

$$\left(x - \frac{3}{2}\right)^2 + (y - (-4))^2 = \left(\frac{\sqrt{29}}{2}\right)^2$$

$$\left(x - \frac{3}{2}\right)^2 + (y + 4)^2 = \frac{29}{4} \qquad \text{Simplify}$$

$$x^2 - 3x + \frac{9}{4} + y^2 + 8y + 16 = \frac{29}{4} \qquad \text{Compute squares}$$

$$x^2 + y^2 - 3x + 8y + 11 = 0 \qquad \text{Simplify}$$

49. The center of circle is the midpoint of the diameter. Therefore,

$$\text{center} = \left(\frac{1+3}{2}, \frac{5-1}{2}\right) = (2, 2)$$

The radius is half the length of the diameter. Thus,

$$\text{radius} = \frac{1}{2}\sqrt{(3-1)^2 + (-1-5)^2} = \frac{\sqrt{40}}{2} = \frac{2\sqrt{10}}{2} = \sqrt{10}$$

The equation of the circle is:

$$(x-2)^2 + (y-2)^2 = \left(\sqrt{10}\right)^2$$

$$(x-2)^2 + (y-2)^2 = 10 \qquad \text{Simplify}$$

$$x^2 - 4x + 4 + y^2 - 4y + 4 = 10 \qquad \text{Compute squares}$$

$$x^2 + y^2 - 4x - 4y - 2 = 0 \qquad \text{Simplify}$$

51. Since the x-axis is tangent to the circle, the length of the vertical line segment originating at the center and terminating at the x-axis equals the radius. The length of this segment is the distance from $(-3, -1)$ to $(-3, 0)$, which is

$\sqrt{(-3-(-3))^2 + (0-(-1))^2} = 1$. The equation of the circle is:

$$(x - (-3))^2 + (y - (-1))^2 = 1$$

$$(x + 3)^2 + (y + 1)^2 = 1 \qquad \text{Simplify}$$

$$x^2 + 6x + 9 + y^2 + 2y + 1 = 1 \qquad \text{Compute squares}$$

$$x^2 + y^2 + 6x + 2y + 9 = 0 \qquad \text{Simplify}$$

53. We need to find a, b, and c so that the three points $(2, 3)$, $(3, 2)$, and $(-4, -5)$ all satisfy the equation $x^2 + y^2 + ax + by + c = 0$. Plugging in the first point, we get $2^2 + 3^2 + a \cdot 2 + b \cdot 3 + c = 0$, or equivalently, $13 + 2a + 3b + c = 0$. Call this equation (1).

Chapter 6 Review

Plugging in the second point, we get $3^2 + 2^2 + a \cdot 3 + b \cdot 2 + c = 0$, or equivalently, $13 + 3a + 2b + c = 0$. Call this equation(2).
Setting equation (1) equal to equation (2) we have:
$13 + 2a + 3b + c = 13 + 3a + 2b + c$
$2a + 3b = 3a + 2b$ Subtract $13 + c$ from both sides
$a = b$ Subtract $2a + 2b$ from both sides
So substituting $a = b$ into equation (1) or (2), yields $c = -13 - 5a$.
Plugging in the third point, we get $(-4)^2 + (-5)^2 + a \cdot (-4) + b \cdot (-5) + c = 0$, or equivalently, $41 - 4a - 5b + c = 0$. However $a = b$, so the equation above becomes $41 - 9a + c = 0$, or $c = 9a - 41$. Now we set the two equations for c equal to each other: $-13 - 5a = 9a - 41$
$28 = 14a$ Add $5a + 41$ to both sides
$2 = a$ Divide both sides by 7
Therefore, $a = b = 2$ and $c = -13 - 5(2) = -23$. Thus, the equation of the circle is $x^2 + y^2 + 2x + 2y - 23 = 0$.

55. See the proof in the back of textbook.

Chapter 6 Review

1. $(2x^3)(5x^4) = 10x^{3+4}$ Apply first law of exponents
$ = 10x^7$ Simplify

2. $(3mn^5)(7m^8n) = 21m^{1+8}n^{5+1}$ Apply first law of exponents
$ = 21m^9n^6$ Simplify

3. $(-a^2b^3)(4ab) = -4a^{2+1}b^{3+1}$ Apply first law of exponents
$ = -4a^3b^4$ Simplify

4. $(5s^3t^2)(-2s^4t) = -10s^{3+4}t^{2+1}$ Apply first law of exponents
$ = -10s^7t^3$ Simplify

5. $\dfrac{3u^4v}{6uv^6} = \dfrac{u^{4-1}}{2v^{6-1}}$ Apply second law of exponents
$\phantom{\dfrac{3u^4v}{6uv^6}} = \dfrac{u^3}{2v^5}$ Simplify

Chapter 6 Review

6. $\dfrac{-24wz^5}{6wz^7} = \dfrac{-4w^{1-1}}{z^{7-5}}$ Apply second law of exponents

$= -\dfrac{4}{z^2}$ Simplify

7. $\dfrac{54r^2s^7t^5}{-18r^3st^3} = -\dfrac{3s^{7-1}t^{5-3}}{r^{3-2}}$ Apply second law of exponents

$= -\dfrac{3s^6t^2}{r}$ Simplify

8. $\dfrac{48a^3bc^2}{36ab^4c^3} = \dfrac{4a^{3-1}}{3b^{4-1}c^{3-2}}$ Apply second law of exponents

$= \dfrac{4a^2}{3b^3c}$ Simplify

9. $\left(-3x^3\right)^2 = (-3)^2\left(x^3\right)^2$ Apply fourth law of exponents

$= 9x^{3(2)}$ Apply third law of exponents

$= 9x^6$ Simplify

10. $\left(-4y^4\right)^3 = (-4)^3\left(y^4\right)^3$ Apply fourth law of exponents

$= -64y^{4(3)}$ Apply third law of exponents

$= -64y^{12}$ Simplify

11. $\left(\dfrac{-3m^2n}{2m^3}\right)^3 = \left(\dfrac{-3n}{2m^{3-2}}\right)^3$ Apply second law of exponents

$= \dfrac{(-3n)^3}{(2m)^3}$ Apply fifth law of exponents

$= \dfrac{(-3)^3 n^3}{(2)^3 m^3}$ Apply fourth law of exponents

$= -\dfrac{27n^3}{8m^3}$ Simplify

Chapter 6 Review

12. $\left(\dfrac{7vw^8}{21v^9w^3}\right)^4 = \left(\dfrac{w^{8-3}}{3v^{9-1}}\right)^4$ Apply second law of exponents

$= \dfrac{\left(w^5\right)^4}{\left(3v^8\right)^4}$ Apply fifth law of exponents

$= \dfrac{\left(w^5\right)^4}{(3)^4\left(v^8\right)^4}$ Apply fourth law of exponents

$= \dfrac{w^{5(4)}}{81v^{8(4)}}$ Apply third law of exponents

$= \dfrac{w^{20}}{81v^{32}}$ Simplify

13. $(-3)^{-4} = \dfrac{1}{(-3)^4} = \dfrac{1}{81}$

14. $4^{-3} = \dfrac{1}{4^3} = \dfrac{1}{64}$

15. $\left(\dfrac{1}{3}\right)^{-2} = 3^2 = 9$

16. $\dfrac{3}{5^{-2}} = 3 \cdot 5^2 = 75$

17. $(3m)^{-5} = \dfrac{1}{(3m)^5} = \dfrac{1}{3^5 m^5} = \dfrac{1}{243m^5}$

18. $-7y^{-8} = -\dfrac{7}{y^8}$

19. $a^{-1} + a^{-2} = \dfrac{1}{a} + \dfrac{1}{a^2}$

20. $\dfrac{3q^{-9}}{r^{-2}} = \dfrac{3r^2}{q^9}$

Chapter 6 Review

21. $6c^{-7} \cdot 3^{-1}c^4 = \dfrac{6c^4}{3c^7} = \dfrac{2}{c^{7-4}} = \dfrac{2}{c^3}$

22. $\dfrac{11z^{-7}}{3^{-2}z^{-5}} = \dfrac{11(3^2)z^5}{z^7} = \dfrac{99}{z^{7-5}} = \dfrac{99}{z^2}$

23. $(2d^{-2}k^3)^{-4} = \dfrac{1}{(2d^{-2}k^3)^4} = \dfrac{1}{2^4(d^{-2})^4(k^3)^4} = \dfrac{1}{16d^{-2(4)}k^{3(4)}} = \dfrac{1}{16d^{-8}k^{12}} = \dfrac{d^8}{16k^{12}}$

24. $\dfrac{2w^3(w^{-2})^{-3}}{5w^{-5}} = \dfrac{2w^3(w^{-2(-3)})}{5w^{-5}} = \dfrac{2w^3 w^6 w^5}{5} = \dfrac{2w^{3+6+5}}{5} = \dfrac{2w^{14}}{5}$

25. $\left(\dfrac{3}{n^{-3}}\right)\dfrac{5}{3}n^{-1} + \left(\dfrac{3}{n^{-3}}\right)2n^{-2} - \left(\dfrac{3}{n^{-3}}\right)n^{-3}$ Divide each term by $\dfrac{1}{3}n^{-3}$

$= 5\dfrac{n^{-1}}{n^{-3}} + \dfrac{6n^{-2}}{n^{-3}} - 3$

$= 5n^{-1-(-3)} + 6n^{-2-(-3)} - 3$ Apply second law of exponents

$= 5n^2 + 6n - 3$ Simplify

$\dfrac{5}{3}n^{-1} + 2n^{-2} - n^{-3} = \dfrac{1}{3n^3}(5n^2 + 6n - 3)$

26. $\left(\dfrac{7}{p^{-1}}\right)\dfrac{3}{7}p - \left(\dfrac{7}{p^{-1}}\right)2 + \left(\dfrac{7}{p^{-1}}\right)\dfrac{2}{7}p^{-1}$ Divide each term by $\dfrac{1}{7}p^{-1}$

$= \dfrac{3p}{p^{-1}} - \dfrac{14}{p^{-1}} + 2$

$= 3p^{1-(-1)} - 14p + 2$ Apply second law of exponents

$= 3p^2 - 14p + 2$ Simplify

$\dfrac{3}{7}p - 2 + \dfrac{2}{7}p^{-1} = \dfrac{1}{7p}(3p^2 - 14p + 2)$

27. $\dfrac{7}{7} + \dfrac{7(5n)^{1/3}}{7}$ Divide each term by 7

$= 1 + (5n)^{1/3}$ Simplify

$7 + 7(5n)^{1/3} = 7\left(1 + (5n)^{1/3}\right)$

229

Chapter 6 Review

28. $\dfrac{12}{4} - \dfrac{8y^{1/4}}{4}$ Divide each term by 4

$= 3 - 2y^{1/4}$ Simplify

$12 - 8y^{1/4} = 4\left(3 - 2y^{1/4}\right)$

29. $\dfrac{3xy^{1/2}}{x^{1/2}} - \dfrac{2(xy)^{1/2}}{x^{1/2}}$ Divide each term by $x^{1/2}$

$= \dfrac{3xy^{1/2}}{x^{1/2}} - \dfrac{2x^{1/2}y^{1/2}}{x^{1/2}}$ Apply fourth law

$= 3x^{1-1/2}y^{1/2} - 2y^{1/2}$ Apply second law

$= 3x^{1/2}y^{1/2} - 2y^{1/2}$ Simplify

$3xy^{1/2} - 2(xy)^{1/2} = x^{1/2}\left(3x^{1/2}y^{1/2} - 2y^{1/2}\right)$

30. $\dfrac{3c(2d)^{1/2}}{d^{1/2}} + \dfrac{(3cd)^{1/2}}{d^{1/2}}$ Divide each term by $\sqrt{d}$ or $d^{1/2}$

$= \dfrac{3\sqrt{2}cd^{1/2}}{d^{1/2}} + \dfrac{\sqrt{3}c^{1/2}d^{1/2}}{d^{1/2}}$ Apply fourth law of exponents

$= 3\sqrt{2}c + \sqrt{3c}$ Simplify

$3c(2d)^{1/2} + (3cd)^{1/2} = \sqrt{d}\left(3\sqrt{2}c + \sqrt{3c}\right)$

31. $\dfrac{q^{-1/3}}{q^{-1/3}} - \dfrac{q^{1/3}}{q^{-1/3}}$ Divide each term by $q^{-1/3}$

$= 1 - q^{1/3 - (-1/3)}$ Apply second law of exponents

$= 1 - q^{2/3}$ Simplify

$q^{-1/3} - q^{1/3} = \dfrac{1}{q^{1/3}}\left(1 - q^{2/3}\right)$

32. $\dfrac{5(x+2)^{3/4}}{(x+2)^{-3/4}} - \dfrac{(x+2)^{-3/4}}{(x+2)^{-3/4}}$ Divide each term by $(x+2)^{-3/4}$

$= 5(x+2)^{3/4 - (-3/4)} - 1$ Apply second law of exponents

$= 5(x+2)^{3/2} - 1$

$5(x+2)^{3/4} - (x+2)^{-3/4} = \dfrac{1}{(x+2)^{3/4}}\left(5(x+2)^{3/2} - 1\right)$

33. $\dfrac{9t}{3\sqrt{t}} + \dfrac{\sqrt{27t}}{3\sqrt{t}} = \dfrac{9t}{3\sqrt{t}} + \dfrac{3\sqrt{3}\sqrt{t}}{3\sqrt{t}}$

$\qquad\qquad\quad = \dfrac{3t}{t^{1/2}} + \sqrt{3}$ Divide each term by $3\sqrt{t}$

$\qquad\qquad\quad = 3t^{1-(1/2)} + \sqrt{3}$ Apply second law of exponents

$\qquad\qquad\quad = 3t^{1/2} + \sqrt{3}$ Simplify

$\qquad\qquad\quad = 3\sqrt{t} + \sqrt{3}$

$9t + \sqrt{27t} = 3\sqrt{t}\left(3\sqrt{t} + \sqrt{3}\right)$

34. $\dfrac{6w}{2\sqrt{w}} + \dfrac{\sqrt{8w}}{2\sqrt{w}} = \dfrac{6w}{2\sqrt{w}} + \dfrac{2\sqrt{2}\sqrt{w}}{2\sqrt{w}}$ Divide each term by $2\sqrt{w}$

$\qquad\qquad\quad = \dfrac{3w}{w^{1/2}} + \sqrt{2}$

$\qquad\qquad\quad = 3w^{1-(1/2)} + \sqrt{2}$ Apply second law of exponents

$\qquad\qquad\quad = 3w^{1/2} + \sqrt{2}$ Simplify

$\qquad\qquad\quad = 3\sqrt{w} + \sqrt{2}$

$6w + \sqrt{8w} = 2\sqrt{w}\left(3\sqrt{w} + \sqrt{2}\right)$

35. $25m^{1/2} = 25\sqrt{m}$

36. $8n^{1/3} = 8\sqrt[3]{n}$

37. $(13d)^{2/3} = \left((13d)^2\right)^{1/3} = \sqrt[3]{(13d)^2}$

38. $6x^{2/5}y^{3/5} = 6\left(x^2\right)^{1/5}\left(y^3\right)^{1/5} = 6\left(x^2 y^3\right)^{1/5} = 6\sqrt[5]{x^2 y^3}$

39. $(3q)^{-3/4} = \dfrac{1}{(3q)^{3/4}} = \dfrac{1}{\left((3q)^3\right)^{1/4}} = \dfrac{1}{\sqrt[4]{(3q)^3}}$

40. $7(uv)^{3/2} = 7\left((uv)^3\right)^{1/2} = 7\sqrt{(uv)^3} = 7\sqrt{u^3 v^3}$

41. $\left(a^2 + b^2\right)^{0.5} = \left(a^2 + b^2\right)^{1/2} = \sqrt{a^2 + b^2}$

Chapter 6 Review

42. $(16-x^2)^{0.25} = (16-x^2)^{1/4} = \sqrt[4]{16-x^2}$

43. $\dfrac{7}{\sqrt{5y}} = \dfrac{7\sqrt{5y}}{\sqrt{5y}\sqrt{5y}} = \dfrac{7\sqrt{5y}}{5y}$

44. $\dfrac{6d}{\sqrt{2d}} = \dfrac{6d\sqrt{2d}}{\sqrt{2d}\sqrt{2d}} = \dfrac{6d\sqrt{2d}}{2d} = 3\sqrt{2d}$

45. $\sqrt{\dfrac{3r}{11s}} = \dfrac{\sqrt{3r}}{\sqrt{11s}} = \dfrac{\sqrt{3r}\sqrt{11s}}{\sqrt{11s}\sqrt{11s}} = \dfrac{\sqrt{33rs}}{11s}$

46. $\sqrt{\dfrac{26}{2m}} = \dfrac{\sqrt{26}}{\sqrt{2m}} = \dfrac{\sqrt{26}\sqrt{2m}}{\sqrt{2m}\sqrt{2m}} = \dfrac{2\sqrt{13m}}{2m} = \dfrac{\sqrt{13m}}{m}$

47. $\dfrac{-3}{\sqrt{a}+2} = \dfrac{-3(\sqrt{a}-2)}{(\sqrt{a}+2)(\sqrt{a}-2)} = \dfrac{-3\sqrt{a}+6}{a-4}$

48. $\dfrac{-3}{\sqrt{z}-4} = \dfrac{-3(\sqrt{z}+4)}{(\sqrt{z}-4)(\sqrt{z}+4)} = \dfrac{-3\sqrt{z}-12}{z-16}$

49. $\dfrac{2x-\sqrt{3}}{x-\sqrt{3}} = \dfrac{(2x-\sqrt{3})(x+\sqrt{3})}{(x-\sqrt{3})(x+\sqrt{3})} = \dfrac{2x^2+2x\sqrt{3}-x\sqrt{3}-3}{x^2-3} = \dfrac{2x^2+x\sqrt{3}-3}{x^2-3}$

50. $\dfrac{m-\sqrt{3}}{5m+2\sqrt{3}} = \dfrac{(m-\sqrt{3})(5m-2\sqrt{3})}{(5m+2\sqrt{3})(5m-2\sqrt{3})}$

$= \dfrac{5m^2-2m\sqrt{3}-5m\sqrt{3}+6}{25m^2-12} = \dfrac{5m^2-7m\sqrt{3}+6}{25m^2-12}$

51. $\dfrac{2}{\sqrt[3]{4}} = \dfrac{2\sqrt[3]{2}}{\sqrt[3]{4}\sqrt[3]{2}} = \dfrac{2\sqrt[3]{2}}{\sqrt[3]{8}} = \dfrac{2\sqrt[3]{2}}{2} = \sqrt[3]{2}$

52. $\dfrac{9}{\sqrt[4]{3}} = \dfrac{9\sqrt[4]{27}}{\sqrt[4]{3}\sqrt[4]{27}} = \dfrac{9\sqrt[4]{27}}{\sqrt[4]{81}} = \dfrac{9\sqrt[4]{27}}{3} = 3\sqrt[4]{27}$

53. $\sqrt[4]{\dfrac{q}{8w^3}} = \dfrac{\sqrt[4]{q}}{\sqrt[4]{8w^3}} = \dfrac{\sqrt[4]{q}\sqrt[4]{2w}}{\sqrt[4]{8w^3}\sqrt[4]{2w}} = \dfrac{\sqrt[4]{2qw}}{\sqrt[4]{16w^4}} = \dfrac{\sqrt[4]{2qw}}{2w}$

Chapter 6 Review

54. $\sqrt[3]{\dfrac{5}{49t}} = \dfrac{\sqrt[3]{5}}{\sqrt[3]{49t}} = \dfrac{\sqrt[3]{5}\sqrt[3]{7t^2}}{\sqrt[3]{49t}\sqrt[3]{7t^2}} = \dfrac{\sqrt[3]{35t^2}}{\sqrt[3]{343t^3}} = \dfrac{\sqrt[3]{35t^2}}{7t}$

55. $x - 3\sqrt{x} + 2 = 0$
$x + 2 = 3\sqrt{x}$ Add $3\sqrt{x}$ to both sides
$(x+2)^2 = (3\sqrt{x})^2$ Square both sides
$x^2 + 4x + 4 = 9x$
$x^2 - 5x + 4 = 0$ Subtract $9x$ from both sides
$(x-1)(x-4) = 0$ Factor
$x = 1$ or $x = 4$ Set each factor equal to 0
Check:
$1 - 3\sqrt{1} + 2 = 3 - 3 = 0$ Yes; 1 is a solution
$4 - 3\sqrt{4} + 2 = 6 - 3\cdot 2 = 0$ Yes; 4 is a solution
Thus, the solutions of the original equation are 1 and 4.

56. $\sqrt{x+1} + \sqrt{x+8} = 7$
$\sqrt{x+1} = 7 - \sqrt{x+8}$ Subtract $\sqrt{x+8}$ from both sides
$(\sqrt{x+1})^2 = (7 - \sqrt{x+8})^2$ Square both sides
$x + 1 = 49 - 14\sqrt{x+8} + x + 8$
$14\sqrt{x+8} = 56$ Rearrange and combine like terms
$\sqrt{x+8} = 4$ Divide both sides by 14
$(\sqrt{x+8})^2 = 4^2$ Square both sides
$x + 8 = 16$
$x = 8$ Subtract 8 from both sides
Check:
$\sqrt{8+1} + \sqrt{8+8} = 3 + 4 = 7$ Yes; 8 is a solution
Thus, the solution of the original equation is 8.

Chapter 6 Review

57. $(x+7)^{1/2} + x^{1/2} = 7$

$(x+7)^{1/2} = 7 - x^{1/2}$ Subtract $x^{1/2}$ from both sides

$\left((x+7)^{1/2}\right)^2 = \left(7 - x^{1/2}\right)^2$ Square both sides

$x + 7 = 49 - 14x^{1/2} + x$

$14x^{1/2} = 42$ Rearrange and combine like terms

$x^{1/2} = 3$ Divide both sides by 14

$\left(x^{1/2}\right)^2 = 3^2$ Square both sides

$x = 9$

Check:

$(9+7)^{1/2} + 9^{1/2} = 4 + 3 = 7$ Yes; 9 is a solution

Thus, the solution to the original equation is 9.

58. $(y-3)^{1/2} + (y+4)^{1/2} = 7$

$(y-3)^{1/2} = 7 - (y+4)^{1/2}$ Subtract $(y+4)^{1/2}$ from both sides

$\left((y-3)^{1/2}\right)^2 = \left(7 - (y+4)^{1/2}\right)^2$ Square both sides

$y - 3 = 49 - 14(y+4)^{1/2} + y + 4$

$14(y+4)^{1/2} = 56$ Rearrange and combine like terms

$(y+4)^{1/2} = 4$ Divide both sides by 14

$\left((y+4)^{1/2}\right)^2 = 4^2$ Square both sides

$y + 4 = 16$

$y = 12$ Subtract 4 from both sides

Check:

$(12-3)^{1/2} + (12+4)^{1/2} = 3 + 4 = 7$ Yes; 12 is a solution

Thus, the solution to the original equation is 12.

Chapter 6 Review

59. $\sqrt[3]{x+1} = 2$

$\left(\sqrt[3]{x+1}\right)^3 = 2^3$ Cube both sides

$x + 1 = 8$
$x = 7$ Subtract 1 from both sides
Check:
$\sqrt[3]{7+1} = \sqrt[3]{8} = 2$ Yes; 7 is a solution
Thus, the solution to the original equation is 7.

60. $x^{2/3} + 2 = 6$

$x^{2/3} = 4$ Subtract 2 from both sides

$\left(x^{2/3}\right)^3 = 4^3$ Cube both sides

$x^2 = 64$
$\sqrt{x^2} = \sqrt{64}$ Take the square root of both sides
$x = \pm 8$
Check:
$8^{2/3} + 2 = 4 + 2 = 6$ Yes; 8 is a solution
$(-8)^{2/3} + 2 = 4 + 2 = 6$ Yes; -8 is a solution
Thus, the solutions to the original equation are 8 and -8.

61. $(x-1)^{-3/2} = \dfrac{1}{8}$

$\dfrac{1}{(x-1)^{3/2}} = \dfrac{1}{8}$ Rewrite without negative exponents

$8 = (x-1)^{3/2}$ Fundamental property of proportions

$\sqrt[3]{8} = \sqrt[3]{(x-1)^{3/2}}$ Take the cube root of both sides

$2 = (x-1)^{1/2}$

$2^2 = \left((x-1)^{1/2}\right)^2$ Square both sides

$4 = x - 1$
$5 = x$ Add 1 to both sides
Check:

$(5-1)^{-3/2} = 4^{-3/2} = \dfrac{1}{4^{3/2}} = \dfrac{1}{8}$ Yes; 5 is a solution

Thus, the solution to the original equation is 5.

235

Chapter 6 Review

62. $(2x+1)^{-1/2} = \dfrac{1}{3}$

$\dfrac{1}{(2x+1)^{1/2}} = \dfrac{1}{3}$ Rewrite without negative exponents

$3 = (2x+1)^{1/2}$ Fundamental property of proportions

$3^2 = \left((2x+1)^{1/2}\right)^2$ Square both sides

$9 = 2x+1$

$8 = 2x$ Subtract 1 from both sides

$4 = x$ Divide both sides by 2

Check:

$(2(4)+1)^{-1/2} = 9^{-1/2} = \dfrac{1}{9^{1/2}} = \dfrac{1}{3}$ Yes; 4 is a solution

Thus, the solution to the original equation is 4.

63. $t = \sqrt{\dfrac{2v}{g}}$

$t^2 = \dfrac{2v}{g}$ Square both sides

$gt^2 = 2v$ Multiply both sides by g

$g = \dfrac{2v}{t^2}$ Divide both sides by t^2

64. $q - 1 = 2\sqrt{\dfrac{r^2 - 1}{3}}$

$\dfrac{q-1}{2} = \sqrt{\dfrac{r^2 - 1}{3}}$ Divide both sides by 2

$\dfrac{q^2 - 2q + 1}{4} = \dfrac{r^2 - 1}{3}$ Square both sides

$\dfrac{3q^2 - 6q + 3}{4} = r^2 - 1$ Multiply both sides by 3

$\dfrac{3q^2 - 6q + 7}{4} = r^2$ Add 1 to both sides

$\pm \dfrac{\sqrt{3q^2 - 6q + 7}}{2} = r$ Take the square root of both sides

65. $R = \dfrac{1 + \sqrt{p^2 + 1}}{2}$

$2R - 1 = \sqrt{p^2 + 1}$ Isolate $\sqrt{p^2 + 1}$

$4R^2 - 4R + 1 = p^2 + 1$ Square both sides

$4R^2 - 4R = p^2$ Subtract 1 from both sides

$\pm 2\sqrt{R^2 - R} = p$ Take the square root of both sides

66. $q = \sqrt[3]{\dfrac{1 + r^2}{2}}$

$q^3 = \dfrac{1 + r^2}{2}$ Cube both sides

$2q^3 - 1 = r^2$ Isolate r^2

$\pm\sqrt{2q^3 - 1} = r$ Take the square root of both sides

67. The speed of light is 186,000 mi/s and 1 mi = 5280 ft, therefore, the speed of light in ft/s is $186000 \cdot 5280 = 9.8208 \times 10^8$ ft/s. The time it takes light to travel 1 ft is $\dfrac{1}{9.8208 \times 10^8} = 1.018 \times 10^{-9}$ s or 0.000000001018 s.

68. The amount of time it would take is $\dfrac{5 \times 10^{12}}{20} = 2.5 \times 10^{11}$ hours or 250,000,000,000 hours. There are 8,760 hours in one year; therefore, 2.5×10^{11} hours equals $\dfrac{2.5 \times 10^{11}}{8760} = 28{,}538{,}812.79$ years.

69. a. $57{,}267{,}400 = 5.72674 \times 10^7$ square miles; $6{,}100{,}000{,}000 = 6.1 \times 10^9$ people.

 b. The number of people to square foot will be $\dfrac{6.1 \times 10^9}{5.72674 \times 10^7} = 107$.

70. a. $92{,}956{,}000 = 9.2956 \times 10^7$ miles; $186{,}000 = 1.86 \times 10^5$ miles per second.

 b. Sunlight will reach Earth in $\dfrac{9.2956 \times 10^7}{1.86 \times 10^5} = 499.76$ seconds, or 8.33 minutes.

Chapter 6 Review

71. Solve for m in the formula $m = \dfrac{M}{\sqrt{1-\dfrac{v^2}{c^2}}}$, where $M = 80$ kg and $v = 0.7c$.

$$m = \dfrac{80}{\sqrt{1-\dfrac{(0.7c)^2}{c^2}}} = \dfrac{80}{\sqrt{1-\dfrac{0.49c^2}{c^2}}} = \dfrac{80}{\sqrt{1-0.49}} = 112$$

Thus, the original mass of the man is 112 kg.

72. The cylinder with smallest surface area for a given volume has radius = height $= \sqrt[3]{\dfrac{V}{\pi}}$. So if the volume is 60 cubic inches, then the radius and the height are $\sqrt[3]{\dfrac{60}{\pi}} = 2.67$ inches.

73. a. To find the membership in 1990, we compute $M(20)$, since 1990 is 20 years after 1970.

$M(20) = 30(20)^{3/4} = 284$
So, the membership was 284 in 1990.

b. $810 = 30t^{3/4}$
$27 = t^{3/4}$
$(27)^{4/3} = \left(t^{3/4}\right)^{4/3}$
$81 = t$
The year will be $1970 + 81 = 2051$.

74. a. To find the number of heron in 1995, we compute $P(5)$, since 1995 is 5 years after 1990.

$P(5) = 36(5)^{-2/3} = 12$
So, there where 12 heron in 1995.

b. $40 = 360t^{-2/3}$
$\dfrac{1}{9} = t^{-2/3}$
$\left(\dfrac{1}{9}\right)^{-3/2} = \left(t^{-2/3}\right)^{-3/2}$
$27 = t$
The year will be $1990 + 27 = 2017$.

75. a. Compute q with $m = 100$ and $w = 1600$.
$$q = 0.6(100)^{1/4}(1600)^{3/4} = 0.6(100)^{1/4}(100)^{3/4}(16)^{3/4}$$
$$= 0.6(100)^{1/4+3/4} 8 = 480$$
Thus, they can produce 480 saddlebags.

b. Solve for w with $m = 100$ and $q = 200$.
$$200 = 0.6(100)^{1/4} w^{3/4}$$
$$\frac{200}{0.6(100)^{1/4}} = w^{3/4}$$
$$\left(\frac{200}{0.6(100)^{1/4}}\right)^{4/3} = \left(w^{3/4}\right)^{4/3} = w$$
$$498 = w$$
Thus, they need 498 hours of labor.

76. a. Compute S with $w = 60$ and $h = 40$.
$$S = 8.5(40)^{0.35}(60)^{0.55} = 293.85$$
Therefore, the child's surface area is 293.85 square inches.

b. Solve for w with $S = 397$ and $h = 50$.
$$397 = 8.5(50)^{0.35} w^{0.55}$$
$$\frac{397}{8.5(50)^{0.35}} = w^{0.55}$$
$$\left(\frac{397}{8.5(50)^{0.35}}\right)^{1/0.55} = \left(w^{0.55}\right)^{1/0.55} = w$$
$$89.96 = w$$
Therefore, the child weighs 89.96 pounds.

77. The car traveling east will be $50t$ miles from the intersection t hours after passing the intersection. The car traveling north will be $5 + 40t$ miles from the intersection t hours after the eastbound car passes the intersection. By the Pythagorean theorem, the distance between the two cars t hours after the eastbound car passes the intersection is $\sqrt{(50t)^2 + (5 + 40t)^2}$ miles. The cars will be 200 miles apart when
$$200 = \sqrt{(50t)^2 + (5 + 40t)^2}$$

Chapter 6 Review

$40000 = 2500t^2 + 25 + 400t + 1600t^2$ Square both sides

$0 = 4100t^2 + 400t - 39975$ Combine like terms

$t = \dfrac{-400 \pm \sqrt{400^2 - 4 \cdot 4100 \cdot (-39975)}}{2(4100)}$ Apply quadratic formula

$t = -3.17$ or 3.07

A negative answer doesn't make sense, therefore, the cars are 200 miles apart after 3.07 hours.

78. Let x represent the distance in feet from the base of the second antenna to where the guy wire is anchored to the ground. By the Pythagorean theorem, the length of the guy wire from the second antenna to the anchored point is $\sqrt{x^2 + 25^2}$ feet. Since the distance between the two antennae is 75 feet, the length of the guy wire from the first antenna to the anchored point is $\sqrt{(75-x)^2 + 20^2}$ feet. The guy wire has total length 90 feet, therefore the sum of the lengths $\sqrt{x^2 + 25^2}$ and $\sqrt{(75-x)^2 + 20^2}$ is 90. Solve for x in the following:

$\sqrt{x^2 + 25^2} + \sqrt{(75-x)^2 + 20^2} = 90$

$\sqrt{x^2 + 625} + \sqrt{6025 - 150x + x^2} = 90$ Simplify

$\sqrt{6025 - 150x + x^2} = 90 - \sqrt{x^2 + 625}$ Rearrange

$6025 - 150x + x^2 = 8100 - 180\sqrt{x^2 + 625} + x^2 + 625$ Square both sides

$\sqrt{x^2 + 625} = 15 + \dfrac{5}{6}x$ Isolate $\sqrt{x^2 + 625}$

$x^2 + 625 = 225 + 25x + \dfrac{25}{36}x^2$ Square both sides

$11x^2 - 900x + 14400 = 0$ Combine like terms

$x = \dfrac{900 \pm \sqrt{900^2 - 4(11)(14400)}}{2 \cdot 11}$ Apply quadratic formula

$x = 60$ or 21.82

Therefore, we may anchor the guy wire either 60 or 21.82 feet from the second antenna.

79. Compute the length of each side.

length of $\overline{AB} = \sqrt{(5-(-1))^2 + (4-2)^2} = \sqrt{36+4} = \sqrt{40} = 2\sqrt{10}$

length of $\overline{BC} = \sqrt{(1-5)^2 + (-4-4)^2} = \sqrt{16+64} = \sqrt{80} = 4\sqrt{5}$

length of $\overline{AC} = \sqrt{(1-(-1))^2 + (-4-2)^2} = \sqrt{4+36} = \sqrt{40} = 2\sqrt{10}$

Thus, the perimeter equals $2\sqrt{10} + 2\sqrt{10} + 4\sqrt{5} = 4\sqrt{10} + 2\sqrt{5}$. Since $\overline{AB}^2 + \overline{AC}^2 = (2\sqrt{10})^2 + (2\sqrt{10})^2 = 40 + 40 = 80 = \overline{BC}^2$, $\triangle ABC$ is a right triangle.

80. Compute the midpoint of each side.

$m\overline{AB} = \left(\dfrac{-1+5}{2}, \dfrac{2+4}{2}\right) = (2,3)$

$m\overline{BC} = \left(\dfrac{5+1}{2}, \dfrac{4+(-4)}{2}\right) = (3,0)$

$m\overline{AC} = \left(\dfrac{-1+1}{2}, \dfrac{2+(-4)}{2}\right) = (0,-1)$

Compute the length of each side

length from $m\overline{AB}$ to $m\overline{BC} = \sqrt{(3-2)^2 + (0-3)^2} = \sqrt{1+9} = \sqrt{10}$

length from $m\overline{BC}$ to $m\overline{AC} = \sqrt{(0-3)^2 + (-1-0)^2} = \sqrt{9+1} = \sqrt{10}$

length from $m\overline{AB}$ to $m\overline{AC}$ $\sqrt{(0-2)^2 + (-1-3)^2} = \sqrt{4+16} = \sqrt{20} = 2\sqrt{5}$

Therefore, the perimeter is $\sqrt{10} + \sqrt{10} + 2\sqrt{5} = 2\sqrt{10} + 2\sqrt{5}$.

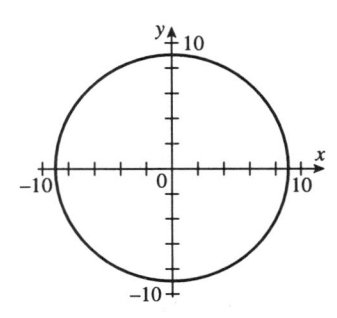

81.

Chapter 6 Review

82. To put in standard form:
$3x^2 + 3y^2 = 12$
$x^2 + y^2 = 4$ Divide both sides by 3

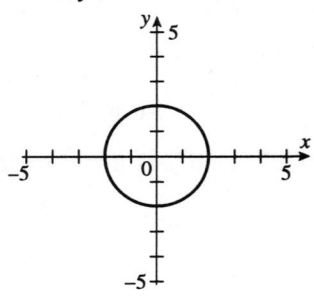

83.

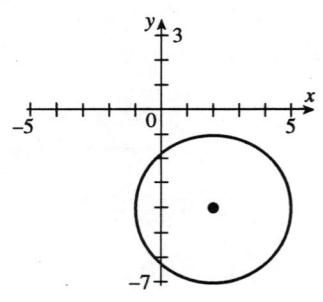

84. To put in standard form:
$x^2 + y^2 + 6y = 0$
$x^2 + y^2 + 6y + 9 = 9$ Add 9 to both sides to complete the square
$(x-0)^2 + (y+3)^2 = 3^2$ Factor

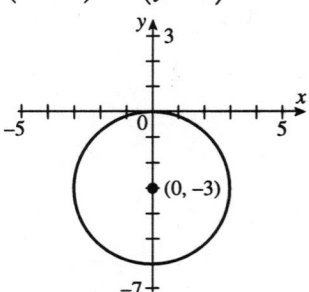

(0, –3)

242

Chapter 6 Review

85. The equation is:
$$(x-5)^2 + (y-(-2))^2 = (4\sqrt{2})^2$$
$$(x-5)^2 + (y+2)^2 = 32$$

86. The radius is equal to the distance from the center to any point on the circle. Therefore, the radius is:
$$\text{radius} = \sqrt{(2-7)^2 + (3-(-1))^2} = \sqrt{25+16} = \sqrt{41}$$
The equation is:
$$(x-7)^2 + (y-(-1))^2 = (\sqrt{41})^2$$
$$(x-7)^2 + (y+1)^2 = 41$$

87. The center is at the midpoint of a diameter.
$$\text{center} = \left(\frac{-2+4}{2}, \frac{3+5}{2}\right) = (1,4)$$
The radius is half the length of the diameter.
$$\text{radius} = \frac{1}{2}\sqrt{(4-(-2))^2 + (5-3)^2} = \frac{1}{2}\sqrt{36+4} = \frac{\sqrt{40}}{2} = \frac{2\sqrt{10}}{2} = \sqrt{10}$$
The equation is:
$$(x-1)^2 + (y-4)^2 = (\sqrt{10})^2$$
$$(x-1)^2 + (y-4)^2 = 10$$

88. We need to find a, b, and c so that (6, 2), (–1, 1), and (0, –6) are roots of the equation $x^2 + y^2 + ax + by + c = 0$. Plugging in (0, –6) yields:
$$0^2 + (-6)^2 + a(0) + b(-6) + c = 0$$
$$36 - 6b + c = 0$$
Therefore $c = 6b - 36$. Plugging in (–1, 1) gives
$$(-1)^2 + 1^2 + a(-1) + b(1) + c = 0$$
$$2 - a + b + c = 0$$
However, $c = 6b - 36$, so:
$$2 - a + b + (6b - 36) = 0$$
$$7b - 34 = a$$
Plugging in (6, 2), we get:
$$6^2 + 2^2 + a(6) + b(2) + c = 0$$
$$40 + 6a + 2b + c = 0$$

Chapter 6 Review

However, $c = 6b - 36$ and $a = 7b - 34$ so:
$40 + 6(7b - 34) + 2b + (6b - 36) = 0$
$50b - 200 = 0$
$b = 4$
And $a = 7(4) - 34 = -6$
$c = 6(4) - 36 = -12$
So the equation of the circle is
$x^2 + y^2 - 6x + 4y - 12 = 0$
$x^2 - 6x + 9 + y^2 + 4y + 4 - 12 + 12 = 25$ Add 25 to both sides
$(x - 3)^2 + (y + 2)^2 = 25$ Factor

Chapter 7

Section 7.1

1. a.

t	P(t)
0	300
1	600
2	1200
3	2400
4	4800
5	9600
6	19,200
7	38,400
8	76,800

b. $P(t) = 300(2)^t$

c. Refer to the graph in the back of the textbook.

d. $P(8) = 300(2)^8 = 76,800$

5 days $= \frac{5}{7}$ weeks and

$P\left(\frac{5}{7}\right) = 300(2)^{5/7} = 492$

3. a.

t	P(t)
0	20,000
6	50,000
12	125,000
18	312,500
24	781,250

b. $P(t) = 20,000(2.5)^{t/6}$

c. Refer to the graph in the back of the textbook.

d. $P(4) = 20,000(2.5)^{4/6} = 36,840$
$P(20) = 20,000(2.5)^{20/6}$
$= 424,128$

5. a.

t	A(t)
0	$4000.00
1	$4320.00
2	$4665.60
3	$5038.85

b. $P(t) = 4000(1.08)^t$

c. Refer to the graph in the back of the textbook.

d. $P(2) = 4000(1.08)^2 = \$4665.60$
$P(10) = 4000(1.08)^{10} = \$8635.70$

7. a.

t	P(t)
0	$20,000.00
5	$25,525.63
10	$32,577.89
15	$41,578.56
20	$53,065.95

b. $P(t) = 20,000(1.05)^t$

c. Refer to the graph in the back of the textbook.

d. $P(12) = \$35,917.13$
$P(27) = \$74,669.13$

9. a.

t	S(t)
0	1500
2	1882
4	2360
6	2961
8	3714
10	4659
12	5844

b. $S(t) = 1500(1.12)^t$

Chapter 7, Section 7.1

    c. Refer to the graph in the back of the textbook.

    d. $S(6) = 2961$
        $S(12) = 5844$

11. a.

t	$P(t)$
0	250,000
2	187,500
4	140,625
6	105,469
8	79,102

    b. $P(t) = 250,000(0.75)^{t/2}$

    c. Refer to the graph in the back of the textbook.

    d. $P(3) = 162,380$
        $P(8) = 79,102$

13. a.

d	$L(d)$
0	100%
10	67%
20	44%
30	30%
40	20%
50	13%

    b. $L(d) = (0.85)^{d/4}$

    c. Refer to the graph in the back of the textbook.

    d. $L(20) = 44\%$
        $L(45) = 16\%$

15. a.

t	$P(t)$
0	50
20	43
40	36
60	31
80	26
100	22

    b. $P(t) = 50(0.992)^t$

    c. Refer to the graph in the back of the textbook.

    d. $P(10) = 46.1$ pounds
        $P(100 \text{ years}) = P(100)$
                             $= 22.4$ pounds

17. For A, $1.30 = (1 + r)^6$, so
$r = 1.30^{1/6} - 1 = 0.0447$
For B, $1.20 = (1 + r)^4$, so
$r = 1.20^{1/4} - 1 = 0.0466$
Species B multiplies faster.

19. a. $L(t) = mt + L_0$, $m = \dfrac{9-6}{2} = 1.5$,
$L_0 = 6$, so $L(t) = 1.5t + 6$.
Refer to the graph in the back of the textbook.

    b. $E(t) = E_0 a^t$, $E_0 = 6$,
$E_0 a^2 = 9 \Rightarrow a = \sqrt{\dfrac{9}{E_0}} = 1.5^{1/2}$,
so $E(t) = 6(1.5)^{t/2}$.
Refer to the graph in the back of the textbook.

21. a.

d	$W(d)$
1	2¢
2	4¢
3	8¢
4	16¢
5	32¢

    b. $W(d) = 2^d$

    c. $W(15) = \$327.68$
        $W(30) = \$10,737,418.24$

Chapter 7, Section 7.2

23. $11{,}196{,}700 = 9{,}579{,}700(1+r)^{10}$, so
$$r = \left(\frac{11{,}196{,}700}{9{,}579{,}700}\right)^{1/10} - 1$$
$= 0.0157 = 1.57\%$

25. a. $20{,}000 = 10{,}000(1+r)^{20}$, so
$$r = \left(\frac{20{,}000}{10{,}000}\right)^{1/20} - 1$$
$= 0.0353 = 3.53\%$

   b. $700{,}000 = 350{,}000(1+r)^{20}$, so
$$r = \left(\frac{700{,}000}{350{,}000}\right)^{1/20} - 1$$
$= 0.0353 = 3.53\%$

   c. No

   d. From the results of a – c, 3.53%.

27. a. The number of letters that will eventually go out with your name on top of the list is $6^6 = 46{,}656$, which would you $466,560.

   b. Assuming all letters move at the same speed, the number of copies will be 6 sent to you and others on your level, plus 6^2 to those on the next level, ..., plus 6^7 to those on the level from which you receive money:
   $6^7 + 6^6 + 6^5 + 6^4 + 6^3 + 6^2 + 6$
   $= 335{,}922$.

Section 7.2

1. 3 [^] [2nd] [√] 2 [ENTER]
   yields 4.728804388

3. 4 [^] [(] [2nd] π [−] 1 [)] [ENTER]
   yields 19.47005841

5. [(−)] 0.6 [^] [(] 2 [2nd] [√] 3 [)] [ENTER]
   yields −0.1704093375

7. 6 [^] [(−)] [2nd] [√] 5 [ENTER]
   yields 0.0181970466

9. 2.8 [×] 9 [^] [2nd] [√] 7 [ENTER]
   yields 937.2304752

11. 8 [−] 4 [^] [2nd] [√] 13 [ENTER]
    yields −140.1692797

13.

x	5^x
−1	$\frac{1}{5}$
0	1
1	5
2	25

Refer to the graph in the back of the textbook.

15.

t	3^{-t}
−2	9
−1	3
0	1
1	$\frac{1}{3}$
2	$\frac{1}{9}$

Refer to the graph in the back of the textbook.

17.

z	-4^z
−1	$-\frac{1}{4}$
0	−1
1	−4
2	−16

Refer to the graph in the back of the textbook.

Chapter 7, Section 7.2

19.

x	$\left(\frac{1}{10}\right)^x$
-2	100
-1	10
0	1
1	$\frac{1}{10}$

Refer to the graph in the back of the textbook.

21.

x	$\left(\frac{1}{2}\right)^{-x}$
-2	$\frac{1}{4}$
-1	$\frac{1}{2}$
0	1
1	2
2	4

Refer to the graph in the back of the textbook.

23. Refer to the graph in the back of the textbook.

25. Refer to the graph in the back of the textbook.

27. Refer to the graph in the back of the textbook.

29. Refer to the graph in the back of the textbook.

31. Refer to the graph in the back of the textbook.

33. Refer to the graph in the back of the textbook.

35. $2^x = 32 = 2^5$, so $x = 5$.

37. $5^{x+2} = 25^{4/3} = (5^2)^{4/3} = 5^{8/3}$, so $x + 2 = \frac{8}{3}$ and $x = \frac{8}{3} - 2 = \frac{2}{3}$.

39. $3^{2x-1} = \frac{\sqrt{3}}{9} = 3^{1/2} \cdot 3^{-2} = 3^{-3/2}$, so $2x - 1 = -\frac{3}{2}$ and $x = \frac{\left(-\frac{3}{2}+1\right)}{2} = -\frac{1}{4}$.

41. $4 \cdot 2^{x-3} = 8^{-2x}$,
$2^2 \cdot 2^{x-3} = (2^3)^{-2x}$, $2^{x-1} = 2^{-6x}$,
so $x - 1 = -6x$, $7x = 1$, and $x = \frac{1}{7}$.

43. $27^{4x+2} = 81^{x-1}$,
$(3^3)^{4x+2} = (3^4)^{x-1}$,
$3^{12x+6} = 3^{4x-4}$, so
$12x + 6 = 4x - 4$, $8x = -10$, and
$x = -\frac{10}{8} = -\frac{5}{4}$.

45. $10^{x^2-1} = 1000 = 10^3$, $x^2 - 1 = 3$,
$x^2 = 4$, and $x = \pm\sqrt{4} = \pm 2$.

47. a. $N(t) = 100(8)^{t/4}$
b.

248

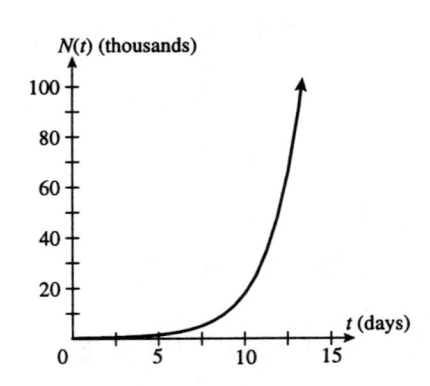

Chapter 7, Section 7.2

c. $51{,}200 = 100(8)^{t/4}$
$\left(\dfrac{51{,}200}{100}\right)^4 = 512^4 = 8^t$
$(8^3)^4 = 8^{12} = 8^t$
$t = 12$

49. a. $N(t) = 26(2)^{t/6}$

b.

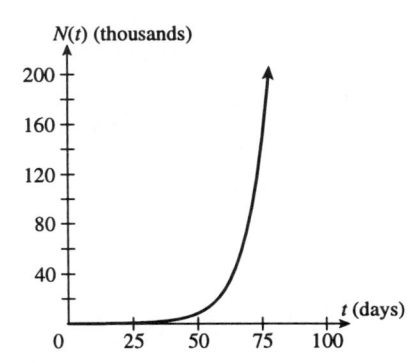

N(t) (thousands) vs t (days)

c. $106{,}496 = 26(2)^{t/6}$
$\left(\dfrac{106{,}496}{26}\right)^6 = 4096^6 = 2^t$
$(2^{12})^6 = 2^{72} = 2^t$
$t = 72$

51. a. $V(t) = 700(0.7)^{t/2}$

b.

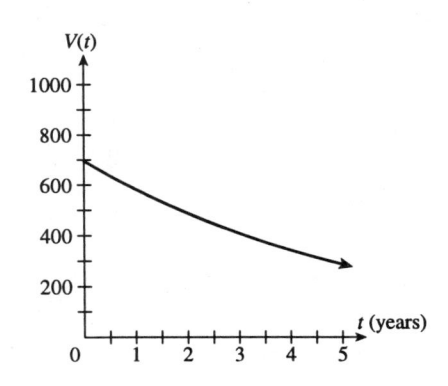

V(t) vs t (years)

c. $343 = 700(0.7)^{t/2}$
$\left(\dfrac{343}{700}\right)^2 = \left(\dfrac{49}{100}\right)^2 = \left(\dfrac{7}{10}\right)^4$
$= (0.7)^t$
$t = 4$

53. Graph Y = 2 ^ X
Start with   Xmin = 0
Xmax = 2
Ymin = 0
Ymax = 4
Then use TRACE and ZOOM to find that
Y = 3 when
X = 1.58.

55. Graph Y = 5 ^ X
Start with   Xmin = 0
Xmax = 1
Ymin = 0
Ymax = 3
Then use TRACE and ZOOM to find that
Y = 2 when
X = 0.43.

57. Graph Y = 3 ^ (X − 1)
Start with   Xmin = 0
Xmax = 3
Ymin = 0
Ymax = 5
Then use TRACE and ZOOM to find that
Y = 4 when
X = 2.26.

59. Graph Y = 4 ^ (−X)
Start with   Xmin = −2
Xmax = 2
Ymin = 0
Ymax = 10
Then use TRACE and ZOOM to find that
Y = 7 when
X = −1.40.

Chapter 7, Section 7.3

61.
a. Yes: $y = 3(2)^x$
b. No: P does not increase by the same factor for each increment of t.
c. No: N does not increase by the same factor for each increment of x.
d. Yes: $R = 405\left(\dfrac{1}{3}\right)^p$

63.
a. $P_0 = f(0) = 300$
b. $f(1) = 300a^1 = 600;\ a = 2$
c. $f(x) = 300(2)^x$

65.
a. $S_0 = S(0) = 150$
b. $S(1) = 150a^1 = 82;$
$a = \dfrac{82}{150} \approx 0.55$
c. $S(d) = 150(0.55)^d$

67.

x	$f(x) = x^2$	$g(x) = 2^x$
-2	4	$\dfrac{1}{4}$
-1	1	$\dfrac{1}{2}$
0	0	1
1	1	2
2	4	4
3	9	8
4	16	16
5	25	32
6	36	64

Refer to the graph in the back of the textbook.

Section 7.3

1. $\log_7(49) = \log_7(7^2) = 2$

3. $\log_4(64) = \log_4(4^3) = 3$

5. $\log_3 \sqrt{3} = \log_3(3^{1/2}) = \dfrac{1}{2}$

7. $\log_5 \dfrac{1}{5} = \log_5(5^{-1}) = -1$

9. $\log_4 4 = \log_4(4^1) = 1$

11. $\log_{10} 1 = \log_{10}(10^0) = 0$

13. $\log_8 8^5 = 5$

15. $\log_{10} 10^{-4} = -4$

17. $\log_{10} 10{,}000 = \log_{10}(10^4) = 4$

19. $\log_{10} 0.1 = \log_{10}(10^{-1}) = -1$

21. $16^{\log_{16} 256} = 16^w$, or $256 = 16^w$

23. $b^{\log_b 9} = b^{-2}$, or $9 = b^{-2}$

25. $10^{\log_{10} A} = 10^{-2.3}$, or $A = 10^{-2.3}$

27. $4^{\log_4 36} = 4^{2q-1}$, or $36 = 4^{2q-1}$

29. $u^{\log_u v} = u^w$, or $v = u^w$

31. $\log_8(8^{-1/3}) = \log_8\left(\dfrac{1}{2}\right)$,
or $-\dfrac{1}{3} = \log_8 \dfrac{1}{2}$

33. $\log_t(t^{3/2}) = \log_t 16$, or $\dfrac{3}{2} = \log_t 16$

35. $\log_{0.8} 0.8^{1.2} = \log_{0.8} M$,
or $1.2 = \log_{0.8} M$

Chapter 7, Section 7.3

37. $\log_x x^{5t} = \log_x(W-3)$,
or $5t = \log_x(W-3)$

39. $\log_3 3^{-0.2t} = \log_3(2N_0)$,
or $-0.2t = \log_3(2N_0)$

41. $8 = b^3$; $b = 8^{1/3} = 2$

43. $x = 4^3 = 64$

45. $y = \log_2 \frac{1}{2} = \log_2(2^{-1}) = -1$

47. $10 = b^{1/2}$; $b = 10^2 = 100$

49. $3x - 1 = 2^5 = 32$; $x = \frac{32+1}{3} = 11$

51. $\log_7 x = \frac{7-5}{3} = \frac{2}{3}$; $x = 7^{2/3} = \sqrt[3]{49}$

53. Since $\log_2 16 = 4$ and $\log_2 32 = 5$, $4 < \log_2 25 < 5$. Graphing $f(x) = \log_2 x$ for $16 < x < 32$ and reading off the $f(x)$ value off $x = 25$ yields $\log_2 25 = 4.64$.

55. Since $\log_{10} 10 = 1$ and $\log_{10} 100 = 2$, $1 < \log_{10} 50 < 2$. Graphing $f(x) = \log_{10} x$ for $10 < x < 100$ and reading off the $f(x)$ value for $x = 50$ yields $\log_{10} 50 = 1.70$.

57. Since $\log_8 1 = 0$ and $\log_8 8 = 1$, $0 < \log_8 5 < 1$. Graphing $f(x) = \log_8 x$ for $1 < x < 8$ and reading off the $f(x)$ value for $x = 5$ yields $\log_8 5 = 0.77$.

59. Since $\log_3 27 = 3$ and $\log_3 81 = 4$, $3 < \log_3 67.9 < 4$. Graphing $f(x) = \log_3 x$ for $27 < x < 81$ and reading off the $f(x)$ value for $x = 67.9$ yields $\log_3 67.9 = 3.84$.

61. [LOG] 54.3 [ENTER]
yields 1.7348

63. [LOG] 2344 [ENTER]
yields 3.3700

65. [LOG] 0.073 [ENTER]
yields -1.1367

67. [LOG] 0.6942 [ENTER]
yields -0.1585

69. $x = \log_{10} 200 = 2.30$

71. $-3x = \log_{10} 5$; $x = -\frac{\log_{10} 5}{3} = -0.23$

73. $10^{0.2x} = \frac{80}{25} = \frac{16}{5}$; $0.2x = \log_{10} \frac{16}{5}$;
$x = \frac{\log_{10} \frac{16}{5}}{0.2} = 2.53$

75. $2(10^{1.4x}) = 23.8$; $10^{1.4x} = 11.9$;
$1.4x = \log_{10} 11.9$; $x = 0.77$

77. $3(10^{-1.5x}) = 31.8$; $10^{-1.5x} = 10.6$;
$-1.5x = \log_{10} 10.6$; $x = -0.68$

79. $1 - 10^{-0.2x} = \frac{13}{16}$; $10^{-0.2x} = \frac{3}{16}$;
$-0.2x = \log_{10} \frac{3}{16}$; $x = 3.63$

81. $a_E = 29{,}028 \text{ ft} \cdot \frac{1 \text{ mi}}{5280 \text{ ft}}$
$= 5.4977 \text{ mi}$;
$P(a_E) = 30(10)^{-(0.09)(5.4977)}$
$= 9.60 \text{ in. of mercury}$

83. $20.2 = 30(10)^{-0.09a}$;
$-0.09a = \log_{10}\left(\frac{20.2}{30}\right) = -0.1718$;
$a = \frac{-0.1718}{-0.09} = 1.91 \text{ mi}$

Chapter 7, Section 7.4

85. $30(10)^{-0.09a} = \frac{1}{2} \cdot 30(10)^0 = 15;$

$10^{-0.09a} = \frac{1}{2};$

$a = \dfrac{\log_{10}\frac{1}{2}}{-0.09} = 3.34$ mi.

87. a. $P(10)$

$= 15{,}717{,}000(10)^{(0.0104)(10)}$

$= 19{,}969{,}613$

b. $P(20) = 25{,}372{,}873$
$P(30) = 32{,}238{,}116$
$P(40) = 40{,}960{,}915$

c. From a., $P(10)$ is just about 20,000,000, so the population will reach 20 million around 1970.

d. $30{,}000{,}000$
$= 15{,}717{,}000(10^{0.0104t});$

$0.0104t = \log_{10}\dfrac{30{,}000{,}000}{15{,}717{,}000}$

$= 0.2808$

$t = \dfrac{0.2808}{0.0104} = 27$ years, or 1987

e. On a graphing calculator, use
Xmin = 0
Xmax = 40
Ymin = 0
Ymax = 50,000,000

Section 7.4

1.

x	$f(x) = 2^x$	$f^{-1}(x) = \log_2 x$
-3	$\frac{1}{8}$	Undefined
-2	$\frac{1}{4}$	Undefined
-1	$\frac{1}{2}$	Undefined
0	1	Undefined
1	2	0
2	4	1
3	8	1.58

Refer to the graph in the back of the textbook.

3.

x	$f(x) = \left(\dfrac{1}{3}\right)^x$	$f^{-1}(x) = \log_{1/3} x$
-2	9	Undefined
-1	3	Undefined
0	1	Undefined
1	$\frac{1}{3}$	0
2	$\frac{1}{9}$	-0.63
3	$\frac{1}{27}$	-1
4	$\frac{1}{81}$	-1.26

Refer to the graph in the back of the textbook.

5. $\log_{10} 487 = 2.688$

7. $\log_{10} 2.16 = 0.334$

9. $\log_{10} -7$ is undefined, since there is no real number a such that 10^a is negative.

11. $6 \log_{10} 28 = 8.683$

c.

(graph)

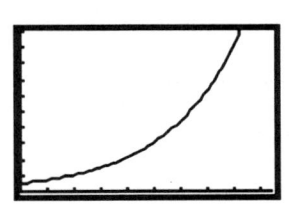

d. Time to double is always 19.08 years.

71. a. $c(t) = 0.7(0.8)^t$

b. $0.4 = 0.7(0.8)^t$;

$$t = \frac{\log_{10}\left(\frac{0.4}{0.7}\right)}{\log_{10} 0.8} = 2.5 \text{ hours}$$

c. Refer to the graph in the back of the textbook.

73. a. $\frac{1}{2} = (0.946)^{t_h}$;

$$t_h = \frac{\log_{10} \frac{1}{2}}{\log_{10} 0.946} = 12.49 \text{ hours}$$

b. Three-fourths will be gone in two half-lives, or 24.97 hours. Seven-eighths will be gone in three half-lives, or 37.46 hours.

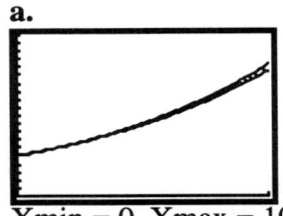

75. a.

(graph)

Xmin = 0, Xmax = 10, Ymin = 0, and Ymax = 20000

$Y_1 = 5000(1+0.12)^t$;

$Y_2 = 5000\left(1+\frac{0.12}{4}\right)^{4t}$

b. $5000(1+0.12)^{10} = \$15,529.24$;

$5000\left(1+\frac{0.12}{4}\right)^{4 \cdot 10} = \$16,310.19$

77. a.

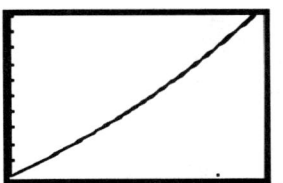

(graph)

Xmin = 0, Xmax = 12, Ymin = 1000, and Ymax = 2000

$Y_1 = 1000\left(1+\frac{0.06}{2}\right)^{2t}$;

$Y_2 = 1000\left(1+\frac{0.06}{4}\right)^{4t}$

The two functions nearly coincide.

b. $1000\left(1+\frac{0.06}{2}\right)^{2t} = 2000$;

$2t \log_{10} 1.03 = \log_{10} 2$; $t = 11.72$;
The investment passes 2000 at the next compounding, at the 12 year mark.

$1000\left(1+\frac{0.06}{4}\right)^{4t} = 2000$;

$4t \log_{10} 1.015 = \log_{10} 2$;
$t = 11.64$; The investment passes 2000 at the next compounding, at the $11\frac{3}{4}$ year mark.

Chapter 7, Section 7.6

79. $1900 = 1000\left(1 + \frac{r}{12}\right)^{12 \cdot 5}$;

$\left(\frac{1900}{1000}\right)^{1/60} = 1 + \frac{r}{12}$;

$r = 12\left[\left(\frac{1900}{1000}\right)^{1/60} - 1\right] = 0.129$

or 12.9%

81. $3 = \left(1 + \frac{0.10}{365}\right)^{365t}$;

$t = \frac{\log_{10} 3}{365 \log_{10}\left(1 + \frac{0.10}{365}\right)} = 10.988$

= 10 years, 361 days

83. $A = A_0(10^{kt} - 1)$; $10^{kt} = \frac{A}{A_0} + 1$;

$kt = \log_{10}\left(\frac{A}{A_0} + 1\right)$;

$t = \frac{\log_{10}\left(\frac{A}{A_0} + 1\right)}{k}$

85. $w = pv^q$; $\frac{w}{p} = v^q$; $q = \log_v \frac{w}{p}$

87. $t = T \log_{10}\left(1 + \frac{A}{k}\right)$;

$\frac{t}{T} = \log_{10}\left(1 + \frac{A}{k}\right)$; $10^{t/T} = 1 + \frac{A}{k}$;

$A = k(10^{t/T} - 1)$

89. $\log_b 4 + \log_b 8 \stackrel{?}{=} \log_b 64 - \log_b 2$

$\log_b(4 \cdot 8) \stackrel{?}{=} \log_b\left(\frac{64}{2}\right)$

$\log_b 32 = \log_b 32$

91. $2\log_b 6 - \log_b 9 \stackrel{?}{=} 2\log_b 2$

$\log_b 6^2 - \log_b 9 \stackrel{?}{=} \log_b 2^2$

$\log_b \frac{36}{9} \stackrel{?}{=} \log_b 4$

$\log_b 4 = \log_b 4$

93. $\frac{1}{2}\log_b 12 - \frac{1}{2}\log_b 3 \stackrel{?}{=} \frac{1}{3}\log_b 8$

$\frac{1}{2}\log_b \frac{12}{3} \stackrel{?}{=} \log_b 8^{1/3}$

$\log_b 4^{1/2} \stackrel{?}{=} \log_b 8^{1/3}$

$\log_b 2 = \log_b 2$

95. Suppose $x = y = 1$. Then
$\log_{10} x + \log_{10} y = 0 + 0 = 0$
$= \log_{10} 1 \neq \log_{10}(1 + 1) = \log_{10} 2$.

97. a. $x = b^m$; $y = b^n$

  b. $\log_b(xy) = \log_b(b^m b^n)$

  c. $\log_b(b^m b^n) = \log_b b^{m+n}$

  d. $\log_b b^{m+n} = m + n$

  e. $m + n = \log_b x + \log_b y$

99. a. $x = b^m$

  b. $\log_b(x^p) = \log_b(b^m)^p$

  c. $\log_b(b^m)^p = \log_b b^{mp}$

  d. $\log_b b^{mp} = mp$

  e. $mp = p \log_b x$

Section 7.6

1. [LOG] 3.9 [ENTER]
yields 1.3610

3. [LOG] 16 [ENTER]
yields 2.7726

5. [LOG] 0.3 [ENTER]
yields −1.2040

Chapter 7, Section 7.6

7. Since $e^0 = 1$, $\ln 1 = 0$

9. [2nd] [e] 0.4 [ENTER]
yields 1.4918

11. [2nd] [e] 2.34 [ENTER]
yields 10.3812

13. [2nd] [e] [(-)] 1.2 [ENTER]
yields 0.3012

15. [2nd] [e] [(-)] 0.4 [ENTER]
yields 0.6703

17. $x = \ln 1.9 = 0.642$

19. $x = \ln 45 = 3.807$

21. $x = \ln 0.3 = -1.204$

23. $x = e^{1.42} = 4.137$

25. $x = e^{0.63} = 1.878$

27. $x = e^{-2.6} = 0.0743$

29. $e^{1.2x} = \dfrac{6.21}{2.3}$; $x = \dfrac{\ln\left(\frac{6.21}{2.3}\right)}{1.2} = 0.8277$

31. $\ln\left(\dfrac{7.74}{1.72}\right) = 0.2x$;

$x = \dfrac{\ln\left(\frac{7.74}{1.72}\right)}{0.2} = 7.5204$

33. $e^{0.3x} = \left(\dfrac{6.4+1.8}{20}\right)$;

$x = \dfrac{\ln\left(\frac{6.4+1.8}{20}\right)}{0.3} = -2.9720$

35. $e^{1.2x} = \dfrac{46.52 - 24.2}{3.1}$;

$x = \dfrac{\ln\left(\frac{46.52-24.2}{3.1}\right)}{1.2} = 1.6451$

37. $e^{-1.3x} = \dfrac{16.24 + 21.7}{0.7}$;

$x = \dfrac{\ln\left(\frac{16.24+21.7}{0.7}\right)}{-1.3} = -3.0713$

39. $\ln y = kt$; $t = \dfrac{\ln y}{k}$

41. $e^{-t} = 1 - \dfrac{y}{k}$;

$t = -\ln\left(1 - \dfrac{y}{k}\right) = \ln\left(\dfrac{k}{k-y}\right)$

43. $e^{T/T_0} = k + 10$; $k = e^{T/T_0} - 10$

45. $N(t) = 100 \cdot 2^t = 100(e^{\ln 2})^t$
$= 100 e^{(\ln 2)t} = 100 e^{0.6931t}$

47. $N(t) = 1200(0.6)^t = 1200(e^{\ln 0.6})^t$
$= 1200 e^{(\ln 0.6)t} = 1200 e^{-0.5108t}$

49. $N(t) = 10(1.15)^t = 10(e^{\ln 1.15})^t$
$= 10 e^{(\ln 1.15)t} = 10 e^{0.1398t}$

51. a.

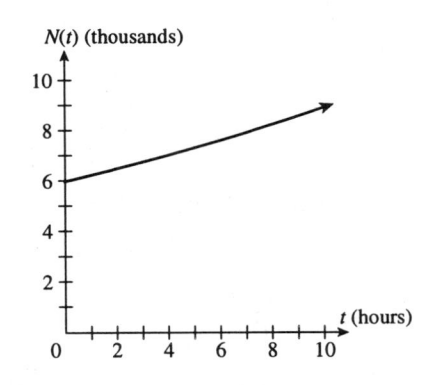

b. $N(10) = 6000 e^{0.4} = 8951$

Chapter 7, Section 7.6

c. $10{,}000 = 6000e^{0.04t}$;
$e^{0.04t} = \dfrac{10{,}000}{6000} = \dfrac{5}{3}$;
$0.04t = \ln\dfrac{5}{3}$;
$t = \dfrac{\ln\frac{5}{3}}{0.04} = 12.8$ hours

53. a. $1000e^{-(0.1)(0.6)} = 941.8$ lumens

b. $800 = 1000e^{-0.1t}$; $0.8 = e^{-0.1t}$;
$t = \dfrac{\ln 0.8}{-0.1} = 2.2$ cm

55. a.

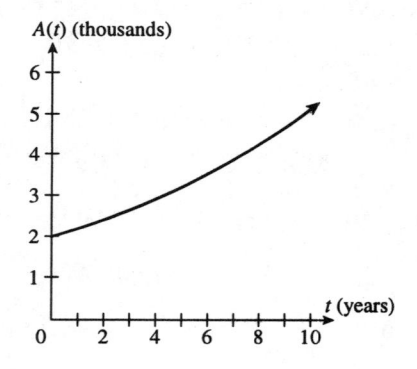

$A(t) = 2000e^{0.095t}$

b. $A(7) = 2000e^{(0.095)(7)} = \$3888.98$

c. $5000 = 2000e^{0.095t}$;
$e^{0.095t} = \dfrac{5000}{2000} = 2.5$;
$t = \dfrac{\ln 2.5}{0.095} = 9.6$ years

57. $809.92 = 600e^{r \cdot 4}$; $e^{4r} = \dfrac{809.92}{600}$;
$r = \dfrac{\ln\left(\dfrac{809.92}{600}\right)}{4} = 0.075 = 7.5\%$

59. a. $0.788 = e^{-0.000124t}$;
$t = \dfrac{\ln 0.788}{-0.000124} = 1921$ years

b. $0.5 = e^{-0.000124t_h}$;
$t_h = \dfrac{\ln 0.5}{-0.000124} = 5589.9$ years

61. a. $P_0 = P(1970) = 20{,}000$

b. $e^{10k} = \dfrac{35{,}000}{20{,}000} = \dfrac{7}{4}$;
$k = \dfrac{\ln\left(\dfrac{7}{4}\right)}{10} = 0.05596$

c. $P(t) = 20{,}000e^{0.05596t}$

d. $P(30) = 20{,}000e^{0.05596 \cdot 30}$
$= 107{,}182$

63. $0.5 = e^{k \cdot 8}$; $k = \dfrac{\ln 0.5}{8} = -0.0866$;
$N = N_0 e^{-0.0866t}$ where t is in days

65. a. $8^x = 20$

b. $\log_{10} 8^x = \log_{10} 20$

c. $x \log_{10} 8 = \log_{10} 20$;
$x = \dfrac{\log_{10} 20}{\log_{10} 8} = 1.4406$

67. Replacing the 20 with Q yields
$\log_8 Q = \dfrac{\log_{10} Q}{\log_{10} 8}$

69. $\ln Q = \log_e Q = \dfrac{\log_{10} Q}{\log_{10} e}$
$= 2.3 \log_{10} Q$

Chapter 7 Review

1. a.

Years After 1974	Number of Degrees
0	8
5	$\frac{3}{2} \cdot 8 = 12$
10	$\frac{3}{2} \cdot 12 = 18$
15	$\frac{3}{2} \cdot 18 = 27$
20	$\frac{3}{2} \cdot 27 = 40.5$

b. $N(t) = 8(1.5)^{t/5}$

c. Refer to the graph in the back of the textbook.

d. $N(10) = 8(1.5)^{10/5} = 18$
$N(21) = 8(1.5)^{21/5} = 43.9 \approx 44$

2. a.

Years After 1975	Price (dollars)
0	0.25
1	$1.1 \times 0.25 = 0.275$
2	$1.1 \times 0.275 = 0.3025$
3	$1.1 \times 0.3025 = 0.3328$
4	$1.1 \times 0.3328 = 0.3661$
5	$1.1 \times 0.3661 = 0.4027$

b. $P(t) = 0.25(1.1)^t$

c. Refer to the graph in the back of the textbook.

d. $P(10) = 0.25(1.1)^{10} = \$0.65$
$P(25) = 0.25(1.1)^{25} = \$2.71$

Chapter 7 Review

3. a.

Hours After 8 a.m.	Amount in body (mg)
0	100
1	0.85 × 100 = 85
2	0.85 × 85 = 72.25
3	0.85 × 72.25 = 61.4125
4	0.85 × 61.4125 = 52.2006
5	0.85 × 52.2006 = 44.3705

b. $A(t) = 100(0.85)^t$

c. Refer to the graph in the back of the textbook.

d. $A(4) = 100(0.85)^4 = 52.20$ mg
$A(10) = 100(0.85)^{10} = 19.69$ mg

4. a.

Weeks After the Series	Sales (dollars)
0	200,000
1	0.70 × 200,000 = 140,000
2	0.70 × 140,000 = 98,000
3	0.70 × 98,000 = 68,600
4	0.70 × 68,600 = 48,020
5	0.70 × 48,020 = 33,614

b. $S(t) = 200,000(0.7)^t$

c. Refer to the graph in the back of the textbook.

d. $S(4) = 200,000(0.7)^4 = \$48,020$
$S(6) = 200,000(0.7)^6 = \$23,529.80$

5. 7 [^] [2nd] [√] 2 [ENTER]
yields 15.673

6. 3 [^] [(] [2nd] [π] [+] 1 [)] [ENTER]
yields 94.633

7. 0.2 [×] 5 [^] [2nd] [√] 3 [ENTER]
yields 3.248

262

Chapter 7 Review

8. 6 ⊟ 2 ⌃ 2nd √ 7 ENTER
    yields −0.258

9. Refer to the graph in the back of the textbook.

10. Refer to the graph in the back of the textbook.

11. Refer to the graph in the back of the textbook.

12. Refer to the graph in the back of the textbook.

13. $3^{x+2} = 9^{1/3} = (3^2)^{1/3}$; $x + 2 = 2 \cdot \frac{1}{3}$;
    $x = \frac{2}{3} - 2 = -\frac{4}{3}$

14. $2^{x-1} = 8^{-2x} = (2^3)^{-2x}$;
    $x - 1 = 3(-2x) = -6x$; $7x = 1$; $x = \frac{1}{7}$

15. $4^{2x+1} = 8^{x-3}$; $(2^2)^{2x+1} = (2^3)^{x-3}$;
    $2(2x + 1) = 3(x - 3)$; $x = -11$

16. $3^{x^2-4} = 27 = 3^3$; $x^2 - 4 = 3$;
    $x^2 = 7$; $x = \pm\sqrt{7}$

17. $\log_2 16 = \log_2 2^4 = 4$

18. $\log_4 2 = \log_4 4^{1/2} = \frac{1}{2}$

19. $\log_3 \frac{1}{3} = \log_3 3^{-1} = -1$

20. $\log_7 7 = \log_7 7^1 = 1$

21. $\log_{10} 10^{-3} = -3$

22. $\log_{10} 0.0001 = \log_{10} 10^{-4} = -4$

23. $3 = 2^{x-2}$

24. $q = n^{p-1}$

25. $-2 = \log_{0.3}(x + 1)$

26. $0.3t = \log_4 (3N_0)$

27. $y = \log_3 \frac{1}{3} = \log_3 3^{-1} = -1$

28. $x = 3^4 = 81$

29. $16 = b^2$; $b = 16^{1/2} = 4$
    (bases are positive)

30. $3x - 1 = 2^3 = 8$; $x = \frac{8+1}{3} = 3$

31. $10^{1.3x} = \frac{20.4}{4} = 5.1$;
    $x = \frac{\log_{10} 5.1}{1.3} = 0.544$

32. $10^{0.5x} = \frac{127 + 17.3}{2} = 72.15$;
    $x = \frac{\log_{10} 72.15}{0.5} = 3.716$

33. $10^{-0.7x} = \frac{9 - 6.1}{3}$;
    $x = \frac{\log_{10}\left(\frac{9-6.1}{3}\right)}{-0.7} = 0.021$

34. $10^{-1.2x} = -\frac{30}{40} + 1 = \frac{1}{4}$;
    $x = \frac{\log_{10} \frac{1}{4}}{-1.2} = 0.502$

35. $\frac{1}{2.3}(\log_{10} 12,000 - \log_{10} 9,000)$
    $= 0.0543$

36. $\frac{1}{0.4}\sqrt{\frac{\log_{10} 48}{1.2}} = 2.959$

37. $1.2 \log_{10}\left(\frac{6400}{6400 - 2000}\right) = 0.195$

38. $\frac{1}{0.3}\left(\frac{\log_{10} 180}{\log_{10} 460}\right) = 2.823$

Chapter 7 Review

39. $\log_b\left(\dfrac{xy^{1/3}}{z^2}\right) = \log_b(xy^{1/3}) - \log_b z^2$
$= \log_b x + \log_b y^{1/3} - \log_b z^2$
$= \log_b x + \dfrac{1}{3}\log_b y - 2\log_b z$

40. $\log_b \sqrt{\dfrac{L^2}{2R}} = \log_b\left(\dfrac{L^2}{2R}\right)^{1/2}$
$= \dfrac{1}{2}\log_b \dfrac{L^2}{2R}$
$= \dfrac{1}{2}(\log_b L^2 - \log_b 2R)$
$= \dfrac{1}{2}[2\log_b L - (\log_b 2 + \log_b R)]$
$= \log_b L - \dfrac{1}{2}(\log_b 2 + \log_b R)$

41. $\log_{10}\left(x\sqrt[3]{\dfrac{x}{y}}\right) = \log_{10}\dfrac{x^{4/3}}{y^{1/3}}$
$= \log_{10} x^{4/3} - \log_{10} y^{1/3}$
$= \dfrac{4}{3}\log_{10} x - \dfrac{1}{3}\log_{10} y$

42. $\log_{10}\sqrt{(s-a)(s-g)^2}$
$= \log_{10}(s-a)^{1/2}(s-g)$
$= \log_{10}(s-a)^{1/2} + \log_{10}(s-g)$
$= \dfrac{1}{2}\log_{10}(s-a) + \log_{10}(s-g)$

43. $\dfrac{1}{3}(\log_{10} x - 2\log_{10} y)$
$= \dfrac{1}{3}(\log_{10} x - \log_{10} y^2)$
$= \dfrac{1}{3}\log_{10}\dfrac{x}{y^2} = \log_{10}\sqrt[3]{\dfrac{x}{y^2}}$

44. $\dfrac{1}{2}\log_{10}(3x) - \dfrac{2}{3}\log_{10} y$
$= \log_{10}\sqrt{3x} - \log_{10} y^{2/3}$
$= \log_{10}\dfrac{\sqrt{3x}}{\sqrt[3]{y^2}}$

45. $\dfrac{1}{3}\log_{10} 8 - 2(\log_{10} 8 - \log_{10} 2)$
$= \log_{10} 8^{1/3} - 2\left(\log_{10}\dfrac{8}{2}\right)$
$= \log_{10} 2 - \log_{10} 4^2$
$= \log_{10}\dfrac{2}{16} = \log_{10}\dfrac{1}{8}$

46. $\dfrac{1}{2}(\log_{10} 9 + 2\log_{10} 4) + 2\log_{10} 5$
$= \dfrac{1}{2}(\log_{10} 9 + \log_{10} 16) + \log_{10} 25$
$= \log_{10}(9 \cdot 16)^{1/2} + \log_{10} 25$
$= \log_{10}(12 \cdot 25) = \log_{10} 300$

47. $\log_3 x + \log_3 4 = 2$; $\log_3 4x = 2$;
$4x = 3^2 = 9$; $x = \dfrac{9}{4}$

48. $\log_2(x+2) - \log_2 3 = 6$;
$\log_2 \dfrac{x+2}{3} = 6$; $\dfrac{x+2}{3} = 2^6 = 64$;
$x = 3 \cdot 64 - 2 = 190$

49. $\log_{10}(x-1) + \log_{10}(x+2) = 3$;
$\log_{10}[(x-1)(x+2)]$
$= \log_{10}(x^2 + x - 2) = 3$;
$x^2 + x - 2 = 10^3 = 1000$;
$x^2 + x - 1002 = 0$;
$x = \dfrac{-1 \pm \sqrt{1+4008}}{2} = 31.16$
or -32.16; since x must be greater than one for $\log_{10}(x-1)$ to be defined,
$x = 31.16$.

50. $\log_{10}(x+2) - \log_{10}(x-3) = 1$;
$\log_{10} \frac{x+2}{x-3} = 1$; $\frac{x+2}{x-3} = 10^1 = 10$;
$x + 2 = 10x - 30$; $x = \frac{32}{9}$

51. $\log_{10} 3^{x-2} = \log_{10} 7$;
$(x-2)\log_{10} 3 = \log_{10} 7$;
$x = \frac{\log_{10} 7}{\log_{10} 3} + 2 = 3.77$

52. $\log_{10}(4 \cdot 2^{1.2x}) = \log_{10} 64$;
$\log_{10} 4 + 1.2x \log_{10} 2 = \log_{10} 64$;
$x = \frac{\log_{10} 64 - \log_{10} 4}{1.2 \log_{10} 2} = 3\frac{1}{3}$

53. $\log_{10} 1200 = \log_{10} 24 + \log_{10} 6^{-0.3x}$;
$\log_{10} 1200 = \log_{10} 24 - 0.3x \log_{10} 6$;
$x = \frac{\log_{10} 1200 - \log_{10} 24}{-0.3 \log_{10} 6} = -7.278$

54. $\log_{10} 0.08 = \log_{10} 12 + \log_{10} 3^{-1.5x}$;
$\log_{10} 0.08 = \log_{10} 12 - 1.5x \log_{10} 3$;
$x = \frac{\log_{10} 0.08 - \log_{10} 12}{-1.5 \log_{10} 3} = 3.041$

55. $\frac{N}{N_0} = 10^{kt}$; $\log_{10} \frac{N}{N_0} = kt$;
$t = \frac{1}{k} \log_{10} \frac{N}{N_0}$

56. $\frac{Q - R_0}{R} = \log_{10} kt$;
$10^{(Q-R_0)/R} = kt$; $t = \frac{1}{k} \cdot 10^{(Q-R_0)/R}$

57. a. $P(80) = 3800 \cdot 2^{-80/20} = 238$

b. $120 = 3800 \cdot 2^{-t/20}$;
$2^{-t/20} = \frac{120}{3800} = \frac{6}{190}$;

$-\frac{t}{20} = \frac{\log_{10} \frac{6}{190}}{\log_{10} 2}$;

$t = -20 \cdot \frac{\log_{10} \frac{6}{190}}{\log_{10} 2} = 100$ years,
or in 2010.

58. a. $N(9) = 8000 \cdot 3^{9/4} = 94{,}757$

b. $2{,}000{,}000 = 8000 \cdot 3^{t/4}$;
$3^{t/4} = \frac{2{,}000{,}000}{8000} = 250$;
$\frac{t}{4} = \frac{\log_{10} 250}{\log_{10} 3}$;
$t = 4 \cdot \frac{\log_{10} 250}{\log_{10} 3} = 20.1$ years.
2001 will be the hot year in which production exceeds 2 million.

59. a. Inflation is like interest compounded continuously. The nominal annual rate r that yields an effective annual rate of 6% is given by $e^{r \cdot 1} = 1.06$;
$r = \ln 1.06 = 0.05827$. Then the price in 10 months is
$90 e^{0.05827(10/12)} = \$94.48$.

b. $120 = 90 e^{0.05827 t}$;
$t = \frac{1}{0.05827} \ln\left(\frac{120}{90}\right) = 4.94$ years
or 4 years, 11 months, 7 days

60. a. Inflation is like interest compounded continuously. The nominal annual rate r that yields an effective annual rate of 8% is given by $e^{r \cdot 1} = 1.08$;
$r = \ln 1.08 = 0.07696$. Then the price in 20 months is
$1200 e^{0.07696(20/12)} = \$1364.23$.

Chapter 7 Review

b. $1500 = 1200e^{0.07696t}$;

$t = \dfrac{1}{0.07696} \ln \dfrac{1500}{1200}$

$= 2.899$ years or
34 months, 24 days

61. $x = \ln 4.7 = 1.548$

62. $x = \ln 0.5 = -0.693$

63. $x = e^{6.02} = 411.58$

64. $x = e^{-1.4} = 0.247$

65. $e^{0.6x} = \dfrac{4.73}{1.2}$;

$x = \dfrac{1}{0.6} \ln \dfrac{4.73}{1.2} = 2.286$

66. $e^{-1.2x} = \dfrac{1.75}{0.3}$;

$x = \dfrac{1}{-1.2} \ln \dfrac{1.75}{0.3} = -1.470$

67. $e^{-kt} = \dfrac{y-6}{12}$;

$t = -\dfrac{1}{k} \ln\left(\dfrac{y-6}{12}\right) = \dfrac{1}{k} \ln\left(\dfrac{12}{y-6}\right)$

68. $\ln(k+10) = \dfrac{N-N_0}{4}$;

$k = e^{(N-N_0)/4} - 10$

69. $N(t) = 600(0.4)^t = 600(e^{\ln 0.4})^t$

$= 600e^{-0.9163t}$

70. $N(t) = 100(1.06)^t = 100(e^{\ln 1.06})^t$

$= 100e^{0.0583t}$

Chapter 8

Section 8.1

1. Binomial (two terms); 3 (highest exponent is 3)
3. Monomial (one term); 4 (exponent is 4)
5. Trinomial (three terms); 2 (exponent is 2);
7. Trinomial (three terms); 3 (highest exponent is 3)
9. b (because of the $\frac{2}{x}$ term); c (because of the $\sqrt{z}$ term)
11. a (because the polynomial is the denominator of a fraction); b (because of the 2^v term); c (because the polynomial appears under a radical); $\frac{m^4}{12}$, on the other hand, can be written as $\frac{1}{12}m^4$.
13. a. $2^3 - 3 \cdot 2^2 + 2 + 1 = -1$
    b. $(-2)^3 - 3(-2)^2 - 2 + 1 = -21$
    c. $(2b)^3 - 3(2b)^2 + 2b + 1 = 8b^3 - 12b^2 + 2b + 1$
15. a. $\left(\frac{1}{2}\right)^2 + 3\left(\frac{1}{2}\right) + 1 = \frac{11}{4}$
    b. $\left(-\frac{1}{3}\right)^2 + 3\left(-\frac{1}{3}\right) + 1 = \frac{1}{9}$
    c. $(-w)^2 + 3(-w) + 1 = w^2 - 3w + 1$
17. a. $3(1.8)^4 - 2(1.8)^2 + 3 = 28.0128$
    b. $3(-2.6)^4 - 2(-2.6)^2 + 3 = 126.5728$
    c. $3(k-1)^4 - 2(k-1)^2 + 3 = 3k^4 - 12k^3 + 16k^2 - 8k + 4$
19. a. $(-1)^6 - (-1)^5 = 2$
    b. $(-2)^6 - (-2)^5 = 96$
    c. $\left(\frac{m}{3}\right)^6 - \left(\frac{m}{3}\right)^5 = \frac{m^6}{729} - \frac{m^5}{243}$
21. $y(y^2 - 2y + 3) + 2(y^2 - 2y + 3) = y^3 - 2y^2 + 3y + 2y^2 - 4y + 6 = y^3 - y + 6$

Chapter 8, Section 8.1

23. $3x(4x^2 + x - 2) - 2(4x^2 + x - 2) = 12x^3 + 3x^2 - 6x - 8x^2 - 2x + 4$
    $= 12x^3 - 5x^2 - 8x + 4$

25. $[x(x-1) - 2(x-1)](x-3) = (x^2 - 3x + 2)(x-3) = x(x^2 - 3x + 2) - 3(x^2 - 3x + 2)$
    $= x^3 - 6x^2 + 11x - 6$

27. $2a^2(3a^2 + 2a - 1) - 3a(3a^2 + 2a - 1) + (3a^2 + 2a - 1) = 6a^4 - 5a^3 - 5a^2 + 5a - 1$

29. $(y-2)(y+2) = y^2 - 4$; $(y+4)(y+1) = y^2 + 5y + 4$;
    $(y^2 - 4)(y^2 + 5y + 4) = y^2(y^2 + 5y + 4) - 4(y^2 + 5y + 4) = y^4 + 5y^3 - 20y - 16$

31. The highest exponent comes from multiplying the x^2 and $3x^2$ terms.
    $x^2 \cdot 3x^2 = 3x^4$. Answer: 4.

33. The highest exponent comes from multiplying the x, $2x$, and x^3 terms.
    $x \cdot 2x \cdot x^3 = 2x^5$. Answer: 5.

35. The highest exponent comes from multiplying the $3x^2$, x^3, and $-2x^2$ terms.
    $3x^2 \cdot x^3 \cdot (-2x^2) = -6x^7$. Answer: 7.

37. Since the degree is 4, there will be 4 factors.

39. Since the degree is 6, there will be 6 factors.

41. $(x+y)^3 = (x+y)(x+y)^2 = (x+y)(x^2 + 2xy + y^2)$
    $= x(x^2 + 2xy + y^2) + y(x^2 + 2xy + y^2) = x^3 + 2x^2y + xy^2 + x^2y + 2xy^2 + y^3$
    $= x^3 + 3x^2y + 3xy^2 + y^3$

43. $(x+y)(x^2 - xy + y^2) = x(x^2 - xy + y^2) + y(x^2 - xy + y^2)$
    $= x^3 - x^2y + xy^2 + x^2y - xy^2 + y^3 = x^3 + y^3$

45. $x(x^2 + x + 1) - (x^2 + x + 1) = x^3 - 1$

47. $2x(4x^2 - 2x + 1) + (4x^2 - 2x + 1) = 8x^3 + 1$

49. $3a(9a^2 + 6ab + 4b^2) - 2b(9a^2 + 6ab + 4b^2) = 27a^3 - 8b^3$

51. Using the formula for $x^3 + y^3$, with $y = 3$, $x^3 + 27 = (x+3)(x^2 - 3x + 9)$

53. Using the formula for $x^3 - y^3$, but with $2x$ in place of x,
    $(2x)^3 - y^3 = (2x - y)[(2x)^2 + (2x)y + y^2] = (2x - y)(4x^2 + 2xy + y^2)$

55. $8b^3 = (2b)^3$; $a^3 - (2b)^3 = (a - 2b)(a^2 + 2ab + 4b^2)$

57. $x^3y^3 = (xy)^3$, $1 = 1^3$; $[(xy)^3 - 1] = (xy - 1)(x^2y^2 + xy + 1)$

59. $27a^3 = (3a)^3$, $64b^3 = (4b)^3$; $[(3a)^3 + (4b)^3] = (3a + 4b)(9a^2 - 12ab + 16b^2)$

61. $125a^3b^3 = (5ab)^3$, $1 = 1^3$; $[(5ab)^3 - 1^3] = (5ab - 1)(25a^2b^2 + 5ab + 1)$

Chapter 8, Section 8.1

63. a. $A(x) = 2lw + 2wh + 2lh = 2x^2 + 32x$

 b. $A(18) = 2(18)^2 + 32(18) = 1224$ sq. in.

65. a. $A(x) = (3x)(2x) - \pi\left(\frac{x}{2}\right)^2 - \pi x^2 = \left(6 - \frac{5\pi}{4}\right)x^2$

 b. $A(8) = \left(6 - \frac{5\pi}{4}\right)8^2 = 132.67$ sq. in.

67. a. $V(r, h) = \pi r^2 h + \frac{1}{2}\left(\frac{4}{3}\right)\pi r^3 = \pi r^2 h + \frac{2}{3}\pi r^3$

 b. The total height of the silo is $h + r$, so $h = 5r - r = 4r$

 $V(r, 4r) = \pi r^2(4r) + \frac{2}{3}\pi r^3 = \frac{14}{3}\pi r^3$

69. a. $500(1 + r)^2$; $500(1 + r)^3$; $500(1 + r)^4$

 b. $500(1 + r)^2 = 500(1 + 2r + r^2) = 500r^2 + 1000r + 500$
 $500(1 + r)^3 = 500(1 + 3r + 3r^2 + r^3) = 500r^3 + 1500r^2 + 1500r + 500$
 $500(1 + r)^4 = 500(1 + 4r + 6r^2 + 4r^3 + r^4)$
 $= 500r^4 + 2000r^3 + 3000r^2 + 2000r + 500$

 c. $500(1.08)^2 = \$583.20$; $500(1.08)^3 = \$629.86$; $500(1.08)^4 = \$680.24$; Yes

71. a. $l = 16 - 2x$, $w = 12 - 2x$, $h = x$

 b. $V(x) = lwh = x(16 - 2x)(12 - 2x)$

 c. $x > 0$, obviously, and $2x$ cannot exceed 12. So $0 < x < 6$.

 d. $V(1) = (1)(14)(10) = 140$
 $V(2) = (2)(12)(8) = 192$
 $V(3) = (3)(10)(6) = 180$

 e. Refer to the graph in the back of the textbook.

 f. Use TRACE and ZOOM to find that at the high point, $x = 2.26$ in. and $y = V(2.26) = 194.07$ cu. in.

73. a. When
 $3x^2 - \frac{1}{3}x^3 = x^2\left(3 - \frac{x}{3}\right) = 0$,
 $x = 0$ or $x = 9$.

 b. Refer to the graph in back of the textbook.

 c. $f(2) = 3(2)^2 - \left(\frac{1}{3}\right)(2)^3$
 $= \frac{28}{3}$ points

 d. At the highest point on the graph, $f(x) = 36$ points.

Chapter 8, Section 8.2

- e. The graph indicates, and computation confirms, that $f(3) = 3^2(3 - 1) = 18$. Administer 3 milliliters. (There is also a higher solution, but there's no point in administering more medication if less will do.)

75. a. Refer to the graph in back of the textbook.
    b. $P(0) = 900$; $P(15) = 11{,}145$; $P(34) = 15{,}078$
    c. $P(1) - P(0) = 2241 - 900 = 1341$
       $P(16) - P(15) = 11{,}316 - 11{,}145 = 171$
       $P(35) - P(34) = 15{,}705 - 15{,}078 = 627$
    d. The graph is flattest at $t = 21$, or in 1981.

77. a. For $x = 0$, $y = 20$ cm
    b. Graphing y versus x using a window with Xmin = −150, Xmax = 0, and then ZOOMing in on the point where $y = 0$, shows that $x = -100$ there. The ramp runs from −100 cm to 0 cm and is 100 cm long.

79. a. Refer to the graph in the back of the textbook.
    b. $H(0) = 864$ min.
    c. $H(-14) = 859.8$ min.
    d. $14 \cdot 60 = 840$, and the graph is above 840 for $-34 \le t \le 34$.
    e. $13 \cdot 60 = 780$, and the graph is below 780 for $t > 66$ or $t < -66$.

Section 8.2

1-9. Refer to the graph in the back of the textbook. If the coefficient of x^3 is positive, the y-values head toward $+\infty$ as $x \to +\infty$ and toward $-\infty$ as $x \to -\infty$. If the coefficient of x^3 is negative, the y-values head toward $-\infty$ as $x \to +\infty$ and toward $+\infty$ as $x \to -\infty$.

11. b. and c. are the same.

13-19. Refer to the graph in the back of the textbook. If the coefficient of x^4 is positive, the y-values head toward $+\infty$ as $x \to +\infty$ and toward $-\infty$ as $x \to -\infty$. If the coefficient of x^4 is negative, the y-values head toward $-\infty$ as $x \to +\infty$ and toward $+\infty$ as $x \to -\infty$.

21. $f(x)$ has zeros at −3 and 1.
23. $G(x)$ has zeros at 2 and −2.
25. $h(x)$ has zeros at 0, −2, and 2.
27. $P(x)$ has zeros at −4, −1, and 1.
29. $q(x)$ has zeros at −3 and −1.
31. a. x-intercepts at −2, −1, 3.
    b. $P(x) = (x + 1)(x + 2)(x - 3)$
33. a. x-intercepts at −2, 0, 1, and 2.
    b. $R(x) = x(x - 2)(x - 1)(x + 2)$
35. a. x-intercepts at −2, 1, and 4.
    b. $p(x) = (x - 4)(x - 1)(x + 2)$
37. a. x-intercepts at −2, 2, and 3
    b. $r(x) = (x + 2)^2(x - 2)(x - 3)$
39. a. $P(x) = x^2(x^2 + 4)$, so the only zero is $x = 0$.
    b. Refer to the graph in the back of the textbook.

41. a. $P(x) = x(x^2 - 8)$
$= x(x + 2\sqrt{2})(x - 2\sqrt{2})$,
so the zeros are $0, \pm 2\sqrt{2}$.

  b. Refer to the graph in the back of the textbook.

43. a. $f(x)k = x^2(x^2 + 4x + 4)$,
$= x^2(x + 2)^2$
so the zeros are 0 and −2.

  b. Refer to the graph in the back of the textbook.

45. a. $g(x) = x(16 - x^2)$
$= x(4 + x)(4 - x)$,
so the zeros are $0, \pm 4$.

  b. Refer to the graph in the back of the textbook.

47. a. $k(x) = (x^2 - 2)(x^2 - 8)$,
so the zeros are $\pm\sqrt{2}$ and $\pm 2\sqrt{2}$.

  b. Refer to the graph in the back of the textbook.

49. a. $r(x) = (x + 1)(x - 1)(x + 3)^2$,
so the zeros are −1, 1, and −3.

  b. Refer to the graph in the back of the textbook.

51. The function is a third-degree polynomial with zeros at −2, 1, and 4, so it is $(x + 2)(x - 1)(x - 4)$.

53. The function is a third-degree polynomial with zeros at −3 and 2. The curve "kisses" the x-axis at −3, so the zero there is twofold. The function is $(x + 3)^2(x - 2)$.

55. The function is a fourth-degree polynomial with zeros at −2 and 2. The curve has an inflection point at 2, so the zero there is threefold. The function is $(x + 2)(x - 2)^3$.

57. a. An added constant term raises or lowers the graph. Here the graph is shifted up 3 units.

  b. An added constant term raises or lowers the graph. Here the graph is shifted down 5 units.

  c. Replacing x with $x - h$ shifts the graph right h units. Here the graph is shifted right 2 units.

  d. Replacing x with $x + h$ shifts the graph left h units. Here the graph is shifted left 3 units.

59. a. An added constant term raises or lowers the graph. Here the graph is shifted up 6 units.

  b. An added constant term raises or lowers the graph. Here the graph is shifted down 2 units.

  c. Replacing x with $x - h$ shifts the graph right h units. Here the graph is shifted right 4 units.

  d. Replacing x with $x + h$ shifts the graph left h units. Here the graph is shifted left 2 units.

Section 8.3

1. $x + 3$ equals zero for $x = -3$, so that represents a vertical asymptote. The degree of the denominator exceeds that of the numerator, so $y = 0$ is a horizontal asymptote. Refer to the graph in the back of the textbook.

3. $x^2 - 5x + 4$ equals zero for $x = 1$ and $x = 4$, so those represent two vertical asymptotes. The degree of the denominator exceeds that of the numerator, so $y = 0$ is a horizontal asymptote. Refer to the graph in the back of the textbook.

Chapter 8, Section 8.3

5. $x + 3$ equals zero for $x = -3$, so that represents a vertical asymptote. The degree of the numerator and denominator are equal, and the ratio of lead coefficients is 1, so $y = 1$ is a horizontal asymptote. Refer to the graph in the back of the textbook.

7. $x + 2$ equals zero for $x = -2$, so that represents a vertical asymptote. The numerator and denominator have equal degree and the ratio of lead coefficients is 1, so $y = 1$ is a horizontal asymptote. Refer to the graph in the back of the textbook.

9. $x^2 - 4$ equals zero for $x = \pm 2$, so those represent two vertical asymptotes. The degree of the denominator exceeds that of the numerator, so $y = 0$ is a horizontal asymptote. Refer to the graph in the back of the textbook.

11. $x^2 + 5x + 4 = (x + 4)(x + 1)$ equals zero for $x = -1$ and $x = -4$, so those represent vertical asymptotes. The denominator's degree exceeds the numerator's, so $y = 0$ is a horizontal asymptote. Refer to the graph in the back of the textbook.

13. $x^2 - 4$ equals zero for $x = \pm 2$, so those represent two vertical asymptotes. The numerator's and denominator's degrees are equal, with a leading coefficient ratio of 1, so $y = 1$ is a horizontal asymptote. Refer to the graph in the back of the textbook.

15. $(x - 1)^2$ equals zero for $x = 1$, so that represents a vertical asymptote. The denominator's degree exceeds the numerator's so $y = 0$ is a horizontal asymptote. Refer to the graph in the back of the textbook.

17. $x^2 + 3$ is never zero, so there is no vertical asymptote. The denominator's degree exceeds the numerator's, so $y = 0$ is a horizontal asymptote. Refer to the graph in the back of the textbook.

19. a. The duck's ground speed is $50 - v$, so $\frac{150}{50 - v}$.

   b. $\frac{150}{50 - 5} = 3\frac{1}{3}$ hours. The time goes up as v goes up.

   c. $50 - v$ is zero for $v = 50$, so that is a vertical asymptote. The degree of the denominator exceeds that of the numerator, so $y = 0$ is a horizontal asymptote. Refer to the graph in the back of the textbook.

21. a. $\frac{20{,}000 + 8n}{n} = 8 + \frac{20{,}000}{n}$

   b. $18 = 8 + \frac{20{,}000}{n}$; $10n = 20{,}000$; $n = 2000$

   c. Refer to the graph in the back of the textbook.

   d. $12 > 8 + \frac{20{,}000}{n}$; $4n > 20{,}000$; $n > 5000$

Chapter 8, Section 8.3

e. The numerator's and denominator's degrees in $\dfrac{20{,}000 + 8n}{n}$ are equal with leading coefficients ratio of 8. So $y = 8$ is a horizontal asymptote. $8 is what the cost per calculator approaches as n gets large.

23. a. $R(x) = \dfrac{300}{x}(15x) + \dfrac{300}{x}(10)$
$= 4500 + \dfrac{3000}{x}$;
$C(x) = R(x) + S(x)$
$= 4500 + \dfrac{3000}{x} + 6x$

b. Refer to the graph in the back of the textbook.
The curve's minimum is $4768.33.

c. The minimum comes at $x = 22.36$. Since computers come as whole units, the inventory cost will be minimized when $x = 22$. (Then $C = \$4768.36$, which is less than $C(23) = \$4768.43$.)
Orders per year is $\dfrac{300}{22} = 13.6$, so there will be 14 orders per year, one smaller than the rest.

d. Refer to the graph in the back of the textbook.
$y = 6x + 4500$ is a slant asymptote for the cost function.

25. a. $V(x) = lwh = x^2 h = 24x - \dfrac{x^3}{2}$

b. Refer to the graph in the back of the textbook.
At the highest point on the curve, $V(x) = 64$.

c. $V(4) = 64$, so $x = 4$ gives the maximum volume.

d. Refer to the graph in the back of the textbook.
When $x = 4$, $h = \dfrac{24}{4} - \dfrac{4}{2} = 4$; the box with maximum volume measures $4 \times 4 \times 4$.

27. a. Percentages range from 0 to 100, so $0 \leq p < 100$.

b. $C(40) = \dfrac{72 \cdot 40}{100 - 40} = \$48{,}000$
$C(50) = \dfrac{72 \cdot 50}{100 - 50} = \$72{,}000$
$C(75) = \dfrac{72 \cdot 75}{100 - 75} = \$216{,}000$

c. $108 = \dfrac{72p}{100 - p}$;
$10{,}800 = p(72 + 108)$;
$p = \dfrac{10{,}800}{72 + 108} = 60$; 60%

d. Refer to the graph in the back of the textbook.
The curve rises above 1,728 when p goes above 96%.

e. $100 - p$ is zero when $p = 100$. This indicates that immunizing 100% is infinitely expensive and so cannot be done.

29. a. $P(20) = \dfrac{440(332)}{332 - 20} = 468.2$ hertz
$P(-68) = \dfrac{440(332)}{332 + 68}$
$= 365.2$ hertz

Chapter 8, Section 8.4

b. $415 = \dfrac{440(332)}{332-v}$;
$137{,}780 - 415v = 146{,}080$;
$v = \dfrac{146{,}080 - 137{,}780}{-415} = -20$;
the train is receding at 20 m/s.
$553\tfrac{1}{3} = \dfrac{440(332)}{332-v}$;
$183{,}706\tfrac{2}{3} - \left(553\tfrac{1}{3}\right)v = 146{,}080$;
$v = \dfrac{146{,}080 - 183{,}706\tfrac{2}{3}}{-553\tfrac{1}{3}} = 68$;
the train is approaching at over 68 m/s.

c. Refer to the graph in the back of the textbook.
$456.5 = \dfrac{440(332)}{332-v}$;
$151{,}558 - 456.5v = 146{,}080$;
$v = \dfrac{146{,}080 - 151{,}558}{-456.5} = 12$; the train must be approaching at over 12 m/s.

d. $332 - v$ is zero when $v = 332$, so that is a vertical asymptote. 332 m/s is the speed of sound, and as the train's speed approaches the speed of sound, the frequency goes to $+\infty$.

31. $x^2 + 1$ is never zero, so there is no vertical asymptote. The denominator's degree exceeds the numerator's, so $y = 0$ is a horizontal asymptote.
Refer to the graph in the back of the textbook.

Section 8.4

1. Factoring $7c^2d$ out of numerator and denominator yields $-\dfrac{2}{d^2}$.

3. Factoring $-6rst$ out of numerator and denominator yields $\dfrac{2r}{t}$.

5. Factoring out 2 yields $\dfrac{2x+3}{3}$.

7. Factoring out $2a$ yields $\dfrac{3a^2 - 2a}{2}$.

9. $\dfrac{6 - 6t^2}{(t-1)^2} = \dfrac{6(1-t)(1+t)}{(1-t)^2} = \dfrac{6+6t}{1-t}$

11. $\dfrac{2y^2 - 8}{2y + 4} = \dfrac{2(y+2)(y-2)}{2(y+2)} = y - 2$

13. $\dfrac{6 - 2v}{v^3 - 27} = \dfrac{2(3-v)}{(v-3)(v^2+3v+9)}$
$= \dfrac{-2}{v^2 + 3v + 9}$

15. $\dfrac{4x^3 + 36x}{6x^2 + 18x} = \dfrac{2(2x)(x^2+9)}{3(2x)(x^2+3)}$
$= \dfrac{2x^2 + 18}{3x^2 + 9}$

17. $\dfrac{y^2 - 9x^2}{(3x-y)^2} = \dfrac{(y-3x)(y+3x)}{(y-3x)^2}$
$= \dfrac{y+3x}{y-3x}$

19. $\dfrac{2x^2 + x - 6}{x^2 + x - 2} = \dfrac{(2x-3)(x+2)}{(x-1)(x+2)}$
$= \dfrac{2x-3}{x-1}$

Chapter 8, Section 8.4

21. $\dfrac{6x^2+x-12}{-6x^2+17x-12} = \dfrac{(3x-4)(2x+3)}{(3x-4)(-2x+3)} = \dfrac{2x+3}{-2x+3}$

23. $\dfrac{8z^3-27}{4z^2-9} = \dfrac{(2z-3)(4z^2+6z+9)}{(2z+3)(2z-3)} = \dfrac{4z^2+6z+9}{2z+3}$

25. b: $\dfrac{4a^2-2a}{2a-1} = \dfrac{2a(2a-1)}{2a-1} = 2a$

27. None

29. Combining and factoring out $3pn^2$ yields $\dfrac{25p}{n}$.

31. Combining and factoring out $24np$ yields $-\dfrac{np^2}{2}$.

33. Combining and factoring out a^2b^2 yields $\dfrac{5}{ab}$.

35. $\dfrac{5x+25}{2x} \cdot \dfrac{4x}{2x+10} = \dfrac{5(x+5) \cdot 2(2x)}{2x \cdot 2(x+5)}$
$= 5$

37. $\dfrac{4a^2-1}{a^2-16} \cdot \dfrac{a^2-4a}{2a+1} = \dfrac{(2a+1)(2a-1) \cdot a(a-4)}{(a-4)(a+4) \cdot (2a+1)} = \dfrac{a(2a-1)}{a+4} = \dfrac{2a^2-a}{a+4}$

39. $\dfrac{2x^2-x-6}{3x^2+4x+1} \cdot \dfrac{3x^2+7x+2}{2x^2+7x+6} = \dfrac{(2x+3)(x-2) \cdot (3x+1)(x+2)}{(3x+1)(x+1) \cdot (2x+3)(x+2)} = \dfrac{x-2}{x+1}$

41. $\dfrac{3x^4-48}{x^4-4x^2-32} \cdot \dfrac{4x^4-8x^3+4x^2}{2x^4+16x}$
$= \dfrac{3(x^2+4)(x+2)(x-2) \cdot 2x(2x)(x-1)^2}{(x^2+4)(x+2\sqrt{2})(x-2\sqrt{2}) \cdot 2x(x+2)(x^2-2x+4)}$
$= \dfrac{3(x-2)(2x)(x-1)^2}{(x+2\sqrt{2})(x-2\sqrt{2})(x^2-2x+4)} = \dfrac{6x^4-24x^3+30x^2-12x}{x^4-2x^3-4x^2+16x-32}$

43. $\dfrac{4x-8}{3y} \div \dfrac{6x-12}{y} = \dfrac{4x-8}{3y} \cdot \dfrac{y}{6x-12} = \dfrac{4(x-2) \cdot y}{3y \cdot 6(x-2)} = \dfrac{2}{9}$

275

Chapter 8, Section 8.4

45. $\dfrac{a^2-a-6}{a^2+2a-15} \div \dfrac{a^2-4}{a^2+6a+5} = \dfrac{a^2-a-6}{a^2+2a-15} \cdot \dfrac{a^2+6a+5}{a^2-4}$

$= \dfrac{(a-3)(a+2)\cdot(a+1)(a+5)}{(a+5)(a-3)\cdot(a+2)(a-2)} = \dfrac{a+1}{a-2}$

47. $\dfrac{x^3+y^3}{x} \div \dfrac{x+y}{3x} = \dfrac{x^3+y^3}{x} \cdot \dfrac{3x}{x+y} = \dfrac{(x+y)(x^2-xy+y^2)(3x)}{x(x+y)} = 3x^2 - 3xy + 3y^2$

49. $1 \div \dfrac{x^2-1}{x+2} = 1 \cdot \dfrac{x+2}{x^2-1} = \dfrac{x+2}{x^2-1}$

51. $(x^2-5x+4) \div \dfrac{x^2-1}{x^2} = (x^2-5x+4) \cdot \dfrac{x^2}{x^2-1} = (x-1)(x-4) \cdot \dfrac{x^2}{(x+1)(x-1)}$

$= \dfrac{x^3-4x^2}{x+1}$

53. $\dfrac{x^2+3x}{2y} \div 3x = \dfrac{x^2+3x}{2y} \cdot \dfrac{1}{3x} = \dfrac{x(x+3)}{6xy} = \dfrac{x+3}{6y}$

55.

$3rs \overline{\smash{)}\begin{matrix} 6rs - 5 + \frac{2}{rs} \\ 18r^2s^2 - 15rs + 6 \end{matrix}}$

$\underline{-(18r^2s^2)}$
$0 - 15rs$
$\underline{-(-15rs)}$
$0 + 6$

59.

$2y+1 \overline{\smash{)}\begin{matrix} 2y + 5 + \frac{2}{2y+1} \\ 4y^2 + 12y + 7 \end{matrix}}$

$\underline{-(4y^2 + 2y)}$
$10y + 7$
$\underline{-(10y + 5)}$
2

57.

$-3s^2 \overline{\smash{)}\begin{matrix} -5s^8 + 7s^3 - \frac{2}{s^2} \\ 15s^{10} - 21s^5 + 6 \end{matrix}}$

$\underline{-(15s^{10})}$
$\phantom{-3s^2)15s^{10}}0 - 21s^5$
$\phantom{-3s^2)15s^{10}}\underline{-(-21s^5)}$
$\phantom{-3s^2)15s^{10}-21s^5}0 + 6$

61.

$$\begin{array}{r} x^2 + 4x + 9 + \frac{19}{x-2} \\ x-2 \overline{\smash{\big)}\, x^3 + 2x^2 + x + 1} \\ \underline{-(x^3 - 2x^2)} \\ 4x^2 + x \\ \underline{-(4x^2 - 8x)} \\ 9x + 1 \\ \underline{-(9x - 18)} \\ 19 \end{array}$$

63.

$$\begin{array}{r} 4z^3 - 2z^2 + 3z + 1 + \frac{2}{2z+1} \\ 2z+1 \overline{\smash{\big)}\, 8z^4 + 0z^3 + 4z^2 + 5z + 3} \\ \underline{-(8z^4 + 4z^3)} \\ -4z^3 + 4z^2 \\ \underline{-(-4z^3 - 2z^2)} \\ 6z^2 + 5z \\ \underline{-(6z^2 + 3z)} \\ 2z + 3 \\ \underline{-(2z+1)} \\ 2 \end{array}$$

65.

$$\begin{array}{r} x^3 + 2x^2 + 4x + 8 + \frac{15}{x-2} \\ x-2 \overline{\smash{\big)}\, x^4 + 0x^3 + 0x^2 + 0x - 1} \\ \underline{-(x^4 - 2x^3)} \\ 2x^3 + 0x^2 \\ \underline{-(2x^3 - 4x^2)} \\ 4x^2 + 0x \\ \underline{-(4x^2 - 8x)} \\ 8x - 1 \\ \underline{-(8x - 16)} \\ 15 \end{array}$$

67. $(1)^3 - 2(1)^2 + 1 = 1 - 2 + 1 = 0$;
$P(x) = x^3 - 2x^2 + 1$
$= (x-1)(x^2 - x - 1)$;
by the quadratic formula, the other zeros are $\frac{1 \pm \sqrt{1+4}}{2} = \frac{1 \pm \sqrt{5}}{2}$.

69. $(-3)^4 - 3(-3)^3 - 10(-3)^2 + 24(-3)$
$= 81 + 81 - 90 - 72 = 0$;
$P(x) = x^4 - 3x^3 - 10x^2 + 24x$
$= (x+3)(x^3 - 6x^2 + 8x)$
$= (x+3) \cdot x(x-2)(x-4)$;
the other zeros are 0, 2, and 4.

71. a. The function is defined except when the denominator is zero, or when $x = 2$. So, $x < 2$ or $2 < x$.

b. $\dfrac{x^2 - 4}{x - 2} = \dfrac{(x-2)(x+2)}{x-2}$
$= x + 2$

c. Refer to the graph in the back of the textbook.

Chapter 8, Section 8.5

73. **a.** The function is defined except when the denominator is zero, or when $x = \pm 1$. So, $x < -1$ or $-1 < x < 1$ or $1 < x$.

b. $\dfrac{x+1}{x^2-1} = \dfrac{x+1}{(x+1)(x-1)} = \dfrac{1}{x-1}$

c. Refer to the graph in the back of the textbook.

Section 8.5

1. $\dfrac{x}{2} - \dfrac{3}{2} = \dfrac{x-3}{2}$

3. $\dfrac{1}{6}a + \dfrac{1}{6}b - \dfrac{5}{6}c = \dfrac{a}{6} + \dfrac{b}{6} - \dfrac{5c}{6}$
   $= \dfrac{a+b-5c}{6}$

5. $\dfrac{x-1}{2y} + \dfrac{x}{2y} = \dfrac{x-1+x}{2y} = \dfrac{2x-1}{2y}$

7. $\dfrac{3-(x-3)-(x-1)}{x+2y} = \dfrac{-2x+7}{x+2y}$

9. $\dfrac{a+1-(5-3a)}{a^2-2a+1} = \dfrac{4a-4}{(a-1)^2}$
   $= \dfrac{4(a-1)}{(a-1)^2} = \dfrac{4}{a-1}$

11. $6(x+y)^2 = 2 \cdot 3 \cdot (x+y)^2$;
    $4xy^2 = 2 \cdot 2 \cdot x \cdot y^2$;
    LCD $= 2 \cdot 2 \cdot 3 \cdot (x+y)^2 \cdot x \cdot y^2$
    $= 12xy^2(x+y)^2$

13. $a^2 + 5a + 4 = (a+1)(a+4)$;
    LCD $= (a+1)^2(a+4)$

15. $x^2 - x = x(x-1)$; LCD $= x(x-1)^3$

17. $6x^3 = 2 \cdot 3 \cdot x^3$;
    $4x^2 - 4x = 2 \cdot 2 \cdot x \cdot (x-1)$;
    LCD $= 2 \cdot 2 \cdot 3 \cdot x^3 \cdot (x-1)^2$
    $= 12x^3(x-1)^2$

19. $\dfrac{2}{6x} \cdot \dfrac{3}{3} = \dfrac{6}{18x}$

21. $y = \dfrac{y}{1} = \dfrac{y}{1} \cdot \dfrac{xy}{xy} = \dfrac{xy^2}{xy}$

23. $\dfrac{3y}{y+2} \cdot \dfrac{y-3}{y-3} = \dfrac{3y^2 - 9y}{y^2 - y - 6}$

25. $\dfrac{3}{a-b} \cdot \dfrac{a+b}{a+b} = \dfrac{3a+3b}{a^2-b^2}$

27. $\dfrac{x}{2} \cdot \dfrac{3}{3} + \dfrac{2x}{3} \cdot \dfrac{2}{2} = \dfrac{7x}{6}$

29. $\dfrac{5y}{6} \cdot \dfrac{2}{2} - \dfrac{3y}{4} \cdot \dfrac{3}{3} = \dfrac{y}{12}$

31. $\dfrac{x+1}{2x} \cdot \dfrac{3}{3} + \dfrac{2x-1}{3x} \cdot \dfrac{2}{2} = \dfrac{7x+1}{6x}$

33. $\dfrac{5}{x+1} \cdot \dfrac{x-1}{x-1} + \dfrac{3}{x-1} \cdot \dfrac{x+1}{x+1} = \dfrac{8x-2}{x^2-1}$

35. $\dfrac{y}{2y-1} \cdot \dfrac{y+1}{y+1} - \dfrac{2y}{y+1} \cdot \dfrac{2y-1}{2y-1}$
    $= \dfrac{-3y^2 + 3y}{2y^2 + y - 1}$

37. $\dfrac{y-1}{y+1} \cdot \dfrac{2y-3}{2y-3} - \dfrac{y-2}{2y-3} \cdot \dfrac{y+1}{y+1}$
    $= \dfrac{y^2 - 4y + 5}{2y^2 - y - 3}$

39. $\dfrac{7}{5x-10} \cdot \dfrac{3}{3} - \dfrac{5}{3x-6} \cdot \dfrac{5}{5}$
    $= \dfrac{-4}{15x-30}$

Chapter 8, Section 8.5

41. $\dfrac{2}{x^2-x-2} \cdot \dfrac{x+1}{x+1} + \dfrac{2}{x^2+2x+1} \cdot \dfrac{x-2}{x-2} = \dfrac{4x-2}{x^3-3x-2}$

43. $\dfrac{y-1}{y^2-3y} \cdot \dfrac{y+2}{y+2} - \dfrac{y+1}{y^2+2y} \cdot \dfrac{y-3}{y-3} = \dfrac{3y+1}{y^3-y^2-6y}$

45. $\dfrac{x}{1} \cdot \dfrac{x}{x} - \dfrac{1}{x} = \dfrac{x^2-1}{x}$

47. $\dfrac{x}{1} \cdot \dfrac{(x-1)^2}{(x-1)^2} + \dfrac{1}{x-1} \cdot \dfrac{x-1}{x-1} - \dfrac{1}{(x-1)^2} = \dfrac{x^3-2x^2+2x-2}{(x-1)^2}$

49. $\dfrac{y}{1} \cdot \dfrac{y^2-1}{y^2-1} - \dfrac{y^2}{y-1} \cdot \dfrac{y+1}{y+1} + \dfrac{y^2}{y+1} \cdot \dfrac{y-1}{y-1} = \dfrac{y^3-2y^2-y}{y^2-1}$

51. $\dfrac{x}{1} \cdot \dfrac{x+2}{x+2} - \dfrac{1}{1} \cdot \dfrac{x+2}{x+2} + \dfrac{3}{x+2} = \dfrac{x^2+x+1}{x+2}$

53. $\dfrac{1-\frac{2}{3}}{3+\frac{1}{3}} \cdot \dfrac{3}{3} = \dfrac{3-2}{9+1} = \dfrac{1}{10}$

55. $\dfrac{\frac{2}{a}+\frac{3}{2a}}{5+\frac{1}{a}} \cdot \dfrac{2a}{2a} = \dfrac{4+3}{10a+2} = \dfrac{7}{10a+2}$

57. $\dfrac{1+\frac{2}{a}}{1-\frac{4}{a^2}} \cdot \dfrac{a^2}{a^2} = \dfrac{a^2+2a}{a^2-4}$
$= \dfrac{a(a+2)}{(a+2)(a-2)} = \dfrac{a}{a-2}$

59. $\dfrac{x+\frac{x}{y}}{1+\frac{1}{y}} \cdot \dfrac{y}{y} = \dfrac{xy+x}{y+1} = \dfrac{x(y+1)}{y+1} = x$

61. $\dfrac{1}{1-\frac{1}{x}} \cdot \dfrac{x}{x} = \dfrac{x}{x-1}$

63. $\dfrac{y-2}{y-\frac{4}{y}} \cdot \dfrac{y}{y} = \dfrac{y^2-2y}{y^2-4}$
$= \dfrac{y(y-2)}{(y+2)(y-2)} = \dfrac{y}{y+2}$

65. $\dfrac{x+y}{\frac{1}{x}+\frac{1}{y}} \cdot \dfrac{xy}{xy} = \dfrac{x^2y+y^2x}{y+x}$
$= \dfrac{xy(x+y)}{x+y} = xy$

67. $\dfrac{x-\frac{x}{y}}{y+\frac{y}{x}} \cdot \dfrac{xy}{xy} = \dfrac{x^2y-x^2}{xy^2+y^2}$

69. $\dfrac{\frac{4}{x^2}-\frac{4}{z^2}}{\frac{2}{z}-\frac{2}{x}} \cdot \dfrac{x^2z^2}{x^2z^2} = \dfrac{4z^2-4x^2}{2x^2z-2xz^2}$
$= \dfrac{4(z+x)(z-x)}{2xz(x-z)} = \dfrac{-2(x+z)}{xz}$

Chapter 8, Section 8.5

71. $\dfrac{\frac{1}{y+1}}{1-\frac{1}{y^2}} \cdot \dfrac{y^2(y+1)}{y^2(y+1)} = \dfrac{y^2}{y^2(y+1)-(y+1)}$

$= \dfrac{y^2}{y^3+y^2-y-1}$

73. a. Speed $= s+8$; distance $= 25$; $t = \dfrac{25}{s+8}$

b. Speed $= s-8$; distance $= 25$; $t = \dfrac{25}{s-8}$

c. $\dfrac{25}{s+8} \cdot \dfrac{s-8}{s-8} + \dfrac{25}{s-8} \cdot \dfrac{s+8}{s+8} = \dfrac{50s}{s^2-64}$

75. a. Ground speed $= 400+w$; $t = \dfrac{900}{400+w}$

b. Ground speed $= 400-w$; $t = \dfrac{900}{400-w}$

c. Orville has a shorter time and arrives sooner, by
$\dfrac{900}{400-w} - \dfrac{900}{400+w} = \dfrac{1800w}{160{,}000-w^2}$

77. a. $\dfrac{1}{f} = \dfrac{1}{q+60} + \dfrac{1}{q}$

$= \dfrac{q}{q(q+60)} + \dfrac{q+60}{q(q+60)}$

$= \dfrac{2q+60}{q^2+60q}$

b. $f = \dfrac{q^2+60q}{2q+60}$

79. a. First leg: $t = \dfrac{d}{r_1}$;

second leg: $t = \dfrac{d}{r_2}$

b. Total distance $= 2d$;

total time $= \dfrac{d}{r_1} + \dfrac{d}{r_2}$

c. Average speed $= \dfrac{\text{total distance}}{\text{total time}}$

$= \dfrac{2d}{\frac{d}{r_1}+\frac{d}{r_2}}$

d. $\dfrac{2d}{\frac{d}{r_1}+\frac{d}{r_2}} \cdot \dfrac{r_1 r_2}{r_1 r_2} = \dfrac{2d r_1 r_2}{d r_2 + d r_1}$

$= \dfrac{2 r_1 r_2}{r_1 + r_2}$

e. Average speed $= \dfrac{2(70 \cdot 50)}{70+50}$

$= 58\dfrac{1}{3}$ mi/h

Section 8.6

1. $\frac{2}{x+1} = \frac{x}{x+1} + 1$; $2 = x + (x+1)$;
$1 = 2x$; $x = \frac{1}{2}$

3. $\frac{3}{x-2} = \frac{1}{2} + \frac{2x-7}{2x-4}$;
$2(3) = x - 2 + 2x - 7$;
$6 + 7 + 2 = x + 2x$; $15 = 3x$; $x = 5$

5. $\frac{4}{x+2} - \frac{1}{x} = \frac{2x-1}{x^2+2x}$;
$4x - (x+2) = 2x - 1$; $3x - 2x = 1$;
$x = 1$

7. $\frac{x}{x+2} - \frac{3}{x-2} = \frac{x^2+8}{x^2-4}$;
$(x^2 - 2x) - (3x + 6) = x^2 + 8$;
$-5x = 14$; $x = -\frac{14}{5}$

9. $\frac{4}{3x} + \frac{3}{3x+1} + 2 = 0$;
$(12x + 4) + 9x + (18x^2 + 6x) = 0$;
$18x^2 + 27x + 4 = 0$;
$x = \frac{-27 \pm \sqrt{27^2 - 4 \cdot 4 \cdot 18}}{2 \cdot 18}$
$= \frac{-9 \pm 7}{12} = -\frac{1}{6}$ or $-\frac{4}{3}$

11. $\frac{2x}{x-1} - \frac{x+1}{2} = 0$; $4x - (x^2 - 1) = 0$;
$x^2 - 4x - 1 = 0$;
$x = \frac{4 \pm \sqrt{4^2 + 4}}{2} = 2 \pm \sqrt{5}$

13. $\frac{2x}{x-1} - \frac{5}{x^2-x} = \frac{x+1}{x}$;
$2x^2 - 5 = x^2 - 1$; $x^2 = 4$; $x = \pm 2$
No real solution

15. $\frac{9}{x^2+x-2} + \frac{1}{x^2-x} = \frac{4}{x-1}$;
$9x + x + 2 = 4x(x + 2)$;
$10x + 2 = 4x^2 + 8x$;
$4x^2 - 2x - 2 = 0$;
$2(2x + 1)(x - 1) = 0$;
$x = -\frac{1}{2}$ or $x = 1$;
However $x = 1$ is undefined.
Therefore $x = -\frac{1}{2}$.

17. a. Refer to the graph in the back of the textbook.
The graphs cross at
$x = 2\frac{1}{2} = \$2.50$

b. $\frac{160}{x} = 6x + 49$; $160 = 6x^2 + 49x$;
$6x^2 + 49x - 160 = 0$;
$x = \frac{-49 \pm \sqrt{49^2 + 4 \cdot 6 \cdot 160}}{12}$
$= \frac{-49 \pm 79}{12}$;
$x = \frac{5}{2} = 2\frac{1}{2}$
(The negative solution is extraneous.)

19. a. $3200 = lw$; $l = \frac{3200}{w}$

b. $P = 2w + 2l = 2w + \frac{6400}{w}$

c. Refer to the graph in the back of the textbook.
The lowest point is at $w = 56.57$, $P = 226.27$. Choosing $w = 56.57$ minimizes P for the given area, at $P = 226.27$.

Chapter 8, Section 8.6

    d. $240 = 2w + \dfrac{6400}{w}$

    e. $240 = 2w + \dfrac{6400}{w}$;

$$2w^2 - 240w + 6400 = 0:$$
$$w = \dfrac{240 \pm \sqrt{240^2 - 4 \cdot 2 \cdot 6400}}{2 \cdot 2}$$
$$= 60 \pm 20$$
$w = 40$, $l = 80$ (w = width is the shorter dimension)

21. a. Ground speed = $180 + w$;
$$t = \dfrac{500}{180 + w}$$

    b. Ground speed = $180 - w$;
$$t = \dfrac{400}{180 - w}$$

    c. Refer to the graph in the back of the textbook.
The functions cross at $x = 20$, $y = 2.5$. With a wind speed of 20 mi/h, the time for the two trips is the same, namely 2.5 hours.

    d. $\dfrac{500}{180 + w} = \dfrac{400}{180 - w}$

    e. $500 \cdot 180 - 500w = 400 \cdot 180 + 400w$;
$100 \cdot 180 = 900w$; $w = 20$

23. a. Ground speed = $s - 20$;
time = $\dfrac{144}{s - 20}$

    b. Ground speed = $s + 20$;
time = $\dfrac{144}{s + 20}$

    c. Refer to the graph in the back of the textbook.

    d. $\dfrac{144}{s - 20} + \dfrac{144}{s + 20} = 3$

    e. $144(s + 20) + 144(s - 20) = 3(s^2 - 400)$;
$3s^2 - 288s - 1200 = 0$;
$s^2 - 96s - 400 = 0$;
$$s = \dfrac{96 \pm \sqrt{96^2 + 4 \cdot 400}}{2}$$
$= 48 \pm 52$
$s = 100$ (the negative solution is extraneous)

25. $x = \dfrac{4}{y}$; $\left(\dfrac{4}{y}\right)^2 + y^2 = 8$;
$4^2 + y^4 = 8y^2$;
$(y^2)^2 - 8y^2 + 16 = 0$;
$$y^2 = \dfrac{8 \pm \sqrt{8^2 - 4 \cdot 16}}{2} = 4$$
$y = 2$, $x = 2$ or $y = -2$, $x = -2$

27. $x = -\dfrac{4}{y}$; $\left(-\dfrac{4}{y}\right)^2 - 2y^2 = -4$;
$(-4)^2 - 2y^4 = -4y^2$;
$(y^2)^2 - 2y^2 - 8 = 0$;
$$y^2 = \dfrac{2 \pm \sqrt{(-2)^2 + 4 \cdot 8}}{2}; y^2 = 1 \pm 3;$$
$y^2 = 4$ (the negative solution is extraneous); $y = 2$, $x = -2$ or $y = -2$, $x = 2$

29. $x = \dfrac{6}{y}$; $\left(\dfrac{6}{y}\right)^2 - y^2 = 16$;

$6^2 - y^4 = 16y^2$;

$(y^2)^2 + 16y^2 - 36 = 0$;

$y^2 = \dfrac{-16 \pm \sqrt{16^2 + 4 \cdot 36}}{2}$;

$y^2 = \dfrac{-16 \pm \sqrt{400}}{2}$

$= \dfrac{-16 + 20}{2}$

$= \dfrac{4}{2}$

$= 2$

(the negative solution is extraneous)

$y^2 = 2$

$y = \pm\sqrt{2}$

$x = \dfrac{6}{y}$

$= \dfrac{6}{\pm\sqrt{2}}$

$= \pm 3\sqrt{2}$

The solutions are

$(3\sqrt{2},\ \sqrt{2})$ and $(-3\sqrt{2},\ -\sqrt{2})$

Chapter 8 Review

1. Factoring out $2a(a-1)^2$ yields $\dfrac{a}{2(a-1)}$.

2. Factoring out $a(1-2a)$ yields $\dfrac{-a}{4}$.

3. Factoring out 2 yields $\dfrac{2y-3}{3}$.

4. Factoring out $2x^2y$ yields $\dfrac{y^2 - 2x}{2}$.

5. Factoring out $2(x+3)$ yields $\dfrac{x}{x+3}$.

6. Factoring out $2y-x$ yields $\dfrac{2y-x}{2y+x}$.

7. Factoring out $(a-3)$ yields $\dfrac{a-3}{2(a+3)}$.

8. Factoring the numerator and denominator yields
$\dfrac{(2xy+1)^2}{(2xy+1)(2xy-1)} = \dfrac{2xy+1}{2xy-1}$

9. Combining and factoring out $3ab$ yields $10ab$.

10. Combining and factoring out 3 yields $-\dfrac{a^4 b^3}{4}$.

11. Combining and factoring out $2x(2x+3)$ yields $\dfrac{6x}{2x+3}$.

283

Chapter 8 Review

12. $\dfrac{4x^2-9}{3x-3} \cdot \dfrac{x^2-1}{4x-6} = \dfrac{(2x+3)(2x-3)}{3(x-1)} \cdot \dfrac{(x+1)(x-1)}{2(2x-3)} = \dfrac{(2x+3)(x+1)}{2\cdot 3} = \dfrac{2x^2+5x+3}{6}$

13. $\dfrac{a^2-a-2}{a^2-4} \div \dfrac{a^2+2a+1}{a^2-2a} = \dfrac{a^2-a-2}{a^2-4} \cdot \dfrac{a^2-2a}{a^2+2a+1} = \dfrac{(a+1)(a-2)}{(a+2)(a-2)} \cdot \dfrac{a(a-2)}{(a+1)^2}$

    $= \dfrac{a(a-2)}{(a+2)(a+1)} = \dfrac{a^2-2a}{a^2+3a+2}$

14. $\dfrac{a^3-8b^3}{a^2b} \div \dfrac{a^2-4ab+4b^2}{ab^2} = \dfrac{a^3-8b^3}{a^2b} \cdot \dfrac{ab^2}{a^2-4ab+4b^2}$

    $= \dfrac{(a-2b)(a^2+2ab+4b^2)}{a^2b} \cdot \dfrac{ab^2}{(a-2b)^2} = \dfrac{a^2b+2ab^2+4b^3}{a^2-2ab}$

15. $1 \div \dfrac{4x^2-1}{2x+1} = 1 \cdot \dfrac{2x+1}{4x^2-1}$

    $= \dfrac{2x+1}{(2x+1)(2x-1)} = \dfrac{1}{2x-1}$

16. $\dfrac{y^2+2y}{3x} \div 4y = \dfrac{y(y+2)}{3x \cdot 4y} = \dfrac{y+2}{12x}$

17.
$$
\begin{array}{r}
4yz + 2 - \frac{1}{yz} \\
3yz \overline{) 12y^2z^2 + 6yz - 3} \\
\underline{-(12y^2z^2)} \\
0 + 6yz \\
\underline{-(6yz)} \\
0 - 3
\end{array}
$$

18.
$$
\begin{array}{r}
9x^2 - 7 + \frac{4x^2-1}{x^4} \\
4x^4 \overline{) 36x^6 - 28x^4 + 16x^2 - 4} \\
\underline{-(36x^6)} \\
0 - 28x^4 \\
\underline{-(-28x^4)} \\
0 + 16x^2 - 4
\end{array}
$$

$9x^2 - 7 + \dfrac{4x^2-1}{x^4}$

$= 9x^2 - 7 + \dfrac{4}{x^2} - \dfrac{1}{x^4}$

Chapter 8 Review

19.

$$\begin{array}{r} y^2 + 2y - 4 \\ y+1 \overline{\smash{)}y^3 + 3y^2 - 2y - 4} \\ \underline{-(y^3 + y^2)} \\ 2y^2 - 2y \\ \underline{-(2y^2 + 2y)} \\ -4y - 4 \\ \underline{-(-4y - 4)} \\ 0 \end{array}$$

20.

$$\begin{array}{r} x^2 - 2x - 2 - \frac{1}{x-2} \\ x-2 \overline{\smash{)}x^3 - 4x^2 + 2x + 3} \\ \underline{-(x^3 - 2x^2)} \\ -2x^2 + 2x \\ \underline{-(-2x^2 + 4x)} \\ -2x + 3 \\ \underline{-(-2x + 4)} \\ -1 \end{array}$$

21.

$$\begin{array}{r} x^2 + x + \frac{1}{2} - \frac{\frac{1}{2}}{2x-1} \\ 2x-1 \overline{\smash{)}2x^3 + x^2 + 0x - 1} \\ \underline{-(2x^3 - x^2)} \\ 2x^2 + 0x \\ \underline{-(2x^2 - x)} \\ x - 1 \\ \underline{-\left(x - \frac{1}{2}\right)} \\ -\frac{1}{2} \end{array}$$

$$x^2 + x + \frac{1}{2} - \frac{\frac{1}{2}}{2x-1} = x^2 + x + \frac{1}{2} - \frac{1}{4x-2}$$

Chapter 8 Review

22.

$$\begin{array}{r} -y^2 + \frac{y}{3} - \frac{4}{9} + \frac{40}{9(3y+1)} \\ 3y+1 \overline{\smash{\big)} -3y^3 + 0y^2 - y + 4} \\ \underline{-(-3y^3 - y^2)} \\ y^2 - y \\ \underline{-\left(y^2 + \frac{y}{3}\right)} \\ -\frac{4y}{3} + 4 \\ \underline{-\left(-\frac{4y}{3} - \frac{4}{9}\right)} \\ \frac{40}{9} \end{array}$$

23. $\dfrac{x+2-(x-4)}{3x} = \dfrac{6}{3x} = \dfrac{2}{x}$

24. $\dfrac{y-1-(y+1)+y}{y+3} = \dfrac{y-2}{y+3}$

25. $\dfrac{1}{2}a - \dfrac{2}{3}a = \dfrac{a}{2} - \dfrac{2a}{3}$
$= \dfrac{3a - 2 \cdot 2a}{2 \cdot 3} = -\dfrac{a}{6}$

26. $\dfrac{5}{6}b - \dfrac{1}{3}b + \dfrac{3}{4}b = \dfrac{5b}{6} - \dfrac{b}{3} + \dfrac{3b}{4}$
$= \dfrac{10b - 4b + 9b}{12} = \dfrac{15b}{12} = \dfrac{5b}{4}$

27. $\dfrac{3}{2(x-3)} \cdot \dfrac{x+3}{x+3} - \dfrac{4}{(x+3)(x-3)} \cdot \dfrac{2}{2} = \dfrac{3x+1}{2(x+3)(x-3)}$

28. $\dfrac{1}{(y+2)^2} \cdot \dfrac{y-2}{y-2} + \dfrac{3}{(y+2)(y-2)} \cdot \dfrac{y+2}{y+2} = \dfrac{4y+4}{(y+2)^2(y-2)}$

29. $\dfrac{2a+1}{a-3} \cdot \dfrac{a-1}{a-1} - \dfrac{-2}{(a-3)(a-1)} = \dfrac{2a^2 - a + 1}{(a-3)(a-1)}$

30. $\dfrac{a}{1} \cdot \dfrac{(a+1)^2(a-1)}{(a+1)^2(a-1)} - \dfrac{1}{(a+1)^2} \cdot \dfrac{a-1}{a-1} + \dfrac{3}{(a+1)(a-1)} \cdot \dfrac{a+1}{a+1} = \dfrac{a^4 + a^3 - a^2 + a + 4}{(a+1)^2(a-1)}$

31. $\dfrac{\frac{3}{4} - \frac{1}{2}}{\frac{3}{4} + \frac{1}{2}} \cdot \dfrac{4}{4} = \dfrac{3-2}{3+2} = \dfrac{1}{5}$

32. $\dfrac{y - \frac{2y}{x}}{1 + \frac{2}{x}} \cdot \dfrac{x}{x} = \dfrac{xy - 2y}{x+2} = \dfrac{y(x-2)}{x+2}$

33. $\dfrac{x-4}{x-\frac{16}{x}} \cdot \dfrac{x}{x} = \dfrac{x^2-4x}{x^2-16} = \dfrac{x(x-4)}{(x+4)(x-4)} = \dfrac{x}{x+4}$

34. $\dfrac{\frac{1}{x-1}}{1-\frac{1}{x^2}} \cdot \dfrac{x^2(x-1)}{x^2(x-1)} = \dfrac{x^2}{x^2(x-1)-(x-1)} = \dfrac{x^2}{(x^2-1)(x-1)}$

35. $\dfrac{x}{x-2} = \dfrac{2}{x-2}+7$; $x = 2+7(x-2)$; $x-2 = 7(x-2)$; $1 = 7$ unless $x-2 = 0$, but then the fractions in the equation are undefined; there is no solution.

36. $\dfrac{x}{x-3}+\dfrac{9}{x+3} = 1$;
$x(x+3)+9(x-3) = (x-3)(x+3)$;
$x^2+3x+9x-27 = x^2-9$;
$12x = 18$; $x = \dfrac{3}{2}$

37. $\dfrac{2}{x-1}-\dfrac{x+2}{x} = 0$;
$2x-(x+2)(x-1) = 0$;
$2x-(x^2+x-2) = 0$;
$x^2-x-2 = 0$; $(x+1)(x-2) = 0$;
$x = -1$ or $x = 2$

38. $\dfrac{3x}{x+1}-\dfrac{2}{x^2+x} = \dfrac{4}{x}$;
$3x(x)-2 = 4(x+1)$;
$3x^2-4x-6 = 0$;
$x = \dfrac{4\pm\sqrt{(-4)^2+4\cdot 3\cdot 6}}{6}$;
$x = \dfrac{2\pm\sqrt{22}}{3}$

39. $Q(x) = x^3(x^2-4) = x^3(x+2)(x-2)$;
$x = 0, -2,$ or 2

40. $R(x) = x(2x^2+3x-2)$
$= x(2x-1)(x+2)$;
$x = 0, \dfrac{1}{2},$ or -2

41. **a.** $(x-2) = 0$ when $x = 2$;
$(x+1) = 0$ when $x = -1$
b. Refer to the graph in the back of the textbook.

42. **a.** $(x-3) = 0$ when $x = 3$;
$(x+2) = 0$ when $x = -2$
b. Refer to the graph in the back of the textbook.

43. **a.** $x^2 = 0$ when $x = 0$; $(x-1) = 0$ when $x = 1$; $(x+3) = 0$ when $x = -3$
b. Refer to the graph in the back of the textbook.

44. **a.** $(x+1) = 0$ when $x = -1$;
$(x-2) = 0$ when $x = 2$
b. Refer to the graph in the back of the textbook.

45. **a.** $x^3-x^5 = x^3(1-x^2)$
$= x^3(1-x)(1+x)$;
x-intercepts at $0, \pm 1$
b. Refer to the graph in the back of the textbook.

46. **a.** $x^4-9x^2 = x^2(x^2-9)$
$= x^2(x+3)(x-3)$;
x-intercepts at $0, \pm 3$
b. Refer to the graph in the back of the textbook.

Chapter 8 Review

47. a. $x^3 + x^2 - x - 1 = (x-1)(x^2+2x+1) = (x-1)(x+1)^2$; x-intercepts at ± 1
    b. Refer to the graph in the back of the textbook.

48. a. $x^3 - x^2 - 4x + 4 = (x-2)(x^2+x-2) = (x-2)(x+2)(x-1)$; x-intercepts at $1, \pm 2$
    b. Refer to the graph in the back of the textbook.

49. a. $x^3 + x^2 - 2x = x(x^2+x-2) = x(x+2)(x-1)$; x-intercepts at $0, -2, 1$
    b. Refer to the graph in the back of the textbook.

50. a. $x^3 - 2x^2 + x - 2 = (x-2)(x^2+1)$; x-intercept at 2
    b. Refer to the graph in the back of the textbook.

51. a. $x^4 - 7x^2 + 6 = (x^2-1)(x^2-6)$; x-intercepts at $\pm 1, \pm\sqrt{6}$
    b. Refer to the graph in the back of the textbook.

52. a. $x^4 + x^3 - 3x^2 - 3x = x(x^3+x^2-3x-3) = x(x+1)(x^2-3)$; x-intercepts at $0, -1, \pm\sqrt{3}$
    b. Refer to the graph in the back of the textbook.

53. a. $x - 4 = 0$ when $x = 4$, so that is a vertical asymptote. The denominator's degree exceeds the numerator's, so $y = 0$ is a horizontal asymptote. $\frac{1}{0-4} = -\frac{1}{4}$ is a y-intercept.
    b. Refer to the graph in the back of the textbook.

54. a. $x^2 - 3x - 10 = (x-5)(x+2) = 0$ when $x = 5$ or $x = -2$, so those are two vertical asymptotes. The denominator's degree exceeds the numerator's, so $y = 0$ is a horizontal asymptote. $\frac{2}{0-0-10} = -\frac{1}{5}$ is a y-intercept.
    b. Refer to the graph in the back of the textbook.

55. a. $x + 3 = 0$ when $x = -3$, so that is a vertical asymptote. The numerator's and denominator's degrees are equal with a leading coefficient ratio of 1, so $y = 1$ is a horizontal asymptote. $\frac{0-2}{0+3} = -\frac{2}{3}$ is a y-intercept.
    b. Refer to the graph in the back of the textbook.

56. a. $x^2 - 2x - 3 = (x+1)(x-3) = 0$ when $x = -1$ or $x = 3$, so those are two vertical asymptotes. The denominator's degree exceeds the numerator's, so $y = 0$ is a horizontal asymptote. $\frac{0-1}{0-0-3} = \frac{1}{3}$ is a y-intercept. $y = 0$ when $x - 1 = 0$ or $x = 1$, so 1 is an x-intercept.
    b. Refer to the graph in the back of the textbook.

Chapter 8 Review

57. a. $x^2 - 4 = 0$ when $x = \pm 2$, so those are two vertical asymptotes. The numerator's and denominator's degrees are equal with a leading coefficient ratio of 3, so $y = 3$ is a horizontal asymptote. $\frac{3(0)}{0-4} = 0$, so $(0, 0)$ is both a y- and an x-intercept.

   b. Refer to the graph in the back of the textbook.

58. a. $x^2 - 9 = 0$ when $x = \pm 3$, so those are two vertical asymptotes. The numerator's and denominator's degrees are equal with a leading coefficient ratio of 2, so $y = 2$ is a horizontal asymptote.
$\frac{2(0)-2}{0-9} = \frac{2}{9}$ is a y-intercept.
$y = 0$ when $2x^2 - 2 = 0$ or $x = \pm 1$, so those are two x-intercepts.

   b. Refer to the graph in the back of the textbook.

59. $\dfrac{1}{x^3} + \dfrac{1}{y} = \dfrac{y}{x^3 y} + \dfrac{x^3}{x^3 y} = \dfrac{x^3 + y}{x^3 y}$

60. $\dfrac{1}{xy} - \dfrac{x}{y-1} = \dfrac{y-1}{xy(y-1)} - \dfrac{x^2 y}{xy(y-1)}$
$= \dfrac{-x^2 y + y - 1}{xy(y-1)}$

61. $\dfrac{\frac{1}{x} - y}{y - 1} \cdot \dfrac{x}{x} = \dfrac{1 - xy}{xy - x}$

62. $\dfrac{\frac{1}{x} + \frac{1}{y}}{\frac{1}{x}} \cdot \dfrac{xy}{xy} = \dfrac{y + x}{y}$

63. $\dfrac{\frac{1}{x} - \frac{1}{y}}{\frac{1}{x-y}} \cdot \dfrac{xy(x-y)}{xy(x-y)}$
$= \dfrac{y(x-y) - x(x-y)}{xy}$
$= \dfrac{-(x-y)^2}{xy}$

64. $\dfrac{\frac{1}{xy}}{\frac{1}{x} - \frac{1}{y}} \cdot \dfrac{xy}{xy} = \dfrac{1}{y-x}$

65. a. $\dfrac{n(n-1)(n-2)}{6} = \dfrac{n^3 - 3n^2 + 2n}{6}$
$= \dfrac{n^3}{6} - \dfrac{3n^2}{6} + \dfrac{2n}{6} = \dfrac{n^3}{6} - \dfrac{n^2}{2} + \dfrac{n}{3}$

   b. $\frac{1}{6} \cdot 12 \cdot 11 \cdot 10 = 220$

   c. $\frac{1}{6} \cdot 19 \cdot 18 \cdot 17 = 969$;
$\frac{1}{6} \cdot 20 \cdot 19 \cdot 18 = 1140$; $n = 20$

66. a. $n(n-1)(n-2) = n(n^2 - 3n + 2)$
$= n^3 - 3n^2 + 2n$

   b. $21 \cdot 20 \cdot 19 = 7980$

   c. $22 \cdot 21 \cdot 20 = 9240$;
$23 \cdot 22 \cdot 21 = 10{,}626$; $n = 23$

67. a. $A = s^2 = (x + y)^2$

   b. $(x + y)^2 = x^2 + 2xy + y^2$
$= x^2 + xy + xy + y^2$

   c. Refer to the graph in the back of the textbook.

68. a. $A = x^2 - y^2$

   b. $x^2 - y^2 = (x + y)(x - y)$

   c. Refer to the graph in the back of the textbook.

Chapter 8 Review

69. a. $A = \dfrac{x^2}{2} - \dfrac{y^2}{2}$

b. $\dfrac{x^2}{2} - \dfrac{y^2}{2} = \dfrac{1}{2}(x+y)(x-y)$

c. $\dfrac{1}{2}(12)(3) = 18$ sq. ft.

70. a. $A = \pi x^2 - \pi y^2$

b. $\pi(x^2 - y^2) = \pi(x+y)(x-y)$

c. $\pi(11)(6) = 66\pi$

71. a. $V = \pi r^2 h = \pi\left(\dfrac{h}{2}\right)^2 h = \dfrac{\pi h^3}{4}$

b. $\dfrac{\pi(2)^3}{4} = 2\pi \approx 6.28 \text{ cm}^3$;

$\dfrac{\pi(4)^3}{4} = 16\pi = 50.27 \text{ cm}^3$

c. Refer to the graph in the back of the textbook.

$100 = \dfrac{\pi h^3}{4}$; $h = \sqrt[3]{\dfrac{400}{\pi}} \approx 5.03$ cm

72. a. $V = \dfrac{1}{3}\pi r^2 h = \dfrac{1}{3}\pi\left(\dfrac{d}{2}\right)^2\left(3 \cdot \dfrac{d}{2}\right)$

$= \dfrac{\pi d^3}{8}$

b. $\dfrac{\pi(3)^3}{8} = \dfrac{27\pi}{8} \approx 10.60 \text{ in.}^3$;

$\dfrac{\pi(4)^3}{8} = \dfrac{64\pi}{8} \approx 25.13 \text{ in.}^3$

c. Refer to the graph in the back of the textbook.

$5 = \dfrac{\pi d^3}{8}$; $d = \sqrt[3]{\dfrac{40}{\pi}} \approx 2.34$ in.

73. a. $P = R - C = -0.02(x - 800)^2 + 12{,}800 - (8x + 4000) = -0.02x^2 + 24x - 4000$

b. Refer to the graph in the back of the textbook.

c. When $P = -0.02x^2 + 24x - 4000 = 0$,

$x = \dfrac{-24 \pm \sqrt{24^2 - 4(0.02)(4000)}}{2(-0.02)} = 600 \pm 400$;

P is negative when $x < 600 - 400 = 200$ or $x > 600 + 400 = 1000$.

d. R is greatest when $x = 800$, at which point $R = \$12{,}800$; P is greatest when $x = 600$, at which point $P = -0.02(600)^2 + 24(600) - 4000 = \$3200$.

74. a. $P = R - C = -0.03(x - 600)^2 + 10{,}800 - (3x + 3000) = -0.03x^2 + 33x - 3000$

b. Refer to the graph in the back of the textbook.

Chapter 8 Review

c. When $P = -0.03x^2 + 33x - 3000 = 0$,

$$x = \frac{-33 \pm \sqrt{33^2 - 4(0.03)(3000)}}{2(-0.03)} = 550 \pm 450;$$

P is negative when $x < 550 - 450 = 100$ or $x > 550 + 450 = 1000$.

d. R is greatest when $x = 600$, at which point $R = \$10,800$; P is greatest when $x = 550$, at which point $P = -0.03(550)^2 + 33(550) - 3000 = \$6075$.

75. a. $P = R - C = -1.75(x - 200)^2 + 70,000 - (175x + 21,875)$
$= -1.75x^2 + 525x - 21,875$

b. Refer to the graph in the back of the textbook.

c. $P = -1.75x^2 + 525x - 21,875 = 0$ when

$$x = \frac{-525 \pm \sqrt{525^2 - 4(1.75)(21,875)}}{2(-1.75)} = 150 \pm 100$$

P is negative when $x < 150 - 100 = 50$ or $x > 150 + 100 = 250$.

d. R is greatest when $x = 200$, at which point $R = \$70,000$; P is greatest when $x = 150$, at which point $P = -1.75(150)^2 + 525(150) - 21,875 = \$17,500$.

76. a. $P = R - C = -2(x - 120)^2 + 28,800 - (80x + 12,800) = -2x^2 + 400x - 12,800$

b. Refer to the graph in the back of the textbook.

c. $P = -2x^2 + 400x - 12,800 = 0$ when

$$x = \frac{-400 \pm \sqrt{400^2 - 4(2)(12,800)}}{2(-2)} = 100 \pm 60$$

P is negative when $x < 100 - 60 = 40$ or $x > 100 + 60 = 160$.

d. R is greatest when $x = 120$, at which point $R = \$28,800$; P is greatest when $x = 100$, at which point $P = -2(100)^2 + 400(100) - 12,800 = \$7200$.

77. $\dfrac{840}{L} = \dfrac{210}{M}$; $L = M + 180$;
$\dfrac{840}{M + 180} = \dfrac{210}{M}$;
$840M = 210M + 37,800$;
$M = \dfrac{37,800}{630} = 60$; $L = 240$

78. $\dfrac{180}{M} = \dfrac{200}{F}$; $F = M + 5$;
$\dfrac{180}{M} = \dfrac{200}{M + 5}$; $180M + 900 = 200M$;
$M = \dfrac{900}{20} = 45$; $F = 50$

291

Chapter 8 Review

79. $P = 2l + 2w = 26$
$A = lw = 12$
$l = \frac{12}{w}$; $\frac{24}{w} + 2w = 26$;
$2w^2 - 26w + 24 = 0$;
$w^2 - 13w + 12 = (w-1)(w-12) = 0$
$w = 1$ (w is the shorter dimension),
$l = \frac{12}{w} = 12$

80. $P = 2l + 2w = 34$
$A = lw = 70$
$l = \frac{70}{w}$; $\frac{140}{w} + 2w = 34$;
$2w^2 - 34w + 140 = 0$
$w^2 - 17w + 70 = (w-7)(w-10) = 0$
$w = 7$ (w is the shorter dimension),
$l = \frac{70}{w} = 10$

81. $P = 2l + 2w = 18$
$A = lw$
$(l-5)(w+12) = 2A = 2lw$;
$l = \frac{18-2w}{2} = 9-w$;
$(9-w-5)(w+12) = 2(9-w)w$
$(4-w)(w+12) = 18w - 2w^2$
$-w^2 - 8w + 48 = 18w - 2w^2$
$w^2 - 26w + 48 = (w-2)(w-24) = 0$
$w = 2$ (w is the shorter dimension),
$l = 9 - w = 7$

82. $A = lw = 216$
$P = 2l + 2w = 60$
$l = \frac{216}{w}$; $\frac{432}{w} + 2w = 60$;
$2w^2 - 60w + 432 = 0$;
$w^2 - 30w + 216 = (w-12)(w-18) = 0$
$w = 12$ (w is the shorter dimension),
$l = \frac{216}{12} = 18$

83. $\frac{10}{m} + \frac{10}{m+10} = \frac{50}{60}$;
$(10m + 100) + 10m = \frac{5}{6}m(m+10)$;
$\frac{5}{6}m^2 - \frac{70}{6}m - \frac{600}{6} = 0$;
$m^2 - 14m - 120 = 0$;
$m = \frac{14 \pm \sqrt{14^2 + 480}}{2} = 7 \pm 13$
$m = 20$ (the negative solution is extraneous), $e = m + 10 = 30$

84. $t_t + t_l = t$; $\frac{10}{l-20} + \frac{40}{l} = 1\frac{1}{2}$;
$10l + 40l - 800 = \frac{3}{2}l(l-20)$;
$\frac{3}{2}l^2 - 80l + 800 = 0$;
$l = \frac{80 \pm \sqrt{80^2 - 4\left(\frac{3}{2}\right)(800)}}{2\left(\frac{3}{2}\right)}$;
$l = 3\frac{1}{3}$ or 40; $l = 13\frac{1}{3}$ would yield a negative city speed, so $l = 40$,
$c = l - 20 = 20$.

85. $Ar = 32$; $A = \frac{32}{r}$;
$(A + 200)(r - 0.005) = 35$;
$\left(\frac{32}{r} + 200\right)(r - 0.005) = 35$;
$(32 + 200r)(r - 0.005) = 35r$;
$200r^2 - 4r - 0.16 = 0$;
$r = \frac{4 \pm \sqrt{16 + 4(200)(0.16)}}{2(200)} = \frac{1 \pm 3}{100}$
Throwing out the negative solution yields $r = 0.04 = 4\%$,
$A = \frac{32}{0.04} = \$800$

86. $PV = 30$; $P = \frac{30}{V}$;
$(P + 4)(V - 2) = 30$;
$\left(\frac{30}{V} + 4\right)(V - 2) = 30$;
$(30 + 4V)(V - 2) = 30V$;
$4V^2 - 8V - 60 = 0$;
$V^2 - 2V - 15 = (V + 3)(V - 5) = 0$;
$V = -3$ or 5. Throwing out the negative solution (absolute pressure cannot be negative) yields $V = 5$ cu. in., $P = \frac{30}{V} = 6$ lb per sq. in.

87. $x^2 - 2\left(\frac{1}{2}y^2\right) = x^2 - y^2$
$= (x + y)(x - y)$

88. Diameter of $A_1 = 2y + 2x$,
Radius $= y + x$;
$A_1 = \pi(y + x)^2 - \pi y^2 - \pi x^2$
$= \pi y^2 + 2\pi xy + \pi x^2 - \pi y^2 - \pi x^2$
$= 2\pi xy = (\pi y)(2x)$

Chapter 9

Section 9.1

1. $a_n = n - 5$
$a_1 = 1 - 5 = -4$
$a_2 = 2 - 5 = -3$
$a_3 = 3 - 5 = -2$
$a_4 = 4 - 5 = -1$

3. $c_n = \dfrac{n^2 - 2}{2}$
$c_1 = \dfrac{1^2 - 2}{2} = -\dfrac{1}{2}$
$c_2 = \dfrac{2^2 - 2}{2} = 1$
$c_3 = \dfrac{3^2 - 2}{2} = \dfrac{7}{2}$
$c_4 = \dfrac{4^2 - 2}{2} = 7$
The first four terms are
$-\dfrac{1}{2}$, 1, $\dfrac{7}{2}$, 7.

5. $s_n = 1 + \dfrac{1}{n}$
$s_1 = 1 + \dfrac{1}{1} = 2$
$s_2 = 1 + \dfrac{1}{2} = 1.5$
$s_3 = 1 + \dfrac{1}{3} = 1.\overline{3}$
$s_4 = 1 + \dfrac{1}{4} = 1.25$
The first four terms are
2, 1.5, 1.$\overline{3}$, 1.25.

7. $u_n = \dfrac{n(n-1)}{2}$
$u_1 = \dfrac{1(1-1)}{2} = \dfrac{0}{2} = 0$
$u_2 = \dfrac{2(2-1)}{2} = \dfrac{2}{2} = 1$
$u_3 = \dfrac{3(3-1)}{2} = \dfrac{6}{2} = 3$
$u_4 = \dfrac{4(4-1)}{2} = \dfrac{12}{2} = 6$
The first four terms are 0, 1, 3, 6.

9. $w_n = (-1)^n$
$w_1 = (-1)^1 = -1$
$w_2 = (-1)^2 = 1$
$w_3 = (-1)^3 = -1$
$w_4 = (-1)^4 = 1$
The first four terms are -1, 1, -1, 1.

11. $B_n = \dfrac{(-1)^n (n-2)}{n}$
$B_1 = \dfrac{(-1)^1 (1-2)}{1} = 1$
$B_2 = \dfrac{(-1)^2 (2-2)}{2} = 0$
$B_3 = \dfrac{(-1)^3 (3-2)}{3} = -\dfrac{1}{3}$
$B_4 = \dfrac{(-1)^4 (4-2)}{4} = \dfrac{1}{2}$
The first four terms are
1, 0, $-\dfrac{1}{3}$, $\dfrac{1}{2}$.

Chapter 9, Section 9.1

13. $D_n = 1$
$D_1 = 1$
$D_2 = 1$
$D_3 = 1$
$D_4 = 1$
The first four terms are 1, 1, 1, 1.

15. $s_1 = 3$, $s_n = s_{n-1} + 2$
$s_2 = s_1 + 2 = 3 + 2 = 5$
$s_3 = s_2 + 2 = 5 + 2 = 7$
$s_4 = s_3 + 2 = 7 + 2 = 9$
$s_5 = s_4 + 2 = 9 + 2 = 11$
The first five terms are 3, 5, 7, 9, 11.

17. $d_1 = 24$, $d_{n+1} = \frac{-1}{2} d_n$
$d_2 = \frac{-1}{2} d_1 = \frac{-1}{2}(24) = -12$
$d_3 = \frac{-1}{2} d_2 = \frac{-1}{2}(-12) = 6$
$d_4 = \frac{-1}{2} d_3 = \frac{-1}{2}(6) = -3$
$d_5 = \frac{-1}{2} d_4 = \frac{-1}{2}(-3) = \frac{3}{2}$
The first five terms are
24, −12, 6, −3, $\frac{3}{2}$.

19. $t_1 = 1$, $t_{n+1} = (n+1) t_n$
$t_2 = (1+1) t_1 = 2(1) = 2$
$t_3 = (2+1) t_2 = 3(2) = 6$
$t_4 = (3+1) t_3 = 4(6) = 24$
$t_5 = (4+1) t_4 = 5(24) = 120$
The first five terms are 1, 2, 6, 24, 120.

21. $w_1 = 100$, $w_n = 1.10 w_{n-1} + 100$
$w_2 = 1.10 w_1 + 100$
$= 1.10(100) + 100 = 210$
$w_3 = 1.10 w_2 + 100$
$= 1.10(210) + 100 = 331$
$w_4 = 1.10 w_3 + 100$
$= 1.10(331) + 100 = 464.1$
$w_5 = 1.10 w_4 + 100$
$= 1.10(464.1) + 100 = 610.51$
The first five terms are 100, 210, 331, 464.1, 610.51.

23. a. At the end of the first year the car will be worth
$14,000 − (0.15)($14,000)
= $11,900.
At the end of the second year the car will be worth
$11,900 − (0.15)($11,900)
= $10,115.
At the end of the third year the car will be worth
$10,115 − (0.15)($10,115)
=$8597.75
At the end of the fourth year the car will be worth
$8597.75 − (0.15)($8597.75)
=$7308.09.
The first four terms of the sequence are 11,900, 10,115, 8597.75, 7308.09.

b. Each term of the sequence can be found by multiplying the previous term by 0.85.
$v_1 = 11,900 \quad v_{n+1} = 0.85 v_n$

296

Chapter 9, Section 9.1

25. a. The first minute will cost
$1.10 + $0.45 = $1.55.
Two minutes will cost
$1.55 + $0.45 = $2.00
Three minutes will cost
$2.00 + $0.45 = $2.45
Four minutes will cost
$2.45 + $0.45 = $2.90
The first four terms are 1.55, 2.00, 2.45, 2.90.

b. Each term of the sequence can be found by adding 0.45 to the previous term.
$c_1 = 1.55 \quad c_{n+1} = c_n + 0.45$

27. a. After the first month it is worth
$50,000(1.01) − $500 = $50,000.
After two months it is worth
$50,000(1.01) − $500 = $50,000.
After three months it is worth
$50,000(1.01) − $500 = $50,000.
After four months it is worth
$50,000(1.01) − $500 = $50,000.
The first four terms are 50,000, 50,000, 50,000, 50,000.

b. The interest earned and the amount withdrawn each month are equal.
$v_1 = 50{,}000 \quad v_{n+1} = v_n$

29. a. After the first dose there are 10 ml present. After the second dose there are $10 + 10(0.8) = 18$ ml present. After the third dose there are $10 + 18(0.8) = 24.4$ ml present. After the fourth dose there are $10 + 24.4(0.8) = 29.52$ ml present. The first four terms are 10, 18, 24.4, 29.52.

b. Each term of the sequence can be found by multiplying the previous term by 0.8 and adding 10.
$d_1 = 10 \quad d_{n+1} = 0.8 d_n + 10$

31. a. $\binom{3}{2} = \dfrac{3!}{2!\,1!} = 3$

b. $\binom{4}{2} = \dfrac{4!}{2!\,2!} = 6$

c. $L_{n+1} = n + L_n$
$L_1 = 0$
$L_2 = 1 + 0 = 1$
$L_3 = 2 + 1 = 3$
$L_4 = 3 + 3 = 6$
$L_5 = 4 + 6 = 10$
The first five terms are 0, 1, 3, 6, 10.

33. $D_6 = 2^6 - 6 = 58$

35. $x_{26} = \log 26 = 1.415$

37. $z_{20} = 2\sqrt{20} = 8.944$

39. a. $f_1 = 1$
$f_2 = 1$
$f_3 = f_1 + f_2 = 1 + 1 = 2$
$f_4 = f_2 + f_3 = 1 + 2 = 3$
$f_5 = f_3 + f_4 = 2 + 3 = 5$
$f_6 = f_4 + f_5 = 3 + 5 = 8$
$f_7 = f_5 + f_6 = 5 + 8 = 13$
$f_8 = f_6 + f_7 = 8 + 13 = 21$
$f_9 = f_7 + f_8 = 13 + 21 = 34$
$f_{10} = f_8 + f_9 = 21 + 34 = 55$
$f_{11} = f_9 + f_{10} = 34 + 55 = 89$
$f_{12} = f_{10} + f_{11} = 55 + 89 = 144$
$f_{13} = f_{11} + f_{12} = 89 + 144 = 233$
$f_{14} = f_{12} + f_{13}$
$\quad\;\; = 144 + 233 = 377$

Chapter 9, Section 9.2

$f_{15} = f_{13} + f_{14}$
$= 233 + 377 = 610$
$f_{16} = f_{14} + f_{15}$
$= 377 + 610 = 987$

The first 16 terms are 1, 1, 2, 3, 5, 8, 13, 21, 34, 55, 89, 144, 233, 377, 610, 987.

b. $\dfrac{f_2}{f_1} = \dfrac{1}{1} = 1$

$\dfrac{f_3}{f_2} = \dfrac{2}{1} = 2$

$\dfrac{f_4}{f_3} = \dfrac{3}{2} = 1.5$

$\dfrac{f_5}{f_4} = \dfrac{5}{3} = 1.\overline{6}$

$\dfrac{f_6}{f_5} = \dfrac{8}{5} = 1.6$

$\dfrac{f_7}{f_6} = \dfrac{13}{8} = 1.625$

$\dfrac{f_8}{f_7} = \dfrac{21}{13} = 1.615385$

$\dfrac{f_9}{f_8} = \dfrac{34}{21} = 1.619048$

$\dfrac{f_{10}}{f_9} = \dfrac{55}{34} = 1.617647$

$\dfrac{f_{11}}{f_{10}} = \dfrac{89}{55} = 1.6181\overline{18}$

$\dfrac{f_{12}}{f_{11}} = \dfrac{144}{89} = 1.617978$

$\dfrac{f_{13}}{f_{12}} = \dfrac{233}{144} = 1.61805\overline{5}$

$\dfrac{f_{14}}{f_{13}} = \dfrac{377}{233} = 1.618026$

$\dfrac{f_{15}}{f_{14}} = \dfrac{610}{377} = 1.618037$

$\dfrac{f_{16}}{f_{15}} = \dfrac{987}{610} = 1.618033$

Rounded to three decimal places, the values are 1, 2, 1.5, 1.667, 1.600, 1.625, 1.615, 1.619, 1.618, 1.618, 1.618, 1.618, 1.618, 1.618.

$\dfrac{1+\sqrt{5}}{2}$ is approximately equal to 1.61803 which is equal to the limit above.

41. a_n approaches 1.4142, or $\sqrt{2}$.

43. c_n approaches 0.64039, or $\dfrac{1+\sqrt{17}}{8}$.

45. s_n approaches 2.

Section 9.2

1. This sequence is geometric. Each term is obtained from the previous term by multiplying by 3.

3. This sequence is arithmetic. Each term is obtained from the previous term by subtracting 8.

5. This sequence is geometric. Each term is obtained from the previous term by multiplying by −1.

7. This sequence is neither arithmetic or geometric.

9. This sequence is geometric. Each term is obtained from the previous term by multiplying by $\dfrac{1}{3}$.

Chapter 9, Section 9.2

11. This sequence is geometric. Each term is obtained from the previous term by multiplying by $-\dfrac{3}{2}$.

13. $a_n = a + (n-1)d$
$a_1 = 2 + (0)4 = 2$
$a_2 = 2 + (1)4 = 6$
$a_3 = 2 + (2)4 = 10$
$a_4 = 2 + (3)4 = 14$
The first four terms are 2, 6, 10, 14.

15. $a_n = a + (n-1)d$
$a_1 = \dfrac{1}{2} + (0)\dfrac{1}{4} = \dfrac{1}{2}$
$a_2 = \dfrac{1}{2} + (1)\dfrac{1}{4} = \dfrac{3}{4}$
$a_3 = \dfrac{1}{2} + (2)\dfrac{1}{4} = 1$
$a_4 = \dfrac{1}{2} + (3)\dfrac{1}{4} = \dfrac{5}{4}$
The first four terms are $\dfrac{1}{2}, \dfrac{3}{4}, 1, \dfrac{5}{4}$.

17. $a_n = a + (n-1)d$
$a_1 = 2.7 + 0(-0.8) = 2.7$
$a_2 = 2.7 + 1(-0.8) = 1.9$
$a_3 = 2.7 + 2(-0.8) = 1.1$
$a_4 = 2.7 + 3(-0.8) = 0.3$
The first four terms are 2.7, 1.9, 1.1, 0.3.

19. $a_n = ar^{n-1}$
$a_1 = 5(-2)^0 = 5$
$a_2 = 5(-2)^1 = -10$
$a_3 = 5(-2)^2 = 20$
$a_4 = 5(-2)^3 = -40$
The first four terms are 5, −10, 20, −40.

21. $a_n = ar^{n-1}$
$a_1 = 9\left(\dfrac{2}{3}\right)^0 = 9$
$a_2 = 9\left(\dfrac{2}{3}\right)^1 = 6$
$a_3 = 9\left(\dfrac{2}{3}\right)^2 = 4$
$a_4 = 9\left(\dfrac{2}{3}\right)^3 = \dfrac{8}{3}$
The first four terms are 9, 6, 4, $\dfrac{8}{3}$.

23. $a_n = ar^{n-1}$
$a_1 = 60(0.4)^0 = 60$
$a_2 = 60(0.4)^1 = 24$
$a_3 = 60(0.4)^2 = 9.6$
$a_4 = 60(0.4)^3 = 3.84$
The first four terms are 60, 24, 9.6, 3.84.

25. Add 4 to each successive term:
$a_4 = 11 + 4 = 15$
$a_5 = 15 + 4 = 19$
$a_6 = 19 + 4 = 23$
The next three terms are 15, 19, 23.
$a = 3$ and $d = 4$
$a_n = 3 + (n-1)4$

27. Subtract 4 from each successive term:
$a_4 = -9 - 4 = -13$
$a_5 = -13 - 4 = -17$
$a_6 = -17 - 4 = -21$
The next three terms are −13, −17, −21.
$a = -1$ and $d = -4$
$a_n = -1 + (n-1)(-4)$

Chapter 9, Section 9.2

29. Multiply each successive term by 2.
$a_4 = \frac{8}{3}(2) = \frac{16}{3}$

$a_5 = \frac{16}{3}(2) = \frac{32}{3}$

$a_6 = \frac{32}{3}(2) = \frac{64}{3}$

The next three terms are
$\frac{16}{3}, \frac{32}{3}, \frac{64}{3}$.

$a = \frac{2}{3}$ and $r = 2$

$a_n = \frac{2}{3}(2)^{n-1}$

31. Multiply each successive term by $-\frac{1}{2}$.

$a_4 = 1\left(-\frac{1}{2}\right) = -\frac{1}{2}$

$a_5 = \left(-\frac{1}{2}\right)\left(-\frac{1}{2}\right) = \frac{1}{4}$

$a_6 = \frac{1}{4}\left(-\frac{1}{2}\right) = -\frac{1}{8}$

The next three terms are
$-\frac{1}{2}, \frac{1}{4}, -\frac{1}{8}$.

$a = 4$ and $r = -\frac{1}{2}$

$a_n = 4\left(-\frac{1}{2}\right)^{n-1}$

33. $a = 2$ and $d = \frac{1}{2}$

$a_n = 2 + (n-1)\frac{1}{2}$

$a_{12} = 2 + (12-1)\frac{1}{2}$

$a_{12} = \frac{15}{2}$

35. $a = -3$ and $r = -\frac{1}{2}$

$a_n = -3\left(-\frac{1}{2}\right)^{n-1}$

$a_8 = -3\left(-\frac{1}{2}\right)^{8-1}$

$a_8 = \frac{3}{128}$

37. $a_5 = 48$ and $r = 2$
$a_5 = ar^4$
$48 = a \, 2^4$
$a = \frac{48}{2^4}$
$a = 3$

39. $a = \frac{1}{8}$ and $r = 2$

$a_n = ar^{n-1}$

$512 = \frac{1}{8} \cdot 2^{n-1}$

$4096 = 2^{n-1}$

$n = 13$
There are 13 terms in this sequence.

41. $s_1 = 3$ and $d = 2$
$s_n = 3 + (n-1)2$

300

Chapter 9, Section 9.2

43. $x_1 = 0$ and $d = -3$
$x_n = 0 + (n-1)(-3)$
$x_n = -3(n-1)$

45. $d_1 = 24$ and $r = \dfrac{-1}{2}$
$d_n = 24\left(\dfrac{-1}{2}\right)^{n-1}$

47. $w_1 = 1$ and $r = 2$
$w_n = 1(2)^{n-1}$
$w_n = 2^{n-1}$

49. a.

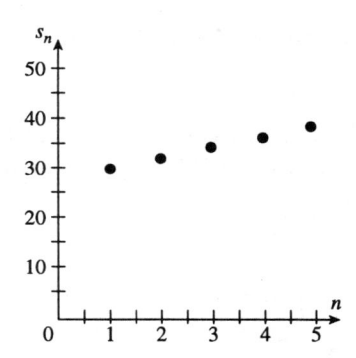

   [graph of S_n vs n]

b. $s_1 = 30$ and $d = 2$
$s_{50} = 30 + (50-1)2$
$s_{50} = 128$
There are 128 seats in the fiftieth row.

51. a.

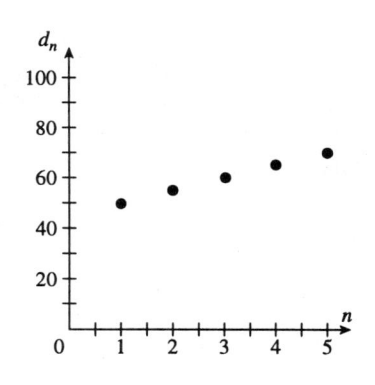

   [graph of d_n vs n]

b. $d_1 = 50$, $d = 5$, $n = \dfrac{70}{5} = 14$
$d_{14} = 50 + (14-1)5$
$d_{14} = 115$
The charge is $115.

53. a.

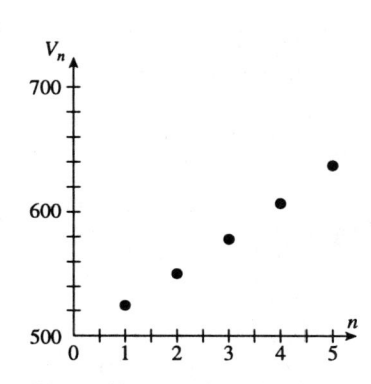

   [graph of V_n vs n]

b. $V_1 = 500(1.05) = 525$
and $r = 1.05$
$V_{18} = 525(1.05)^{18-1}$
$V_{18} = 1203.31$
It will be worth $1203.31.

55. a.

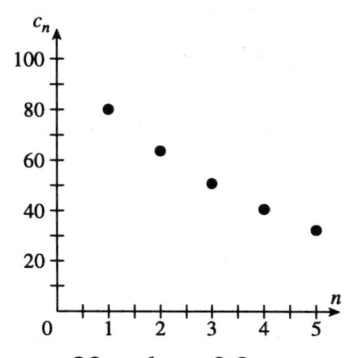

   [graph of c_n vs n]

b. $c_1 = 80$ and $r = 0.8$
$c_{20} = 80(0.8)^{20-1}$
$c_{20} = 1.15$
There will be 1.15 kg remaining.

Chapter 9, Section 9.3

Section 9.3

1. $a_1 = -4 + 3(1) = -1$
   $a_9 = -4 + 3(9) = 23$
   $S_9 = \dfrac{9}{2}(-1+23) = 99$
   The sum is 99.

3. $a_1 = 18 - \dfrac{4}{3}(1) = \dfrac{50}{3}$
   $a_{16} = 18 - \dfrac{4}{3}(16) = -\dfrac{10}{3}$
   $S_{16} = \dfrac{16}{2}\left(\dfrac{50}{3} - \dfrac{10}{3}\right) = \dfrac{320}{3} = 106\dfrac{2}{3}$
   The sum is $106\dfrac{2}{3}$.

5. $a_1 = 1.6 + 0.2(1) = 1.8$
   $a_{30} = 1.6 + 0.2(30) = 7.6$
   $S_{30} = \dfrac{30}{2}(1.8 + 7.6) = 141$
   The sum is 141.

7. $a_1 = 2(-4)^{1-1} = 2$
   $a_6 = 2(-4)^{6-1} = -2048$
   $S_5 = \dfrac{-2048 - 2}{-4 - 1} = 410$
   The sum is 410.

9. $a_1 = -48\left(\dfrac{1}{2}\right)^{1-1} = -48$
   $a_{10} = -48\left(\dfrac{1}{2}\right)^{10-1} = -\dfrac{3}{32}$
   $S_9 = \dfrac{-\dfrac{3}{32} - (-48)}{\dfrac{1}{2} - 1} = -95.8125$
   The sum is -95.8125.

11. $a_1 = 18(1.15)^{1-1} = 18$
    $a_5 = 18(1.15)^{5-1} = 31.482$
    $S_4 = \dfrac{31.482 - 18}{1.15 - 1} = 89.881$
    The sum is 89.881.

13. Arithmetic: Each term is found from the previous term by adding 2.
    $a = 2$ and $d = 2$
    $a_n = 2 + (n-1)2 = 2n$
    $a_1 = 2(1) = 2$
    $a_{50} = 2(50) = 100$;
    100 is the 50th term
    $S_{50} = \dfrac{50}{2}(2 + 100) = 2550$
    The sum is 2550.

15. Geometric: Each term is obtained from previous term by multiplying by 2.
    $a = 2$ and $r = 2$
    $a_n = 2(2)^{n-1} = 2^n$
    $a_1 = 2^1 = 2$
    $a_{10} = 2^{10} = 1024$;
    1024 is the 10th term
    $a_{11} = 2^{11} = 2048$
    $S_{10} = \dfrac{2048 - 2}{2 - 1} = 2046$
    The sum is 2046.

17. Neither: each term cannot be obtained from adding or multiplying the previous term by a constant.
    $1 + 8 + 27 + 64 + 125 + 216 + 343 = 784$
    The sum is 784.

Chapter 9, Section 9.3

19. Arithmetic: each term can be obtained from the previous term by subtracting 3.
$a = 87$ and $d = -3$
$a_n = 87 - 3(n-1) = 90 - 3n$
$a_1 = 90 - 3(1) = 87$
$a_{17} = 90 - 3(17) = 39$;
39 is the 17th term
$S_{17} = \dfrac{17}{2}(87 + 39) = 1071$
The sum is 1071.

21. Geometric: each term is obtained from the previous term by multiplying by $\dfrac{1}{3}$.
$a = 6$ and $r = \dfrac{1}{3}$
$a_n = 6\left(\dfrac{1}{3}\right)^{n-1}$
$a_1 = 6\left(\dfrac{1}{3}\right)^{1-1} = 6$
$a_7 = 6\left(\dfrac{1}{3}\right)^{7-1} = \dfrac{2}{243}$
$\dfrac{2}{243}$ is the 7th term
$a_8 = 6\left(\dfrac{1}{3}\right)^{8-1} = \dfrac{2}{729}$
$S_7 = \dfrac{\dfrac{2}{729} - 6}{\dfrac{1}{3} - 1} = 8.996$
The sum is 8.996.

23. $14 + 16 + 18 + \ldots + 88$
$a = 14$ and $d = 2$
$a_n = 14 + (n-1)2 = 12 + 2n$
$a_1 = 14$
$a_{38} = 12 + 2(38) = 88$; 88 is the 38th term
$S_{38} = \dfrac{38}{2}(14 + 88) = 1938$
The sum is 1938.

25. $1 + 2 + 3 + \ldots + 12$
$a = 1$ and $d = 1$
$a_n = 1 + (n-1)(1) = n$
$a_1 = 1$
$a_{12} = 12$; There are 12 terms in the series
$S_{12} = \dfrac{12}{2}(1 + 12) = 78$
The clock will strike 78 times.

27. a. Drops first time 24 feet, rises first time $24\left(\dfrac{3}{4}\right) = 18$ feet, drops second time 18 feet, rises second time $18\left(\dfrac{3}{4}\right) = 13.5$ feet, drops third time 13.5 feet, rises third time $13.5\left(\dfrac{3}{4}\right) = 10.125$ feet. It will bounce 10.125 feet.

303

Chapter 9, Section 9.3

b. First round down and up:
24 + 18
Second round down and up:
18 + 13.5
Third round down and up:
13.5 + 10.125
Fourth time down only:
10.125
24 + 18 + 18 + 13.5 + 13.5 + 10.125 + 10.125 = 107.25
The ball will travel 107.25 feet.

29. 920,000 + 880,000 + 840,000 + ... + 560,000
$a = 920,000$ and $d = -40,000$
$a_n = 920,000 - 40,000(n-1)$
$a_1 = 920,000$
$a_{10} = 920,000 - 40,000(10-1)$
$= 560,000$
The year 2000 corresponds to the 10th term.
$S_{10} = \frac{10}{2}(920,000 + 560,000)$
$= 7,400,000$
The total revenue should be $7,400,000.

31. 0.10 + 0.15 + 0.20 + ... + 2.55
$a = 0.10$ and $d = 0.05$
$a_n = 0.10 + (n-1)(0.05)$
$= 0.05 + 0.05n$
$a_1 = 0.10$
$a_{50} = 0.05 + 0.05(50) = 2.55$
$S_{50} = \frac{50}{2}(0.10 + 2.55) = 66.25$
It will take 66.25 seconds.

33. 920,000 + 846,400 + 778,688 + ... + 434,388.45
$a = 920,000$ and $r = 0.92$
$a_n = 920,000(0.92)^{n-1}$
$a_1 = 920,000$
$a_{10} = 920,000(0.92)^{10-1}$
$= 434,388.45$
The year 2000 corresponds to the 10th term.
$a_{11} = 920,000(0.92)^{11-1}$
$= 399,637.38$
$S_{10} = \frac{399,637.38 - 920,000}{0.92 - 1}$
$= 6,504,532.78$
The total revenue is $6,504,532.78.

35. 0.1 + 0.12 + 0.144 + ... + 758.36985
$a = 0.1$ and $r = 1.2$
$a_n = 0.1(1.2)^{n-1}$
$a_1 = 0.1$
$a_{51} = 0.1(1.2)^{51-1} = 910.0438$
$S_{50} = \frac{910.0438 - 0.1}{1.2 - 1} = 4549.7$
The performance time is 4549.7 seconds.

37. Since the first $500 was deposited on the day David was born, $n = 2$ on David's 1st birthday, $n = 3$ on David's 2nd birthday, and so on to David's 18th birthday when $n = 19$.
$500 + 525 + 551.25 + \ldots + 1203.31$
$a = 500$ and $r = 1.05$

$a_n = 500(1.05)^{n-1}$

$a_{19} = 500(1.05)^{19-1} = 1203.31$

$a_{20} = 500(1.05)^{20-1} = 1263.475$

$S_{19} = \dfrac{1263.475 - 500}{1.05 - 1} = 15{,}269.50$

There will be $15,269.50 in the account.

39. $0.01 + 0.02 + 0.04 + \ldots + 5{,}368{,}709.12$
$a = 0.01$ and $r = 2$

$a_n = 0.01(2)^{n-1}$

$a_{31} = 0.01(2)^{31-1} = 10{,}737{,}418.24$

$S_{30} = \dfrac{10{,}737{,}418.24 - 0.01}{2 - 1}$

$= 10{,}737{,}418.23$

The total income is $10,737,418.23.

41. a. There are N terms in the series.

b. $S = 1 + 2 + 3 + \ldots + (N-2) + (N-1) + N$
$S = N + (N-1) + (N-2) + \ldots + 3 + 2 + 1$
$2S = (N+1) + (N+1) + (N+1) + \ldots + (N+1) + (N+1)$
Each pair sums to $N + 1$

c. $2S = N(N+1)$
$S = \dfrac{N(N+1)}{2}$

d. $S = \dfrac{N(N+1)}{2}$

43. a. The common ratio is r.

b. $Ar = r + r^2 + r^3 + \ldots + r^N + r^{N+1}$

c. $A = 1 + r + r^2 + \ldots + r^{N-1} + r^N$
$\underline{-Ar = r + r^2 + r^3 + \ldots + r^N + r^{N+1}}$
$A - Ar = 1 - r^{N+1}$

Chapter 9, Section 9.4

    d. $A - Ar = 1 - r^{N+1}$

         $A(1-r) = 1 - r^{N+1}$

         $A = \dfrac{1 - r^{N+1}}{1-r}$

    e. $A = \dfrac{1 - r^{N+1}}{1-r}$

45. a.   Fourth term $= F + (4-1)d$

                 $= F + 3d$

      Ninth term $= F + (9-1)d$

                 $= F + 8d$

  b. $L = N$th term $= F + (N-1)d$

  c. $S = \dfrac{N}{2}[F + F + (N-1)d]$

       $S = \dfrac{N}{2}[2F + (N-1)d]$

Section 9.4

1. $\sum\limits_{i=1}^{4} i^2 = 1^2 + 2^2 + 3^2 + 4^2$

3. $\sum\limits_{j=5}^{7} (j-2) = (5-2) + (6-2) + (7-2) = 3 + 4 + 5$

5. $\sum\limits_{k=1}^{4} k(k+1) = 1(1+1) + 2(2+1) + 3(3+1) + 4(4+1) = 1(2) + 2(3) + 3(4) + 4(5)$

7. $\sum\limits_{m=1}^{4} \dfrac{(-1)^m}{2^m} = \dfrac{(-1)^1}{2^1} + \dfrac{(-1)^2}{2^2} + \dfrac{(-1)^3}{2^3} + \dfrac{(-1)^4}{2^4} = -\dfrac{1}{2} + \dfrac{1}{2^2} - \dfrac{1}{2^3} + \dfrac{1}{2^4}$

9. $\sum\limits_{k=1}^{4} (2k-1) = 1 + 3 + 5 + 7$

Chapter 9, Section 9.4

11. $\sum_{k=1}^{4} 5^{2k-1} = 5 + 5^3 + 5^5 + 5^7$

13. $\sum_{k=1}^{5} k^2 = 1 + 4 + 9 + 16 + 25$

15. $\sum_{k=1}^{5} \frac{k}{k+1} = \frac{1}{2} + \frac{2}{3} + \frac{3}{4} + \frac{4}{5} + \frac{5}{6}$

17. $\sum_{k=1}^{6} \frac{k}{2k-1} = \frac{1}{1} + \frac{2}{3} + \frac{3}{5} + \frac{4}{7} + \frac{5}{9} + \frac{6}{11}$

19. $\sum_{k=1}^{\infty} \frac{2^{k-1}}{k} = \frac{1}{1} + \frac{2}{2} + \frac{4}{3} + \frac{8}{4} + \ldots$

21. Neither.
$\sum_{i=1}^{6} (i^2 + 1) = (1^2+1) + (2^2+1) + (3^2+1) + (4^2+1) + (5^2+1) + (6^2+1)$
$= 2 + 5 + 10 + 17 + 26 + 37 = 97$

The sum is 97.

23. Neither
$\sum_{j=1}^{4} \frac{1}{j} = \frac{1}{1} + \frac{1}{2} + \frac{1}{3} + \frac{1}{4} = \frac{25}{12}$
The sum is $\frac{25}{12}$.

25. Neither
$\sum_{k=1}^{100} 1 = 1 + 1 + 1 + \ldots + 1 = 100(1) = 100$
The sum is 100.

27. Geometric
$\sum_{q=1}^{20} 3^q = 3^1 + 3^2 + 3^3 + \ldots + 3^{20}$
$a = 3$ and $r = 3$
$a^n = 3(3)^{n-1} = 3^n$
$a_1 = 3$
$a_{21} = 3^{21} = 1.046 \times 10^{10}$
$\sum_{q=1}^{20} 3^q = \frac{1.46 \times 10^{10} - 3}{3 - 1}$
$= 5.2302 \times 10^9$
The sum is 5.2302×10^9.

Chapter 9, Section 9.4

29. Arithmetic

$$\sum_{k=1}^{200} k = 1 + 2 + 3 + 4 + \ldots + 200$$

$a = 1$ and $d = 1$
$a_n = 1 + (n-1)(1) = n$
$a_1 = 1$ and $a_{200} = 200$

$$\sum_{k=1}^{200} k = \frac{200}{2}(1 + 200) = 20{,}100$$

The sum is 20,100.

31. Neither

$$\sum_{n=1}^{6} n^3 = 1^3 + 2^3 + 3^3 + 4^3 + 5^3 + 6^3$$
$$= 441$$

The sum is 441.

33. Arithmetic

$$\sum_{n=0}^{30}(3n - 1) = -1 + 2 + 5 + 8 + \ldots + 89$$

$a = -1$ and $d = 3$
$a_n = -1 + (n-1)3 = 3n - 4$
$a_1 = -1$
$a_{31} = 3(31) - 4 = 89$

$$\sum_{n=0}^{30}(3n - 1) = \frac{31}{2}(-1 + 89) = 1364$$

The sum is 1364.

35. Arithmetic

$$\sum_{k=0}^{25}(5 - 2k) = 5 + 3 + 1 + \ldots + (-45)$$

$a = 5$ and $d = -2$
$a_n = 5 - 2(n-1) = 7 - 2n$
$a_1 = 5$ and $a_{26} = -45$

$$\sum_{k=0}^{25}(5 - 2k) = \frac{26}{2}(5 - 45) = -520$$

The sum is −520.

37. Geometric

$$\sum_{j=0}^{10} 2 \cdot 5^j = 2 + 10 + 50 + \ldots$$
$$+ 19{,}531{,}250$$

$a = 2$ and $r = 5$
$a_n = 2(5)^{n-1}$
$a_1 = 2$
$a_{12} = 2(5)^{12-1} = 97{,}656{,}250$

$$\sum_{j=0}^{10} 2 \cdot 5^j = \frac{97{,}656{,}250 - 2}{5 - 1}$$
$$= 24{,}414{,}062$$

The sum is 24,414,062.

39. Geometric

$$\sum_{m=0}^{12} 50(1.08)^m = 50 + 54 + 58.32 + \ldots + 125.90851$$

$a = 50$ and $r = 1.08$
$a_n = 50(1.08)^{n-1}$
$a_1 = 50$
$a_{14} = 50(1.08)^{14-1} = 135.9812$

$$\sum_{m=0}^{12} 50(1.08)^m = \frac{135.9812 - 50}{1.08 - 1} = 1074.765$$

The sum is 1074.765.

41. $a = \dfrac{1}{2}$ and $r = \dfrac{1}{2}$

$S_\infty = \dfrac{\frac{1}{2}}{1 - \frac{1}{2}} = 1$

The sum is 1.

43. $a = 12$ and $r = 0.15$

$S_\infty = \dfrac{12}{1 - 0.15} = 14.12$

The sum is 14.12.

45. $a = 4$ and $r = \dfrac{-3}{5}$

$S_\infty = \dfrac{4}{1 - \left(\frac{-3}{5}\right)} = 2.5$

The sum is 2.5.

47. $a = \dfrac{3}{16}$ and $r = \dfrac{1}{2}$

$S_\infty = \dfrac{\frac{3}{16}}{1 - \frac{1}{2}} = 0.375$

The sum is 0.375.

49. $0.\overline{4} = 0.4 + 0.04 + 0.004 + \ldots$

$a = \dfrac{4}{10}$ and $r = \dfrac{1}{10}$

$S_\infty = \dfrac{\frac{4}{10}}{1 - \frac{1}{10}} = \dfrac{4}{9}$

The decimal is equal to $\dfrac{4}{9}$.

51. $0.\overline{31} = 0.31 + 0.0031$
$\qquad\qquad + 0.000031 + \ldots$

$a = \dfrac{31}{100}$ and $r = \dfrac{1}{100}$

$S_\infty = \dfrac{\frac{31}{100}}{1 - \frac{1}{100}} = \dfrac{31}{99}$

The decimal is equal to $\dfrac{31}{99}$.

53. $2.\overline{410} = 2 + 0.410 + 0.000410$
$\qquad\qquad + 0.000000410 + \ldots$

Calculate the fraction for the repeating decimal.

$a = \dfrac{410}{1000}$ and $r = \dfrac{1}{1000}$

$S_\infty = \dfrac{\frac{410}{1000}}{1 - \frac{1}{1000}} = \dfrac{410}{999}$

$2.\overline{410} = 2 + \dfrac{410}{999} = 2\dfrac{410}{999}$

The decimal is equal to $2\dfrac{410}{999}$.

55. $0.12\overline{8} = 0.12 + 0.008 + 0.0008 + \ldots$

Calculate the fraction for the repeating decimal.

$a = \dfrac{8}{1000}$ and $r = \dfrac{1}{10}$

$S_\infty = \dfrac{\frac{8}{1000}}{1 - \frac{1}{10}} = \dfrac{8}{900} = \dfrac{2}{225}$

$0.12\overline{8} = \dfrac{12}{100} + \dfrac{2}{225} = \dfrac{27}{225} + \dfrac{2}{225}$

$= \dfrac{29}{225}$

The fraction is equal to $\dfrac{29}{225}$.

Chapter 9, Section 9.5

57. $12 + 10.8 + 9.72 + \ldots$
$a = 12$ and $r = 0.9$
$S_\infty = \dfrac{12}{1 - 0.9} = 120$
The bob will move approximately 120 inches.

59. Falls: $6 + 4 + \dfrac{8}{3} + \ldots$
$a = 6$ and $r = \dfrac{2}{3}$
$S_\infty = \dfrac{6}{1 - \frac{2}{3}} = 18$
Moves upwards: $4 + \dfrac{8}{3} + \dfrac{16}{9} + \ldots$
$a = 4$ and $r = \dfrac{2}{3}$
$S_\infty = \dfrac{4}{1 - \frac{2}{3}} = 12$
The ball travels a total distance of approximately $18 + 12 = 30$ feet.

Section 9.5

1. terms $= n + 1 = 50 + 1 = 51$
terms $= n + 1 = 100 + 1 = 101$

3. Sum of exponents $= n$; 100, 50

5. 11 rows:

$n=0$						1					
$n=1$					1		1				
$n=2$					1	2	1				
$n=3$				1	3		3	1			
$n=4$			1	4		6		4	1		
$n=5$		1	5		10		10		5	1	
$n=6$	1	6		15		20		15		6	1
$n=7$	1	7	21	35	35	21	7	1			
$n=8$	1	8	28	56	70	56	28	8	1		
$n=9$	1	9	36	84	126	126	84	36	9	1	
$n=10$	1	10	45	120	210	252	210	120	45	10	1

Chapter 9, Section 9.5

7. $(x+3)^5 = x^5 + 5x^4 3^1 + 10x^3 3^2 + 10x^2 3^3 + 5x^1 3^4 + 3^5$
$= x^5 + 15x^4 + 90x^3 + 270x^2 + 405x + 243$

9. $(x-3)^4 = x^4 + 4x^3(-3)^1 + 6x^2(-3)^2 + 4x^1(-3)^3 + (-3)^4$
$= x^4 - 12x^3 + 54x^2 - 108x + 81$

11. $\left(2x - \dfrac{y}{2}\right)^3 = (2x)^3 + 3(2x)^2\left(-\dfrac{y}{2}\right)^1 + 3(2x)^1\left(-\dfrac{y}{2}\right)^2 + \left(-\dfrac{y}{2}\right)^3$
$= 8x^3 - 6x^2 y + \dfrac{3}{2}xy^2 - \dfrac{1}{8}y^3$

13. $(x^2 - 3)^7 = (x^2)^7 + 7(x^2)^6(-3)^1 + 21(x^2)^5(-3)^2 + 35(x^2)^4(-3)^3 + 35(x^2)^3(-3)^4$
$\qquad + 21(x^2)^2(-3)^5 + 7(x^2)^1(-3)^6 + (-3)^7$
$= x^{14} - 21x^{12} + 189x^{10} - 945x^8 + 2835x^6 - 5103x^4 + 5103x^2 - 2187$

15. $\left(\dfrac{2}{3} - a^2\right)^4 = \left(\dfrac{2}{3}\right)^4 + 4\left(\dfrac{2}{3}\right)^3(-a^2)^1 + 6\left(\dfrac{2}{3}\right)^2(-a^2)^2 + 4\left(\dfrac{2}{3}\right)^1(-a^2)^3 + (-a^2)^4$
$= \dfrac{16}{81} - \dfrac{32}{27}a^2 + \dfrac{8}{3}a^4 - \dfrac{8}{3}a^6 + a^8$

17. $(x+y)^6 = x^6 + 6x^5 y + 15x^4 y^2 + 20x^3 y^3 + 15x^2 y^4 + 6xy^5 + y^6$

19. $(2x-y)^7 = (2x)^7 + 7(2x)^6(-y)^1 + 21(2x)^5(-y)^2 + 35(2x)^4(-y)^3 + 35(2x)^3(-y)^4$
$\qquad + 21(2x)^2(-y)^5 + 7(2x)^1(-y)^6 + (-y)^7$
$= 128x^7 - 448x^6 y + 672x^5 y^2 - 560x^4 y^3 + 280x^3 y^4 - 84x^2 y^5 + 14xy^6$
$\qquad - y^7$

21. $(a-b)^8 = a^8 - 8a^7 b + 28a^6 b^2 - 56a^5 b^3 + 70a^4 b^4 - 56a^3 b^5 + 28a^2 b^6 - 8ab^7 + b^8$

23. $5! = 5 \cdot 4 \cdot 3 \cdot 2 \cdot 1 = 120$

25. $\dfrac{9!}{7!} = \dfrac{9 \cdot 8 \cdot 7 \cdot 6 \cdot 5 \cdot 4 \cdot 3 \cdot 2 \cdot 1}{7 \cdot 6 \cdot 5 \cdot 4 \cdot 3 \cdot 2 \cdot 1} = 72$

27. $\dfrac{5!7!}{12!} = \dfrac{(5 \cdot 4 \cdot 3 \cdot 2 \cdot 1)(7 \cdot 6 \cdot 5 \cdot 4 \cdot 3 \cdot 2 \cdot 1)}{12 \cdot 11 \cdot 10 \cdot 9 \cdot 8 \cdot 7 \cdot 6 \cdot 5 \cdot 4 \cdot 3 \cdot 2 \cdot 1} = \dfrac{1}{792}$

Chapter 9, Section 9.5

29. $\dfrac{8!}{2!(8-2)!} = \dfrac{8!}{2!6!} = \dfrac{8 \cdot 7 \cdot 6 \cdot 5 \cdot 4 \cdot 3 \cdot 2 \cdot 1}{(2 \cdot 1)(6 \cdot 5 \cdot 4 \cdot 3 \cdot 2 \cdot 1)} = 28$

31. $_9C_6 = \dfrac{9!}{(9-6)!6!} = \dfrac{9!}{3!6!} = \dfrac{9 \cdot 8 \cdot 7 \cdot 6!}{3 \cdot 2 \cdot 1 \cdot 6!} = 84$

33. $_{12}C_3 = \dfrac{12!}{(12-3)!3!} = \dfrac{12!}{9!3!} = \dfrac{12 \cdot 11 \cdot 10 \cdot 9!}{3 \cdot 2 \cdot 1 \cdot 9!} = 220$

35. $_{20}C_{18} = \dfrac{20!}{(20-18)!18!} = \dfrac{20!}{2!18!} = \dfrac{20 \cdot 19 \cdot 18!}{2 \cdot 1 \cdot 18!} = 190$

37. $_{14}C_9 = \dfrac{14!}{(14-9)!9!} = \dfrac{14!}{5!9!} = \dfrac{14 \cdot 13 \cdot 12 \cdot 11 \cdot 10 \cdot 9!}{5 \cdot 4 \cdot 3 \cdot 2 \cdot 1 \cdot 9!} = 2002$

39. $x^{13}y^7$ has $n = 20$ and $k = 7$. Thus the coefficient of $x^{13}y^7$ in the expansion of $(x+y)^{20}$ is $_{20}C_7 = 77{,}520$.

41. $n = 12$ and $k = 7$
$_{12}C_7\ a^{12-7}(-2b)^7$
$= 792a^5(-128b^7)$
$= -101{,}376 a^5 b^7$
The coefficient is $-101{,}376$.

43. $n = 10$ and $k = 6$
$_{10}C_6 x^{10-6}(-\sqrt{2})^6$
$= 210 x^4 (8)$
$= 1680 x^4$
The coefficient is 1680.

45. $n = 15$ and $k = 4$
$_{15}C_4\ a^{15-4}(-b)^4$
$= 1365 a^{11} b^4$
The coefficient is 1365.

47. $n = 9$ and $k = 3$
$_9C_3(a^3)^{9-3}(-b^3)^3$
$= 84 a^{18}(-b^9)$
$= -84 a^{18} b^9$
The coefficient is -84.

49. $n = 8$ and $k = 3$
$_8C_3\ x^{8-3}\left(-\dfrac{1}{2}\right)^3$
$= 56 x^5 \left(-\dfrac{1}{8}\right)$
$= -7x^5$
The coefficient is -7.

51. $11^0 = 1$
$11^1 = 11$
$11^2 = 121$
$11^3 = 1331$
$11^4 = 14{,}641$

The digits in the terms of the sequence correspond to the numbers in the first five rows of Pascal's triangle. If $11^n = (10 + 1)^n$ is expanded as a binomial, each term is the product of a number from Pascal's triangle times a power of 10 times a power of 1.

Chapter 9 Review

1. $a_n = \dfrac{n}{n^2 + 1}$

   $a_1 = \dfrac{1}{1^2 + 1} = \dfrac{1}{2}$

   $a_2 = \dfrac{2}{2^2 + 1} = \dfrac{2}{5}$

   $a_3 = \dfrac{3}{3^2 + 1} = \dfrac{3}{10}$

   $a_4 = \dfrac{4}{4^2 + 1} = \dfrac{4}{17}$

   The four terms are $\dfrac{1}{2}, \dfrac{2}{5}, \dfrac{3}{10}, \dfrac{4}{17}$.

2. $b_n = \dfrac{(-1)^{n-1}}{n}$

   $b_1 = \dfrac{(-1)^{1-1}}{1} = 1$

   $b_2 = \dfrac{(-1)^{2-1}}{2} = -\dfrac{1}{2}$

   $b_3 = \dfrac{(-1)^{3-1}}{3} = \dfrac{1}{3}$

   $b_4 = \dfrac{(-1)^{4-1}}{4} = -\dfrac{1}{4}$

   The first four terms are $1, -\dfrac{1}{2}, \dfrac{1}{3}, -\dfrac{1}{4}$.

3. $c_{n+1} = c_n - 3$
   $c_1 = 5$
   $c_2 = c_1 - 3 = 5 - 3 = 2$
   $c_3 = c_2 - 3 = 2 - 3 = -1$
   $c_4 = c_3 - 3 = -1 - 3 = -4$
   $c_5 = c_4 - 3 = -4 - 3 = -7$
   The first five terms are
   $5, 2, -1, -4, -7$.

4. $d_{n+1} = -\dfrac{3}{4} d_n$
   $d_1 = 1$
   $d_2 = -\dfrac{3}{4}(1) = -\dfrac{3}{4}$
   $d_3 = -\dfrac{3}{4}\left(-\dfrac{3}{4}\right) = \dfrac{9}{16}$
   $d_4 = -\dfrac{3}{4}\left(\dfrac{9}{16}\right) = -\dfrac{27}{64}$
   $d_5 = -\dfrac{3}{4}\left(-\dfrac{27}{64}\right) = \dfrac{81}{256}$
   The first five terms are
   $1, -\dfrac{3}{4}, \dfrac{9}{16}, -\dfrac{27}{64}, \dfrac{81}{256}$.

5. a. After one year it is worth
   $1800 - 1800(0.12) = \$1584$.
   After two years it is worth
   $\$1584 - \$1584(0.12) = \$1393.92$.
   After three years it is worth
   $\$1393.92 - \$1393.92(0.12)$
   $= \$1226.6496$.
   After four years it is worth
   $\$1226.6496 - \$1226.6496(0.12)$
   $= \$1079.4516$
   The first four terms are 1584, 1393.92, 1226.6496, 1079.4516.

   b. Each term of the sequence can be found by multiplying the previous term by 0.88.
   $a_1 = 1584$; $a_{n+1} = 0.88 a_n$

Chapter 9 Review

6. a. The first year she will earn $24,000.
The second year she will earn
$24,000 + $24,000(0.06)
= $25,440.
The third year she will earn
$25,440 + $25,440(0.06)
= 26,966.40.
The fourth year she will earn
$26966.40(1.06) = $28584.384.
The first four terms are 24,000, 25,440, 26,966.40, 28,584.384.

b. Each term of the sequence can be found by multiplying the previous term by 1.06.
$a_1 = 24,000; a_{n+1} = 1.06a_n$

7. a. Just after the first dose there are 30 ml present. Just after the second dose there are 30(0.75) + 15 = 37.5 ml present. Just after the third dose there are 37.5(0.75) + 15 = 43.125 ml present. Just after the fourth dose there are 43.125(0.75) + 15 = 47.34375 ml present.
The first four terms are 30, 37.5, 43.125, 47.34375.

b. Each term can be found by multiplying the previous term by 0.75 and adding 15.
$a_1 = 30; a_{n+1} = 0.75a_n + 15$

8. a. After the first meeting she weighs 187 pounds. After the second meeting she weighs 187 − 2 = 185 pounds. After the third meeting she weighs 185 − 2 = 183 pounds. After the fourth meeting she weights 183 − 2 = 181 pounds.
The first four terms are 187, 185, 183, 181.

b. Each term in the sequence can be found by subtracting 2 from the previous term.
$a_1 = 187; a_{n+1} = a_n - 2$

9. $x_7 = (-1)^7(7-2)^2 = (-1)(5)^2 = -25$

10. $y_3 = \sqrt{3^3 - 2} = \sqrt{25} = 5$

11. $a = -4$ and $d = 4$
$a_{10} = -4 + (10-1)4$
$a_{10} = 32$
The tenth term is 32.

12. $s_1 = x - a$ and $d = 2a$
$s_6 = x - a + (6-1)2a$
$s_6 = x + 9a$
The sixth term is $x + 9a$.

13. $a = \dfrac{16}{27}$ and $r = -\dfrac{3}{2}$

$a_8 = \dfrac{16}{27}\left(-\dfrac{3}{2}\right)^{8-1}$

$a_8 = -\dfrac{81}{8}$

The eighth term is $-\dfrac{81}{8}$.

14. Third term: $-\dfrac{2}{3} = ar^2$

sixth term: $\dfrac{16}{81} = ar^5$

Use two equations and two unknowns to solve for a and r.

$a = -\dfrac{3}{2}$ and $r = -\dfrac{2}{3}$

$a_5 = -\dfrac{3}{2}\left(-\dfrac{2}{3}\right)^{5-1}$

$a_5 = -\dfrac{8}{27}$

The fifth term is $-\dfrac{8}{27}$.

Chapter 9 Review

15. $a = -84$ and $d = 10$
$a_{23} = -84 + (23 - 1)10$
$a_{23} = 136$
The twenty-third term is 136.

16. $a = -\dfrac{1}{2}$ and $d = \dfrac{3}{2}$
$a_9 = -\dfrac{1}{2} + (9 - 1)\dfrac{3}{2}$
$a_9 = \dfrac{23}{2}$
The ninth term is $\dfrac{23}{2}$.

17. $a_1 = 8$
$a_{28} = 89 = 8 + (28 - 1)d$
$d = 3$
$a_{21} = 8 + (21 - 1)3$
$a_{21} = 68$
The twenty-first term is 68.

18. $a = 5$ and $d = -3$
$a_n = -37$, solve for n
$-37 = 5 + (n - 1)(-3)$
$n = 15$
It is the fifteenth term.

19. Geometric: Each term is found from the previous term by multiplying by $-\dfrac{1}{2}$.
$a_5 = \dfrac{1}{8}\left(-\dfrac{1}{2}\right) = -\dfrac{1}{16}$
$a_6 = -\dfrac{1}{16}\left(-\dfrac{1}{2}\right) = \dfrac{1}{32}$
$a_7 = \dfrac{1}{32}\left(-\dfrac{1}{2}\right) = -\dfrac{1}{64}$
$a_8 = -\dfrac{1}{64}\left(-\dfrac{1}{2}\right) = \dfrac{1}{128}$
The next four terms are
$-\dfrac{1}{16}, \dfrac{1}{32}, -\dfrac{1}{64}, \dfrac{1}{128}$.
$a = -1$ and $r = -\dfrac{1}{2}$
$a_n = -1\left(-\dfrac{1}{2}\right)^{n-1}$

20. Neither: No constant added or multiplied by the previous terms yields the successive terms.

21. Arithmetic: Each term is found from the pervious term by subtracting 5.
$a_5 = -9 - 5 = -14$
$a_6 = -14 - 5 = -19$
$a_7 = -19 - 5 = -24$
$a_8 = -24 - 5 = -29$
The next four terms are $-14, -19, -24, -29$.
$a = 6$ and $d = -5$
$a_n = 6 + (n - 1)(-5)$
$a_n = 11 - 5n$

22. Geometric: Each term is found from the previous term by multiplying by -4.
$a_5 = (-64)(-4) = 256$
$a_6 = (256)(-4) = -1024$
$a_7 = (-1024)(-4) = 4096$
$a_8 = (4096)(-4) = -16{,}384$
The next four terms are $256, -1024, 4096, -16{,}384$.
$a = 1$ and $r = -4$
$a_n = 1(-4)^{n-1}$
$a_n = (-4)^{n-1}$

23. Geometric: Each term is found from the previous term by multiplying by -2.
$a_5 = (8)(-2) = -16$
$a_6 = (-16)(-2) = 32$
$a_7 = (32)(-2) = -64$
$a_8 = (-64)(-2) = 128$

Chapter 9 Review

The next four terms are −16, 32, −64, 128.
$a = -1$ and $r = -2$
$a_n = (-1)(-2)^{n-1}$

24. Geometric: Each term is found from the previous term by multiplying by $\frac{3}{4}$.

$a_5 = \frac{9}{32}\left(\frac{3}{4}\right) = \frac{27}{128}$

$a_6 = \frac{27}{128}\left(\frac{3}{4}\right) = \frac{81}{512}$

$a_7 = \frac{81}{512}\left(\frac{3}{4}\right) = \frac{243}{2048}$

$a_8 = \frac{243}{2048}\left(\frac{3}{4}\right) = \frac{729}{8192}$

The next four terms are
$\frac{27}{128}, \frac{81}{512}, \frac{243}{2048}, \frac{729}{8192}$.

$a = \frac{2}{3}$ and $r = \frac{3}{4}$

$a_n = \frac{2}{3}\left(\frac{3}{4}\right)^{n-1}$

25. Arithmetic
$a_2 = 3 - 4 = -1$
$a_3 = -1 - 4 = -5$
$a_4 = -5 - 4 = -9$
$a_5 = -9 - 4 = -13$
The next four terms are −1, −5, −9, −13.
$a = 3$ and $d = -4$
$a_n = 3 + (n-1)(-4)$
$a_n = 7 - 4n$

26. Arithmetic
$a_2 = \frac{1}{4} + \frac{1}{2} = \frac{3}{4}$

$a_3 = \frac{3}{4} + \frac{1}{2} = \frac{5}{4}$

$a_4 = \frac{5}{4} + \frac{1}{2} = \frac{7}{4}$

$a_5 = \frac{7}{4} + \frac{1}{2} = \frac{9}{4}$

The next four terms are
$\frac{3}{4}, \frac{5}{4}, \frac{7}{4}, \frac{9}{4}$.

$a = \frac{1}{4}$ and $d = \frac{1}{2}$

$a_n = \frac{1}{4} + (n-1)\left(\frac{1}{2}\right)$

$a_n = \frac{1}{2}n - \frac{1}{4}$

27. Geometric
$a_2 = (12)(-4) = -48$
$a_3 = (-48)(-4) = 192$
$a_4 = (192)(-4) = -768$
$a_5 = (-768)(-4) = 3072$
The next four terms are −48, 192, −768, 3072.
$a = 12$ and $r = -4$
$a_n = 12(-4)^{n-1}$

28. Geometric
$a_2 = 6\left(\frac{1}{3}\right) = 2$

$a_3 = 2\left(\frac{1}{3}\right) = \frac{2}{3}$

$a_4 = \frac{2}{3}\left(\frac{1}{3}\right) = \frac{2}{9}$

$a_5 = \frac{2}{9}\left(\frac{1}{3}\right) = \frac{2}{27}$

Chapter 9 Review

The next four terms are $2, \dfrac{2}{3}, \dfrac{2}{9}, \dfrac{2}{27}$.

$a = 6$ and $r = \dfrac{1}{3}$

$a_n = 6\left(\dfrac{1}{3}\right)^{n-1}$

29. $\displaystyle\sum_{k=2}^{5} k(k-1) = 2(2-1) + 3(3-1) + 4(4-1) + 5(5-1)$
$= 2(1) + 3(2) + 4(3) + 5(4)$

30. $\displaystyle\sum_{j=2}^{\infty} \dfrac{j}{2j-1} = \dfrac{2}{2(2)-1} + \dfrac{3}{2(3)-1} + \dfrac{4}{2(4)-1} + \dfrac{5}{2(5)-1} + \ldots$
$= \dfrac{2}{3} + \dfrac{3}{5} + \dfrac{4}{7} + \dfrac{5}{9} + \ldots$

31. $\displaystyle\sum_{k=1}^{12} (2^k - 1)$

32. $\displaystyle\sum_{k=4}^{15} k^2 x^k$

33. Arithmetic: Each term is obtained from the previous term by adding 3.
$1 + 4 + 7 + 10 + \ldots + 34$
$a = 1$ and $d = 3$
$a_n = 1 + (n-1)3 = 3n - 2$
$a_1 = 1$ and $a_{12} = 34$
$S_{12} = \dfrac{12}{2}(1 + 34) = 210$
The sum is 210.

34. Arithmetic: Each term is obtained from the previous term by adding 0.1.
$1.5 + 1.6 + 1.7 + \ldots + 3.4$
$b_1 = 1.5$ and $d = 0.1$
$b_n = 1.5 + (n-1)0.1 = 1.4 + 0.1n$
$b_1 = 1.5$ and $b_{20} = 3.4$
$S_{20} = \dfrac{20}{2}(1.5 + 3.4) = 49$
The sum is 49.

35. Arithmetic: Each term is obtained from the previous term by adding 3.
$a = 2$ and $d = 3$
$a_n = 2 + (n-1)3 = 3n - 1$
$a_1 = 2$ and $a_6 = 3(6) - 1 = 17$
$S_6 = \dfrac{6}{2}(2 + 17) = 57$
The sum is 57.

Chapter 9 Review

36. Arithmetic: Each term is obtained from the previous term by adding $\frac{2}{3}$.

$a = -\frac{1}{3}$ and $d = \frac{2}{3}$

$a_n = -\frac{1}{3} + (n-1)\left(\frac{2}{3}\right) = \frac{2}{3}n - 1$

$a_1 = -\frac{1}{3}$ and $a_{12} = \frac{2}{3}(12) - 1 = 7$

$S_{12} = \frac{12}{2}\left(-\frac{1}{3} + 7\right) = 40$

The sum is 40.

37. Geometric: Each term is obtained from the previous term by multiplying by $\frac{1}{3}$.

$a = \frac{1}{3}$ and $r = \frac{1}{3}$

$a_n = \frac{1}{3}\left(\frac{1}{3}\right)^{n-1} = \left(\frac{1}{3}\right)^n$

$a_1 = \frac{1}{3}$ and $a_6 = \left(\frac{1}{3}\right)^6 = \frac{1}{729}$

$S_5 = \frac{\frac{1}{729} - \frac{1}{3}}{\frac{1}{3} - 1} = \frac{121}{243}$

The sum is $\frac{121}{243}$.

38. Geometric: Each term is obtained from the previous term by multiplying by 2.
$a = 1$ and $r = 2$
$a_n = 1(2)^{n-1} = 2^{n-1}$
$a_1 = 1$ and $a_7 = 2^{7-1} = 64$
$S_6 = \frac{64 - 1}{2 - 1} = 63$
The sum is 63.

39. Neither

$\sum_{k=1}^{5}(-1)^k(k+1) = (-1)^1(1+1) + (-1)^2(2+1) + (-1)^3(3+1) + (-1)^4(4+1)$
$\qquad + (-1)^5(5+1)$
$= -2 + 3 - 4 + 5 - 6$
$= -4$

The sum is -4.

40. Neither

$$\sum_{k=1}^{4} \frac{k}{k+1} = \frac{1}{1+1} + \frac{2}{2+1} + \frac{3}{3+1} + \frac{4}{4+1}$$

$$= \frac{1}{2} + \frac{2}{3} + \frac{3}{4} + \frac{4}{5}$$

$$= \frac{163}{60}$$

The sum is $\frac{163}{60}$.

41. Geometric: Each term is obtained from the previous term by multiplying by $-\frac{2}{3}$.

$a = -3$ and $r = -\frac{2}{3}$

$$S_\infty = \frac{-3}{1-\left(-\frac{2}{3}\right)} = -\frac{9}{5}$$

The sum is $-\frac{9}{5}$.

42. Geometric: Each term is obtained from the previous term by multiplying by $\frac{1}{3}$.

$a = 3$ and $r = \frac{1}{3}$

$$S_\infty = \frac{3}{1-\frac{1}{3}} = \frac{9}{2}$$

The sum is $\frac{9}{2}$.

43. Drops 1st time 12 ft, rises 1st time

$12\left(\frac{2}{3}\right) = 8$ ft

Drops 2nd time 8 ft, rises 2nd time

$8\left(\frac{2}{3}\right) = \frac{16}{3}$ ft

Drops 3rd time $\frac{16}{3}$ ft, rises 3rd time

$\frac{16}{3}\left(\frac{2}{3}\right) = \frac{32}{9}$ ft

Drops 4th time $\frac{32}{9}$ ft, rises 4th time

$\frac{32}{9}\left(\frac{2}{3}\right) = \frac{64}{27}$ ft

The ball will bounce approximately 2.37 ft.

44. $a = 840$ and $r = 1.02$
$a_n = 840(1.02)^{n-1}$
The year 2000 corresponds to $n = 7$.
$a_7 = 840(1.02)^{7-1} = 945.98$
The taxes will be approximately $945.98.

45. a. $a = 12$ and $d = 6$
$a_n = 12 + (n-1)6 = 6 + 6n$
$a_1 = 12$ and $a_{15} = 6 + 6(15) = 96$
$$S_{15} = \frac{15}{2}(12+96) = 810$$
The sum is 810.

b. $\sum_{n=1}^{15}(6+6n)$

Chapter 9 Review

46. $a = 7$ and $d = 1.3$
$a_n = 7 + (n-1)1.3 = 5.7 + 1.3n$
Solve for n in the equation
$20 = 5.7 + 1.3n$
$n = 11$
The year corresponding to $n = 11$ is 2004.

47. Down: $12 + 9 + 6.75 + ...$
$a = 12$ and $r = \dfrac{3}{4}$
$S_\infty = \dfrac{12}{1-\frac{3}{4}} = 48$
Up: $9 + 6.75 + 5.0625 + ...$
$a = 9$ and $r = \dfrac{3}{4}$
$S_\infty = \dfrac{9}{1-\frac{3}{4}} = 36$
The total distance traveled is approximately $48 + 36 = 84$ ft.

48. $a = 0.01$ and $r = 2$
$a_n = 0.01(2)^{n-1}$
$a_1 = 0.01$ and $a_{31} = 0.01(2)^{31-1}$
$\phantom{a_1 = 0.01 \text{ and } a_{31}} = 10,737,418.24$
There are 30 days in June.
$S_{30} = \dfrac{10,737,418.24 - 0.01}{2-1}$
$\phantom{S_{30}} = 10,737,418.23$
The total amount will be $10,737,418.23.

49. $3.\overline{2} = 3 + 0.2 + 0.02 + 0.002 + ...$
First find the fraction for just the repeating decimal.
$a = \dfrac{2}{10}$ and $r = \dfrac{1}{10}$
$S_\infty = \dfrac{\frac{2}{10}}{1-\frac{1}{10}} = \dfrac{2}{9}$
The fraction is $3 + \dfrac{2}{9} = 3\dfrac{2}{9}$ or $\dfrac{29}{9}$.

50. $0.4\overline{18} = 0.4 + 0.018 + 0.00018 + ...$
First find the fraction for the repeating decimal.
$a = \dfrac{18}{1000}$ and $r = \dfrac{1}{100}$
$S_\infty = \dfrac{\frac{18}{1000}}{1-\frac{1}{100}} = \dfrac{18}{990} = \dfrac{1}{55}$
$0.4\overline{18} = \dfrac{4}{10} + \dfrac{1}{55} = \dfrac{22}{55} + \dfrac{1}{55} = \dfrac{23}{55}$
The fraction is $\dfrac{23}{55}$.

51. $(x-2)^5 = x^5 + 5x^4(-2)^1 + 10x^3(-2)^2 + 10x^2(-2)^3 + 5x^1(-2)^4 + (-2)^5$
$ = x^5 - 10x^4 + 40x^3 - 80x^2 + 80x - 32$

Chapter 9 Review

52. $\left(\dfrac{x}{2}-y\right)^4 = \left(\dfrac{x}{2}\right)^4 + 4\left(\dfrac{x}{2}\right)^3(-y)^1 + 6\left(\dfrac{x}{2}\right)^2(-y)^2 + 4\left(\dfrac{x}{2}\right)^1(-y)^3 + (-y)^4$

$= \dfrac{x^4}{16} - \dfrac{1}{2}x^3y + \dfrac{3}{2}x^2y^2 - 2xy^3 + y^4$

53. $\dfrac{6!}{3!(6-3)!} = \dfrac{6!}{3!3!} = \dfrac{6 \cdot 5 \cdot 4 \cdot 3!}{3 \cdot 2 \cdot 1 \cdot 3!} = 20$

54. $\dfrac{9!}{5!(9-5)!} = \dfrac{9!}{5!4!} = \dfrac{9 \cdot 8 \cdot 7 \cdot 6 \cdot 5!}{4 \cdot 3 \cdot 2 \cdot 1 \cdot 5!} = 126$

55. $_7C_2 = \dfrac{7!}{(7-2)!2!} = \dfrac{7!}{5!2!} = \dfrac{7 \cdot 6 \cdot 5!}{2 \cdot 1 \cdot 5!} = 21$

56. $_{16}C_{14} = \dfrac{16!}{(16-14)!14!} = \dfrac{16!}{2!14!} = \dfrac{16 \cdot 15 \cdot 14!}{2 \cdot 1 \cdot 14!} = 120$

57. $\displaystyle\sum_{k=0}^{5} {}_5C_k = {}_5C_0 + {}_5C_1 + {}_5C_2 + {}_5C_3 + {}_5C_4 + {}_5C_5$

$= \dfrac{5!}{5!0!} + \dfrac{5!}{4!1!} + \dfrac{5!}{3!2!} + \dfrac{5!}{2!3!} + \dfrac{5!}{1!4!} + \dfrac{5!}{0!5!}$
$= 1 + 5 + 10 + 10 + 5 + 1$
$= 32$

58. $\displaystyle\sum_{k=0}^{6} {}_6C_k(1.4)^{6-k}(0.6)^k = (1.4+0.6)^6 = 2^6 = 64$

59. $n = 9$ and $k = 3$

$_9C_3 x^{9-3}(-2y)^3 = 84x^6(-8y^3)$
$= -672x^6y^3$

The coefficient is -672

60. $n = 8$ and $k = 7$

$_8C_7\left(\dfrac{x}{2}\right)^{8-7}(-3)^7 = 8\left(\dfrac{x}{2}\right)(-2187)$
$= -8748x$

The coefficient is -8748.

Chapter 10

Section 10.1

1. The graph of $f(x) = |x| - 4$ is a translation of the basic graph of $y = |x|$ shifted downward four units. Refer to the graph in the back of the textbook.

3. The graph of $g(s) = \sqrt[3]{s-4}$ is a translation of the basic graph of $y = \sqrt[3]{s}$ shifted four units to the right. Refer to the graph in the back of the textbook.

5. The graph of $F(t) = \dfrac{1}{t^2} + 1$ is a translation of the basic graph of $y = \dfrac{1}{t^2}$ shifted upward one unit. Refer to the graph in the back of the textbook.

7. The graph of $G(r) = (r+2)^3$ is a translation of the basic graph of $y = r^3$ shifted two units to the left. Refer to the graph in the back of the textbook.

9. The graph of $H(d) = \sqrt{d} - 3$ is a translation of the basic graph of $y = \sqrt{d}$ shifted downward three units. Refer to the graph in the back of the textbook.

11. The graph of $h(v) = \dfrac{1}{v+6}$ is a translation of the basic graph of $y = \dfrac{1}{v}$ shifted six units to the left. Refer to the graph in the back of the textbook.

13. The graph of $f(x) = 2 + (x-3)^2$ is a translation of the basic graph of $y = x^2$ shifted three units to the right and upward two units. Refer to the graph in the back of the textbook.

15. The graph of $g(z) = \dfrac{1}{z+2} - 3$ is a translation of the basic graph $y = \dfrac{1}{z}$ shifted two units to the left and downward three units. Refer to the graph in the back of the textbook.

17. The graph of $F(u) = \sqrt{u+4} + 4$ is a translation of the basic graph $y = \sqrt{u}$ shifted four units to the left and upward four units. Refer to the graph in the back of the textbook.

19. The graph of $G(t) = |t-5| - 1$ is a translation of the basic graph of $y = |t|$ shifted five units to the right and downward one unit. Refer to the graph in the back of the textbook.

21. The graph of $h(p) = (p+2)^3 - 5$ is a translation of the basic graph of $y = p^3$ shifted two units to the left and downward five units. Refer to the graph in the back of the textbook.

Chapter 10, Section 10.1

23. The graph of $H(w) = \dfrac{1}{(w-1)^2} + 6$ is a translation of the basic graph of $y = \dfrac{1}{w^2}$ shifted one unit to the right and upward six units. Refer to the graph in the back of the textbook.

25. The graph of $f(t) = \sqrt[3]{t-8} - 1$ is a translation of the basic graph $y = \sqrt[3]{t}$ shifted eight units to the right and downward one unit. Refer to the graph in the back of the textbook.

27. The graph of $F(t) = 4t^2$ is an expansion of the basic graph $y = t^2$ by a factor of four. Refer to the graph in the back of the textbook.

29. The graph of $f(x) = \dfrac{1}{3}|x|$ is a compression of the basic graph $y = |x|$ by a factor of $\dfrac{1}{3}$. Refer to the graph in the back of the textbook.

31. The graph of $h(z) = \dfrac{2}{z^2}$ is an expansion of the basic graph $y = \dfrac{1}{z^2}$ by a factor of two. Refer to the graph in the back of the textbook.

33. The graph of $G(v) = -2\sqrt{v}$ is an expansion of the basic graph $y = \sqrt{v}$ by a factor of two, combined with a reflection about the x-axis. Refer to the graph in the back of the textbook.

35. The graph of $g(s) = -\dfrac{1}{2}s^3$ is a compression of the basic graph $y = s^3$ by a factor of $\dfrac{1}{2}$, combined with a reflection about the x-axis. Refer to the graph in the back of the textbook.

37. The graph of $H(x) = \dfrac{1}{3x}$ is a compression of the basic graph $y = \dfrac{1}{x}$ by a factor of $\dfrac{1}{3}$. Refer to the graph in the back of the textbook.

39. The graph is a translation of the basic graph $y = |x|$, shifted one unit to the left and downward two units. Refer to the graph in the back of the textbook.
$y = |x+1| - 2$

41. The graph is a translation of the basic graph $y = \sqrt{x}$ reflected about the x-axis, shifted upward three units. Refer to the graph in the back of the textbook. $y = -\sqrt{x} + 3$

43. The graph is a translation of the basic graph of $y = x^3$ shifted to the right three units and upward one unit. $y = (x-3)^3 + 1$

45. a. $y = x^2 - 4x + 7$
$y = (x^2 - 4x + 4) - 4 + 7$
$y = (x-2)^2 + 3$

b. The graph of $y = (x-2)^2 + 3$ is a translation of the basic graph $y = x^2$ shifted two units to the right and upward three units. Refer to the graph in the back of the textbook.

47. a. $y = x^2 + 2x - 3$

$y = (x^2 + 2x + 1) - 1 - 3$

$y = (x+1)^2 - 4$

b. The graph of $y = (x+1)^2 - 4$ is a translation of the basic graph $y = x^2$ shifted one unit to the left and downward four units.

Section 10.2

1. $x^2 + y^2 = 25$

$x^2 + y^2 = 5^2$

This is the equation for a circle of radius 5 centered at (0, 0). Refer to the graph in the back of the textbook.

3. $4x^2 = 16 - 4y^2$

$4x^2 + 4y^2 = 16$

$x^2 + y^2 = 4$

$x^2 + y^2 = 2^2$

This is the equation for a circle of radius 2 centered at (0, 0). Refer to the graph in the back of the textbook.

5. $\dfrac{x^2}{16} + \dfrac{y^2}{4} = 1$

$\dfrac{x^2}{4^2} + \dfrac{y^2}{2^2} = 1$

This is the equation for an ellipse centered at (0, 0). The vertices are at (4, 0) and (–4, 0), and the covertices are (0, 2) and (0, –2). Refer to the graph in the back of the textbook.

7. $\dfrac{x^2}{10} + \dfrac{y^2}{25} = 1$

$\dfrac{x^2}{\left(\sqrt{10}\right)^2} + \dfrac{y^2}{5^2} = 1$

This is the equation for an ellipse centered at (0, 0). The vertices are at (0, 5) and (0, –5), and the covertices are at $\left(\sqrt{10}, 0\right)$ and $\left(-\sqrt{10}, 0\right)$ or approximately at (3.2, 0) and (–3.2, 0). Refer to the graph in the back of the textbook.

9. $x^2 + \dfrac{y^2}{14} = 1$

$\dfrac{x^2}{1^2} + \dfrac{y^2}{\left(\sqrt{14}\right)^2} = 1$

This is the equation for an ellipse centered at (0, 0). The vertices are at $\left(0, \sqrt{14}\right)$ and $\left(0, -\sqrt{14}\right)$ or approximately at (0, 3.7) and (0, –3.7), and the covertices are at (1, 0) and (–1, 0). Refer to the graph in the back of the textbook.

Chapter 10, Section 10.2

11. $3x^2 + 4y^2 = 36$

$$\frac{x^2}{12} + \frac{y^2}{9} = 1$$

$$\frac{x^2}{\left(\sqrt{12}\right)^2} + \frac{y^2}{3^2} = 1$$

This is the equation for an ellipse centered at (0, 0). The vertices are at $\left(\sqrt{12},\ 0\right)$ and $\left(-\sqrt{12},\ 0\right)$ or approximately at (3.5, 0) and (−3.5, 0), and the covertices are at (0, 3) and (0, −3). Refer to the graph in the back of the textbook.

13. $x^2 = 36 - 9y^2$

$$x^2 + 9y^2 = 36$$

$$\frac{x^2}{36} + \frac{y^2}{4} = 1$$

$$\frac{x^2}{6^2} + \frac{y^2}{2^2} = 1$$

This is the equation for an ellipse centered at (0, 0). The vertices are at (6, 0) and (−6, 0), and the covertices are (0, 2) and (0, −2). Refer to the graph in the back of the textbook.

15. $3y^2 = 30 - 2x^2$

$$2x^2 + 3y^2 = 30$$

$$\frac{x^2}{15} + \frac{y^2}{10} = 1$$

$$\frac{x^2}{\left(\sqrt{15}\right)^2} + \frac{y^2}{\left(\sqrt{10}\right)^2} = 1$$

This is the equation for an ellipse centered at (0, 0). The vertices are at $\left(\sqrt{15},\ 0\right)$ and $\left(-\sqrt{15},\ 0\right)$ or approximately at (3.9, 0) and (−3.9, 0), and the covertices are at $\left(0,\ \sqrt{10}\right)$ and $\left(0,\ -\sqrt{10}\right)$ or approximately at (0, 3.2) and (0, −3.2). Refer to the graph in the back of the textbook.

17. $$\frac{x^2}{25} - \frac{y^2}{9} = 1$$

$$\frac{x^2}{5^2} - \frac{y^2}{3^2} = 1$$

This is the equation for a hyperbola centered at (0, 0). The branches open to the left and the right. The vertices are at (5, 0) and (−5, 0). Refer to the graph in the back of the textbook.

19. $$\frac{y^2}{12} - \frac{x^2}{8} = 1$$

$$\frac{y^2}{\left(\sqrt{12}\right)^2} - \frac{x^2}{\left(\sqrt{8}\right)^2} = 1$$

This is the equation for a hyperbola centered at (0, 0). The branches open upward and downward. The vertices are at $\left(0,\ \sqrt{12}\right)$ and $\left(0,\ -\sqrt{12}\right)$ or approximately at (0, 3.5) and (0, −3.5). Refer to the graph in the back of the textbook.

Chapter 10, Section 10.2

21. $9x^2 - 4y^2 = 36$

$$\frac{x^2}{4} - \frac{y^2}{9} = 1$$

$$\frac{x^2}{2^2} - \frac{y^2}{3^2} = 1$$

This is the equation for a hyperbola centered at (0, 0). The branches open to the left and the right. The vertices are at (2, 0) and (–2, 0). Refer to the graph in the back of the textbook.

23. $y^2 - 9x^2 = 36$

$$\frac{y^2}{36} - \frac{x^2}{4} = 1$$

$$\frac{y^2}{6^2} - \frac{x^2}{2^2} = 1$$

This is the equation for a hyperbola centered at (0, 0). The branches open upward and downward. The vertices are at (0, 6) and (0, –6). Refer to the graph in the back of the textbook.

25. $3x^2 = 4y^2 + 24$

$3x^2 - 4y^2 = 24$

$$\frac{x^2}{8} - \frac{y^2}{6} = 1$$

$$\frac{x^2}{\left(\sqrt{8}\right)^2} - \frac{y^2}{\left(\sqrt{6}\right)^2} = 1$$

This is the equation for a hyperbola centered at (0, 0). The branches open to the left and right. The vertices are at $\left(\sqrt{8},\ 0\right)$ and $\left(-\sqrt{8},\ 0\right)$ or approximately at (2.8, 0) and (–2.8, 0). Refer to the graph in the back of the textbook.

27. $\frac{1}{2}x^2 = y^2 - 12$

$y^2 - \frac{1}{2}x^2 = 12$

$$\frac{y^2}{12} - \frac{x^2}{24} = 1$$

$$\frac{y^2}{\left(\sqrt{12}\right)^2} - \frac{x^2}{\left(\sqrt{24}\right)^2} = 1$$

This is the equation for a hyperbola centered at (0, 0). The branches open upward and downward. The vertices are at $\left(0,\ \sqrt{12}\right)$ and $\left(0,\ -\sqrt{12}\right)$ or approximately at (0, 3.5) and (0, –3.5). Refer to the graph in the back of the textbook.

29. $x^2 = 2y$

$x^2 = 4\left(\frac{1}{2}\right)y$

This is the equation for a parabola with vertex at (0, 0) and opening upward. $p = \frac{1}{2}$, the focus is at $\left(0,\ \frac{1}{2}\right)$, and the directrix is the horizontal line $y = -\frac{1}{2}$. Refer to the graph in the back of the textbook.

Chapter 10, Section 10.2

31. $x = -\dfrac{1}{16}y^2$

$y^2 = -16x$

$y^2 = -4(4)x$

This is the equation for a parabola with vertex at (0, 0) and opening to the left. $p = 4$, the focus is at (−4, 0), and the directrix is the vertical line $x = 4$. Refer to the graph in the back of the textbook.

33. $y^2 = 12x$

$y^2 = 4(3)x$

This is the equation for a parabola with vertex at (0, 0) and opening to the right. $p = 3$, the focus is at (3, 0), and the directrix is the vertical line $x = -3$. Refer to the graph in the back of the textbook.

35. $x^2 + 8y = 0$

$x^2 = -8y$

$x^2 = -4(2)y$

This is the equation for a parabola with vertex at (0, 0) and opening downward. $p = 2$, the focus is at (0, −2), and the directrix is the horizontal line $y = 2$. Refer to the graph in the back of the textbook.

37. $2y^2 - 3x = 0$

$2y^2 = 3x$

$y^2 = \dfrac{3}{2}x$

$y^2 = 4\left(\dfrac{3}{8}\right)x$

This is the equation for a parabola with vertex at (0, 0) and opening to the right. $p = \dfrac{3}{8}$, the focus is at $\left(\dfrac{3}{8},\ 0\right)$, and the directrix is the vertical line $x = -\dfrac{3}{8}$. Refer to the graph in the back of the textbook.

39. $y^2 = 4 - x^2$

$x^2 + y^2 = 4$

$x^2 + y^2 = 2^2$

This is the equation for a circle of radius 2 centered at (0, 0).

41. $4y^2 = x^2 - 8$

$x^2 - 4y^2 = 8$

$\dfrac{x^2}{8} - \dfrac{y^2}{2} = 1$

$\dfrac{x^2}{\left(\sqrt{8}\right)^2} - \dfrac{y^2}{\left(\sqrt{2}\right)^2} = 1$

This is the equation for a hyperbola centered at (0, 0). The branches open to the left and right. The vertices are at $\left(\sqrt{8},\ 0\right)$ and $\left(-\sqrt{8},\ 0\right)$ or approximately at (2.8, 0) and (−2.8, 0).

43. $4x^2 = 12 - 2y^2$

$4x^2 + 2y^2 = 12$

$\dfrac{x^2}{3} + \dfrac{y^2}{6} = 1$

$\dfrac{x^2}{(\sqrt{3})^2} + \dfrac{y^2}{(\sqrt{6})^2} = 1$

This is the equation for an ellipse centered at (0, 0). The vertices are at $(0, \sqrt{6})$ and $(0, -\sqrt{6})$ or approximately at (0, 2.4) and (0, −2.4), and the covertices are at $(\sqrt{3}, 0)$ and $(-\sqrt{3}, 0)$ or approximately at (1.7, 0) and (−1.7, 0).

45. $4x^2 = 6 + 4y$

$x^2 = y + \dfrac{3}{2}$

$x^2 = 4\left(\dfrac{1}{4}\right)\left(y + \dfrac{3}{2}\right)$

This is the equation for a parabola with vertex at $\left(0, -\dfrac{3}{2}\right)$ and opening upward. $p = \dfrac{1}{4}$, the focus is at $\left(0, -\dfrac{5}{4}\right)$, and the directrix is the horizontal line $y = -\dfrac{7}{4}$.

47. $6 + \dfrac{x^2}{4} = y^2$

$y^2 - \dfrac{x^2}{4} = 6$

$\dfrac{y^2}{6} - \dfrac{x^2}{24} = 1$

$\dfrac{y^2}{(\sqrt{6})^2} - \dfrac{x^2}{(\sqrt{24})^2} = 1$

This is the equation for a hyperbola centered at (0, 0). The branches open upward and downward. The vertices are at $(0, \sqrt{6})$ and $(0, -\sqrt{6})$ or approximately at (0, 2.4) and (0, −2.4).

49. $\dfrac{1}{2}x^2 - y = 4$

$\dfrac{1}{2}x^2 = y + 4$

$x^2 = 2(y + 4)$

$x^2 = 4\left(\dfrac{1}{2}\right)(y + 4)$

This is the equation for a parabola with vertex at (0, −4) and opening upward. $p = \dfrac{1}{2}$, the focus is at $\left(0, -\dfrac{7}{2}\right)$, and the directrix is the horizontal line $y = -\dfrac{9}{2}$.

Chapter 10, Section 10.2

51. a. $2a = 16$, $2b = 10$, and the equation is of the form $\dfrac{x^2}{b^2} + \dfrac{y^2}{a^2} = 1$. Therefore, $a^2 = 64$, $b^2 = 25$, and the equation is $\dfrac{x^2}{25} + \dfrac{y^2}{64} = 1$.

b. Substitute $x = 4$ into the equation and solve for y.

$$\dfrac{(4)^2}{25} + \dfrac{y^2}{64} = 1$$

$$\dfrac{16}{25} + \dfrac{y^2}{64} = 1$$

$$\dfrac{y^2}{64} = \dfrac{9}{25}$$

$$y^2 = \dfrac{576}{25}$$

$$y = \pm\dfrac{24}{5}$$

53. Let the center of the ellipse be at the origin. Then the ellipse can be described by the equation $\dfrac{x^2}{a^2} + \dfrac{y^2}{b^2} = 1$ where $2a = 20$ and $b = 7$. Therefore, the equation is $\dfrac{x^2}{100} + \dfrac{y^2}{49} = 1$.

To find the height of the arch at a distance of 8 feet from the peak, substitute $x = 8$ into the equation and solve for y.

$$\dfrac{(8)^2}{100} + \dfrac{y^2}{49} = 1$$

$$\dfrac{64}{100} + \dfrac{y^2}{49} = 1$$

$$\dfrac{y^2}{49} = \dfrac{36}{100}$$

$$y^2 = \dfrac{1764}{100}$$

$$y = \pm\dfrac{42}{10} = \pm\dfrac{21}{5}$$

Therefore the height at 8 feet from the peak is $\dfrac{21}{5}$ ft.

55. Let the center of the ellipse be at the origin. Then the ellipse can be described by the equation $\dfrac{x^2}{b^2} + \dfrac{y^2}{a^2} = 1$ where $2a = 360$ and $b = 50$. Therefore, the equation is $\dfrac{x^2}{2500} + \dfrac{y^2}{32,400} = 1$. To find the distance from the leading edge to the major axis, substitute $y = 165$ into the equation and solve for x.

$$\dfrac{x^2}{2500} + \dfrac{(165)^2}{32,400} = 1$$

$$\dfrac{x^2}{2500} + \dfrac{27,225}{32,400} = 1$$

$$\dfrac{x^2}{2500} + \dfrac{121}{144} = 1$$

$$\dfrac{x^2}{2500} = \dfrac{23}{144}$$

$$x^2 = \dfrac{57,500}{144}$$

$$x \approx \pm 19.98$$

Chapter 10, Section 10.2

Therefore the width of the keel at its widest point is approximately 69.98 cm.

57. Let the vertex of the parabola be at the origin. Then the parabola can be described by an equation of the form $x^2 = 4py$. Substitute $x = 36$ and $y = 3$ into the equation and solve for p.

$(36)^2 = 4p(3)$

$1296 = 12p$

$p = 108$

Since $p = 108$, the focus is 108 in. from the vertex.

59. Let the vertex of the parabola be at the origin. Then the parabola can be described by an equation of the form $x^2 = 4py$. Substitute $x = 30$ and $y = 18$ into the equation and solve for p.

$(30)^2 = 4p(18)$

$900 = 72p$

$p = 12.5$

Since $p = 12.5$, the receiver is 12.5 cm from the vertex.

61. Substitute $y = -360$ into the equation and solve for x.

$$\frac{x^2}{100^2} - \frac{(-360)^2}{150^2} = 1$$

$$\frac{x^2}{10,000} - \left(\frac{-360}{150}\right)^2 = 1$$

$$\frac{x^2}{10,000} - \left(\frac{-12}{5}\right)^2 = 1$$

$$\frac{x^2}{10,000} = \frac{169}{25}$$

$x^2 = 67,600$

$x = \pm 260$

Therefore, the diameter is 520 ft.

63. Substitute $x = 125$ into the equation and solve for y.

$$\frac{(125)^2}{100^2} - \frac{y^2}{150^2} = 1$$

$$\left(\frac{125}{100}\right)^2 - \frac{y^2}{22,500} = 1$$

$$\left(\frac{5}{4}\right)^2 - \frac{y^2}{22,500} = 1$$

$$-\frac{y^2}{22,500} = -\frac{9}{16}$$

$$y^2 = \frac{50,625}{4}$$

$$y = \pm\frac{225}{2} = \pm 112.5$$

Therefore, the distance from the top is $200 - 112.5 = 87.5$ ft.

65. $x^2 - y^2 = 0$

$(x - y)(x + y) = 0$

$y = x$ or $y = -x$

Graph the two lines. Refer to the graph in the back of the textbook.

67. $x^2 - y^2 = 4$

$$\frac{x^2}{2^2} - \frac{y^2}{2^2} = 1$$

This graph is a hyperbola with vertices $(2, 0)$ and $(-2, 0)$.

Chapter 10, Section 10.3

$x^2 - y^2 = 1$

$\dfrac{x^2}{1^2} - \dfrac{y^2}{1^2} = 1$

This graph is a hyperbola with vertices $(1, 0)$ and $(-1, 0)$.

$x^2 - y^2 = 0$

This graph is two lines, $y = x$ and $y = -x$. (See Exercise 65.) Note the asymptotes of the two hyperbolas are $y = x$ and $y = -x$. Refer to the graph in the back of the textbook.

Section 10.3

1. $\dfrac{(x-3)^2}{16} + \dfrac{(y-4)^2}{9} = 1$

   $\dfrac{(x-3)^2}{4^2} + \dfrac{(y-4)^2}{3^2} = 1$

   This is the equation for an ellipse centered at $(3, 4)$. The vertices lie four units to the left and right of the center, at $(-1, 4)$ and $(7, 4)$, and the covertices lie three units above and below the center, at $(3, 7)$ and $(3, 1)$. Refer to the graph in the back of the textbook.

3. $\dfrac{(x+2)^2}{6} + \dfrac{(y-5)^2}{12} = 1$

   $\dfrac{(x+2)^2}{\left(\sqrt{6}\right)^2} + \dfrac{(y-5)^2}{\left(\sqrt{12}\right)^2} = 1$

   This is the equation for an ellipse centered at $(-2, 5)$. The vertices lie $\sqrt{12}$ units above and below the center, at approximately $(-2, 8.5)$ and $(-2, 1.5)$, and the covertices lie $\sqrt{6}$ units to the left and right of the center, at approximately $(-4.4, 5)$

and $(0.4, 5)$. Refer to the graph in the back of the textbook.

5. $\dfrac{x^2}{16} + \dfrac{(y+4)^2}{6} = 1$

   $\dfrac{x^2}{4^2} + \dfrac{(y+4)^2}{\left(\sqrt{6}\right)^2} = 1$

   This is the equation for an ellipse centered at $(0, -4)$. The vertices lie four units to the left and right of the center, at $(-4, -4)$ and $(4, -4)$, and the covertices lie $\sqrt{6}$ units above and below the center, at approximately $(0, -1.6)$ and $(0, -6.4)$. Refer to the graph in the back of the textbook.

7. $9x^2 + 4y^2 - 16y = 20$

   $9x^2 + 4(y^2 - 4y + 4) = 20 + 16$

   $9x^2 + 4(y-2)^2 = 36$

   $\dfrac{x^2}{4} + \dfrac{(y-2)^2}{9} = 1$

   $\dfrac{x^2}{2^2} + \dfrac{(y-2)^2}{3^2} = 1$

   This is the equation for an ellipse centered at $(0, 2)$. The vertices lie three units above and below the center, at $(0, 5)$ and $(0, -1)$, and the covertices lie two units to the left and right of the center, at $(-2, 2)$ and $(2, 2)$. Refer to the graph in the back of the textbook.

9. $9x^2 + 16y^2 - 18x + 96y + 9 = 0$

   $9(x^2 - 2x + 1) + 16(y^2 + 6y + 9)$
   $= -9 + 9 + 144$

332

$9(x-1)^2 + 16(y+3)^2 = 144$

$\dfrac{(x-1)^2}{16} + \dfrac{(y+3)^2}{9} = 1$

$\dfrac{(x-1)^2}{4^2} + \dfrac{(y+3)^2}{3^2} = 1$

This is the equation for an ellipse centered at $(1, -3)$. The vertices lie four units to the left and right of the center, at $(-3, -3)$ and $(5, -3)$, and the covertices lie three units above and below the center, at $(1, 0)$ and $(1, -6)$. Refer to the graph in the back of the textbook.

11. $x^2 + 4y^2 + 4x - 16y + 4 = 0$

$(x^2 + 4x + 4) + 4(y^2 - 4y + 4)$
$\qquad = -4 + 4 + 16$

$(x+2)^2 + 4(y-2)^2 = 16$

$\dfrac{(x+2)^2}{16} + \dfrac{(y-2)^2}{4} = 1$

$\dfrac{(x+2)^2}{4^2} + \dfrac{(y-2)^2}{2^2} = 1$

This is the equation for an ellipse centered at $(-2, 2)$. The vertices lie four units to the left and right of the center, at $(-6, 2)$ and $(2, 2)$, and the covertices lie two units above and below the center, at $(-2, 4)$ and $(-2, 0)$. Refer to the graph in the back of the textbook.

13. $6x^2 + 5y^2 - 12x + 20y - 4 = 0$

$6(x^2 - 2x + 1) + 5(y^2 + 4y + 4)$
$\qquad = 4 + 6 + 20$

$6(x-1)^2 + 5(y+2)^2 = 30$

$\dfrac{(x-1)^2}{5} + \dfrac{(y+2)^2}{6} = 1$

$\dfrac{(x-1)^2}{\left(\sqrt{5}\right)^2} + \dfrac{(y+2)^2}{\left(\sqrt{6}\right)^2} = 1$

This is the equation for an ellipse centered at $(1, -2)$. The vertices lie $\sqrt{6}$ units above and below the center, at approximately $(1, 0.4)$ and $(1, -4.4)$, and the covertices lie $\sqrt{5}$ units to the left and right of the center, at approximately $(-1.2, -2)$ and $(3.2, -2)$. Refer to the graph in the back of the textbook.

15. $8x^2 + y^2 - 48x + 4y + 68 = 0$

$8(x^2 - 6x + 9) + (y^2 + 4y + 4)$
$\qquad = -68 + 72 + 4$

$8(x-3)^2 + (y+2)^2 = 8$

$\dfrac{(x-3)^2}{1} + \dfrac{(y+2)^2}{8} = 1$

$\dfrac{(x-3)^2}{1^2} + \dfrac{(y+2)^2}{\left(\sqrt{8}\right)^2} = 1$

This is the equation for an ellipse centered at $(3, -2)$. The vertices lie $\sqrt{8}$ units above and below the center, at approximately $(3, -0.2)$ and $(3, -4.8)$, and the covertices lie one unit to the left and right of the center, at $(2, -2)$ and $(4, -2)$. Refer to the graph in the back of the textbook.

Chapter 10, Section 10.3

17. $\dfrac{(x-1)^2}{3^2} + \dfrac{(y-6)^2}{2^2} = 1$

$\dfrac{(x-1)^2}{3^2} + \dfrac{(y-6)^2}{2^2} = 1$

$\dfrac{(x-1)^2}{9} + \dfrac{(y-6)^2}{4} = 1$

$4(x-1)^2 + 9(y-6)^2 = 36$

$4(x^2 - 2x + 1) + 9(y^2 - 12y + 36) = 36$

$4x^2 + 9y^2 - 8x - 108y + 292 = 0$

19. The center is between the vertices at (–2, 2). The major axis length is the distance between the vertices, which is 10 units. Therefore, $2a = 10$, so $a = 5$. Also, $2b = 6$, so $b = 3$. The major axis is horizontal.

$\dfrac{(x+2)^2}{5^2} + \dfrac{(y-2)^2}{3^2} = 1$

$\dfrac{(x+2)^2}{25} + \dfrac{(y-2)^2}{9} = 1$

$9(x+2)^2 + 25(y-2)^2 = 225$

$9(x^2 + 4x + 4) + 25(y^2 - 4y + 4) = 225$

$9x^2 + 25y^2 + 36x - 100y - 89 = 0$

21. The center is between the vertices at (–4, 3). The major axis length is the distance between the vertices, which is 12 units. The minor axis length is the distance between the covertices, which is 6 units. Therefore, $2a = 12$, so $a = 6$, and $2b = 6$, so $b = 3$. The major axis is vertical.

$\dfrac{(x+4)^2}{3^2} + \dfrac{(y-3)^2}{6^2} = 1$

$\dfrac{(x+4)^2}{9} + \dfrac{(y-3)^2}{36} = 1$

$4(x+4)^2 + (y-3)^2 = 36$

$4(x^2 + 8x + 16) + (y^2 - 6y + 9) = 36$

$4x^2 + y^2 + 32x - 6y - 37 = 0$

23. $\dfrac{(x-4)^2}{9} - \dfrac{(y+2)^2}{16} = 1$

$\dfrac{(x-4)^2}{3^2} - \dfrac{(y+2)^2}{4^2} = 1$

This is the equation for a hyperbola centered at (4, –2). The transverse axis is parallel to the x-axis. The vertices are three units to the left and right of the center at (1, –2) and (7, –2). The ends of the conjugate axes are four units above and below the center at (4, 2) and (4, –6). Refer to the graph in the back of the textbook.

25. $\dfrac{x^2}{4} - \dfrac{(y-3)^2}{8} = 1$

$\dfrac{x^2}{2^2} - \dfrac{(y-3)^2}{\left(\sqrt{8}\right)^2} = 1$

This is the equation for a hyperbola centered at (0, 3). The transverse axis is parallel to the x-axis. The vertices are two units to the left and right of the center at (–2, 3) and (2, 3). The ends of the conjugate axes are $\sqrt{8}$ units above and below the center at approximately (0, 5.8)

334

27. $\dfrac{(y+2)^2}{6} - \dfrac{(x+2)^2}{10} = 1$

$\dfrac{(y+2)^2}{\left(\sqrt{6}\right)^2} - \dfrac{(x+2)^2}{\left(\sqrt{10}\right)^2} = 1$

and (0, 0.2). Refer to the graph in the back of the textbook.

This is the equation for a hyperbola centered at (−2, −2). The transverse axis is parallel to the y-axis. The vertices are $\sqrt{6}$ units above and below the center at approximately (−2, 0.4) and (−2, −4.4). The ends of the conjugate axes are $\sqrt{10}$ units to the left and right of the center at approximately (−5.2, −2) and (1.2, −2). Refer to the graph in the back of the textbook.

29. $\dfrac{y^2}{6} - \dfrac{(x-3)^2}{15} = 1$

$\dfrac{y^2}{\left(\sqrt{6}\right)^2} - \dfrac{(x-3)^2}{\left(\sqrt{15}\right)^2} = 1$

This is the equation for a hyperbola centered at (3, 0). The transverse axis is parallel to the y-axis. The vertices are $\sqrt{6}$ units above and below the center at approximately (3, 2.4) and (3, −2.4). The ends of the conjugate axes are $\sqrt{15}$ units to the left and right of the center at approximately (−0.9, 0) and (6.9, 0). Refer to the graph in the back of the textbook.

31. $9x^2 - 4y^2 - 36x - 24y - 36 = 0$

$9(x^2 - 4x + 4) - 4(y^2 + 6y + 9)$
$ = 36 + 36 - 36$

$9(x-2)^2 - 4(y+3)^2 = 36$

$\dfrac{(x-2)^2}{4} - \dfrac{(y+3)^2}{9} = 1$

$\dfrac{(x-2)^2}{2^2} - \dfrac{(y+3)^2}{3^2} = 1$

This is the equation for a hyperbola centered at (2, −3). The transverse axis is parallel to the x-axis. The vertices are two units to the left and right of the center at (0, −3) and (4, −3). The ends of the conjugate axes are three units above and below the center at (2, 0) and (2, −6). Refer to the graph in the back of the textbook.

33. $16y^2 - 4x^2 + 32x - 128 = 0$

$16y^2 - 4(x^2 - 8x + 16) = 128 - 64$

$16y^2 - 4(x-4)^2 = 64$

$\dfrac{y^2}{4} - \dfrac{(x-4)^2}{16} = 1$

$\dfrac{y^2}{2^2} - \dfrac{(x-4)^2}{4^2} = 1$

This is the equation for a hyperbola centered at (4, 0). The transverse axis is parallel to the y-axis. The vertices are two units above and below the center at (4, 2) and (4, −2). The ends of the conjugate axes are four units to the left and right of the center at approximately (0, 0) and (8, 0). Refer to the graph in the back of the textbook.

Chapter 10, Section 10.3

35. $4x^2 - 6y^2 - 32x - 24y + 16 = 0$

$4(x^2 - 8x + 16) - 6(y^2 + 4y + 4)$
$= -16 + 64 - 24$

$4(x-4)^2 - 6(y+2)^2 = 24$

$\dfrac{(x-4)^2}{6} - \dfrac{(y+2)^2}{4} = 1$

$\dfrac{(x-4)^2}{(\sqrt{6})^2} - \dfrac{(y+2)^2}{2^2} = 1$

This is the equation for a hyperbola centered at (4, −2). The transverse axis is parallel to the x-axis. The vertices are $\sqrt{6}$ units to the left and right of the center at approximately (1.6, −2) and (6.4, −2). The ends of the conjugate axes are two units above and below the center at (4, 0) and (4, −4). Refer to the graph in the back of the textbook.

37. $12x^2 - 3y^2 + 24y - 84 = 0$

$12x^2 - 3(y^2 - 8y + 16) = 84 - 48$

$12x^2 - 3(y-4)^2 = 36$

$\dfrac{x^2}{3} - \dfrac{(y-4)^2}{12} = 1$

$\dfrac{x^2}{(\sqrt{3})^2} - \dfrac{(y-4)^2}{(\sqrt{12})^2} = 1$

This is the equation for a hyperbola centered at (0, 4). The transverse axis is parallel to the x-axis. The vertices are $\sqrt{3}$ units to the left and right of the center at approximately (−1.7, 4) and (1.7, 4). The ends of the conjugate axes are $\sqrt{12}$ above and below the center at (0, 7.5) and (0, 0.5). Refer to the graph in the back of the textbook.

39. $\dfrac{(y-5)^2}{8^2} - \dfrac{(x+1)^2}{6^2} = 1$

$\dfrac{(y-5)^2}{64} - \dfrac{(x+1)^2}{36} = 1$

$9(y^2 - 10y + 25) - 16(x^2 + 2x + 1)$
$= 576$

$-16x^2 + 9y^2 - 32x - 90y - 367$
$= 0$

41. Since the conjugate axis is horizontal, it lies on the line $y = 1$. Therefore, the transverse axis is vertical and lies on the line $x = -1$. The center is the intersection of the two axes at (−1, 1). The distance from the center to a vertex is 2, so $a = 2$. The distance from the center to one end of the conjugate axis is 4, so $b = 4$.

$\dfrac{(y-1)^2}{2^2} - \dfrac{(x+1)^2}{4^2} = 1$

$\dfrac{(y-1)^2}{4} - \dfrac{(x+1)^2}{16} = 1$

$4(y^2 - 2y + 1) - (x^2 + 2x + 1) = 16$

$-x^2 + 4y^2 - 2x - 8y - 13 = 0$

43. $2x = (y+3)^2$

$(y+3)^2 = 4\left(\dfrac{1}{2}\right)x$

This is the equation for a parabola with vertex at (0, −3) and opening to the right. $p = \dfrac{1}{2}$, the focus is at

Chapter 10, Section 10.3

$\left(\dfrac{1}{2}, -3\right)$, and the directrix is the vertical line $x = -\dfrac{1}{2}$. Refer to the graph in the back of the textbook.

45. $-6(y+4) = (x-3)^2$

$(x-3)^2 = -4\left(\dfrac{3}{2}\right)(y+4)$

This is the equation for a parabola with vertex at $(3, -4)$ and opens downward. $p = \dfrac{3}{2}$, the focus is at $\left(3, -\dfrac{11}{2}\right)$, and the directrix is the horizontal line $y = -\dfrac{5}{2}$. Refer to the graph in the back of the textbook.

47. $(y-4)^2 + 3 = x$

$(y-4)^2 = 4\left(\dfrac{1}{4}\right)(x-3)$

This is the equation for a parabola with vertex at $(3, 4)$ and opens to the right. $p = \dfrac{1}{4}$, the focus is at $\left(\dfrac{13}{4}, 4\right)$, and the directrix is the vertical line $x = \dfrac{11}{4}$. Refer to the graph in the back of the textbook.

49. $y^2 - 6y + 10x + 4 = 0$

$(y^2 - 6y + 9) = -10x - 4 + 9$

$(y-3)^2 = -10x + 5$

$(y-3)^2 = -4\left(\dfrac{5}{2}\right)\left(x - \dfrac{1}{2}\right)$

This is the equation for a parabola with vertex at $\left(\dfrac{1}{2}, 3\right)$ and opens to the left. $p = \dfrac{5}{2}$, the focus is at $(-2, 3)$, and the directrix is the vertical line $x = 3$. Refer to the graph in the back of the textbook.

51. $4x^2 - 4x = 23 - 12y$

$4\left(x^2 - x + \dfrac{1}{4}\right) = -12y + 23 + 1$

$4\left(x - \dfrac{1}{2}\right)^2 = -12y + 24$

$\left(x - \dfrac{1}{2}\right)^2 = -3y + 6$

$\left(x - \dfrac{1}{2}\right)^2 = -4\left(\dfrac{3}{4}\right)(y - 2)$

This is the equation for a parabola with vertex at $\left(\dfrac{1}{2}, 2\right)$ and opens downward. $p = \dfrac{3}{4}$, the focus is at $\left(\dfrac{1}{2}, \dfrac{5}{4}\right)$, and the directrix is the horizontal line $y = \dfrac{11}{4}$. Refer to the graph in the back of the textbook.

Chapter 10, Section 10.3

53. $9y^2 = 6y + 12x - 1$

$9\left(y^2 - \frac{2}{3}y + \frac{1}{9}\right) = 12x - 1 + 1$

$9\left(y - \frac{1}{3}\right)^2 = 12x$

$\left(y - \frac{1}{3}\right)^2 = 4\left(\frac{1}{3}\right)x$

This is the equation for a parabola with vertex at $\left(0, \frac{1}{3}\right)$ and opening to the right. $p = \frac{1}{3}$, the focus is at $\left(\frac{1}{3}, \frac{1}{3}\right)$, and the directrix is the vertical line $x = -\frac{1}{3}$. Refer to the graph in the back of the textbook.

55. p is equal to the distance from the focus to the vertex, which is 1. The parabola opens downward since the focus is below the vertex.

$(x - 1)^2 = -4(1)(y - 2)$

$x^2 - 2x + 1 = -4y + 8$

$x^2 - 2x + 4y - 7 = 0$

57. p is equal to the distance from the focus to the vertex, which is 2. The parabola opens to the right since the focus is to the right of the vertex.

$(y + 2)^2 = 4(2)(x + 4)$

$y^2 + 4y + 4 = 8x + 32$

$y^2 - 8x + 4y - 28 = 0$

59. p is equal to the distance from the vertex to the directrix which is 1. The parabola opens to the left since the directrix is to the right of the vertex.

$(y - 5)^2 = -4(1)(x - 2)$

$y^2 - 10y + 25 = -4x + 8$

$y^2 + 4x - 10y + 17 = 0$

61. $2p$ is equal to the distance from the focus to the directrix, which is 4, so $p = 2$. The vertex is between the focus and directrix at $(3, 0)$. The parabola opens downward since the focus is below the vertex.

$(x - 3)^2 = -4(2)y$

$x^2 - 6x + 9 = -8y$

$x^2 - 6x + 8y + 9 = 0$

63. $y^2 = 4 - x^2$

$x^2 + y^2 = 4$

$x^2 + y^2 = 2^2$

This is the equation for a circle of radius 2 centered at $(0, 0)$.

65. $4y^2 = x^2 - 8$

$x^2 - 4y^2 = 8$

$\frac{x^2}{8} - \frac{y^2}{2} = 1$

$\frac{x^2}{\left(\sqrt{8}\right)^2} - \frac{y^2}{\left(\sqrt{2}\right)^2} = 1$

This is the equation for a hyperbola centered at $(0, 0)$. The transverse axis lies on the x-axis, $a^2 = 8$ and $b^2 = 2$. the vertices are $\sqrt{8}$ units to the left and right of the center at approximately $(-2.8, 0)$ and $(2.8, 0)$.

Chapter 10, Section 10.3

67. $4x^2 = 12 - 2y^2$

$4x^2 + 2y^2 = 12$

$\dfrac{x^2}{3} + \dfrac{y^2}{6} = 1$

$\dfrac{x^2}{(\sqrt{3})^2} + \dfrac{y^2}{(\sqrt{6})^2} = 1$

This is the equation for an ellipse centered at (0, 0). The major axis is vertical, $a^2 = 6$, and $b^2 = 3$. The vertices lie $\sqrt{6}$ units above and below the center, at approximately (0, 2.4) and (0, −2.4), and the covertices lie $\sqrt{3}$ units to the left and right of the center, at approximately (−1.7, 0) and (1.7, 0).

69. $4x^2 = 6 + 4y$

$4x^2 = 4\left(y + \dfrac{3}{2}\right)$

$x^2 = 4\left(\dfrac{1}{4}\right)\left(y + \dfrac{3}{2}\right)$

This is the equation for a parabola with vertex at $\left(0, -\dfrac{3}{2}\right)$ and opening upward. $p = \dfrac{1}{4}$, the focus is at $\left(0, -\dfrac{5}{4}\right)$, and the directrix is the horizontal line $y = -\dfrac{7}{4}$.

71. $6 + \dfrac{x^2}{4} = y^2$

$y^2 - \dfrac{x^2}{4} = 6$

$\dfrac{y^2}{6} - \dfrac{x^2}{24} = 1$

$\dfrac{y^2}{(\sqrt{6})^2} - \dfrac{x^2}{(\sqrt{24})^2} = 1$

This is the equation for a hyperbola centered at (0, 0). The transverse axis lies on the y-axis, $a^2 = 6$ and $b^2 = 24$. The vertices are $\sqrt{6}$ units above and below the center at approximately (0, 2.4) and (0, −2.4).

73. $\dfrac{1}{2}y^2 - x = 4$

$\dfrac{1}{2}y^2 = x + 4$

$y^2 = 4\left(\dfrac{1}{2}\right)(x + 4)$

This is the equation for a parabola with vertex at (−4, 0) and opening to the right. $p = \dfrac{1}{2}$, the focus is at $\left(-\dfrac{7}{2}, 0\right)$, and the directrix is the vertical line $x = -\dfrac{9}{2}$.

Chapter 10, Section 10.4

75. $y = (x-3)^2 + 2$

$(x-3)^2 = y - 2$

$(x-3)^2 = 4\left(\dfrac{1}{4}\right)(y-2)$

This is the equation for a parabola with vertex at (3, 2) and opening upward. $p = \dfrac{1}{4}$, the focus is at $\left(3, \dfrac{9}{4}\right)$, and the directrix is the horizontal line $y = \dfrac{7}{4}$.

77. $(x+1)^2 + (y-4)^2 = 16$

$(x+1)^2 + (y-4)^2 = 4^2$

This is the equation for a circle of radius 4 centered at (−1, 4).

79. $2x^2 + y^2 + 4x = 2$

$2(x^2 + 2x + 1) + y^2 = 2 + 2$

$2(x+1)^2 + y^2 = 4$

$\dfrac{(x+1)^2}{2} + \dfrac{y^2}{4} = 1$

$\dfrac{(x+1)^2}{\left(\sqrt{2}\right)^2} + \dfrac{y^2}{2^2} = 1$

This is the equation for an ellipse centered at (−1, 0). The major axis is vertical, $a^2 = 4$, and $b^2 = 2$. The vertices lie 2 units above and below the center, at approximately (−1, 2) and (−1, −2), and the covertices lie $\sqrt{2}$ units to the left and right of the center, at approximately (−2.4, 0) and (0.4, 0).

81. $x^2 + 6x = 4 - y^2$

$(x^2 + 6x + 9) + y^2 = 4 + 9$

$(x+3)^2 + y^2 = 13$

$(x+3)^2 + y^2 = \left(\sqrt{13}\right)^2$

This is the equation for a circle of radius $\sqrt{13}$ centered at (−3, 0).

Section 10.4

1. Write the equation for f in the form $y = x + 2$. Interchange the variables and solve for y:
$x = y + 2$
$y = x - 2$
Therefore, the inverse function is $g(x) = x - 2$.

3. Write the equation for f in the form $y = 2x$. Interchange the variables and solve for y:
$x = 2y$

$y = \dfrac{x}{2}$

Therefore, the inverse function is $g(x) = \dfrac{x}{2}$.

5. Write the equation for f in the form $y = 2x - 6$. Interchange the variables and solve for y:
$x = 2y - 6$

$2y = x + 6$

$y = \dfrac{x+6}{2}$

Therefore, the inverse function is $g(x) = \dfrac{x+6}{2}$.

Chapter 10, Section 10.4

7. Write the equation for f in the form $y = \dfrac{3-x}{2}$. Interchange the variables and solve for y:
$x = \dfrac{3-y}{2}$
$2x = 3 - y$
$y = 3 - 2x$
Therefore, the inverse function is $g(x) = 3 - 2x$.

9. Write the equation for f in the form $y = x^3 + 1$. Interchange the variables and solve for y:
$x = y^3 + 1$
$y^3 = x - 1$
$y = \sqrt[3]{x-1}$
Therefore, the inverse function is $g(x) = \sqrt[3]{x-1}$.

11. Write the equation for f in the form $y = \sqrt[3]{x}$. Interchange the variables and solve for y:
$x = \sqrt[3]{y}$
$y = x^3$
Therefore, the inverse function is $g(x) = x^3$.

13. Write the equation for f in the form $y = \dfrac{1}{x-1}$. Interchange the variables and solve for y:
$x = \dfrac{1}{y-1}$
$y - 1 = \dfrac{1}{x}$
$y = \dfrac{1}{x} + 1$
$y = \dfrac{x+1}{x}$
Therefore, the inverse function is $g(x) = \dfrac{x+1}{x}$.

15. Write the equation for f in the form $y = \sqrt[3]{x} + 4$. Interchange the variables and solve for y:
$x = \sqrt[3]{y} + 4$
$\sqrt[3]{y} = x - 4$
$y = (x-4)^3$
Therefore, the inverse function is $g(x) = (x-4)^3$.

17. a. Write the equation for f in the form $y = (x-2)^3$. Interchange the variables and solve for y:
$x = (y-2)^3$
$y - 2 = \sqrt[3]{x}$
$y = \sqrt[3]{x} + 2$
Therefore the inverse function is $g(x) = \sqrt[3]{x} + 2$.

b. $f(4) = (4-2)^3 = 2^3 = 8$
$g(8) = \sqrt[3]{8} + 2 = 2 + 2 = 4$

c. $g(-8) = \sqrt[3]{-8} + 2 = -2 + 2 = 0$
$f(0) = (0-2)^3 = (-2)^3 = -8$

Chapter 10, Section 10.4

19. Graph $f(x) = x + 2$ and $g(x) = x - 2$. Refer to the graph in the back of the textbook.

21. Graph $f(x) = 2x$ and $g(x) = \dfrac{x}{2}$. Refer to the graph in the back of the textbook.

23. Graph $f(x) = 2x - 6$ and $g(x) = \dfrac{x+6}{2}$. Refer to the graph in the back of the textbook.

25. Graph $f(x) = \dfrac{3-x}{2}$ and $g(x) = 3 - 2x$. Refer to the graph in the back of the textbook.

27. Graph $f(x) = x^3 + 1$ and $g(x) = \sqrt[3]{x-1}$. Refer to the graph in the back of the textbook.

29. Graph $f(x) = \sqrt[3]{x}$ and $g(x) = x^3$. Refer to the graph in the back of the textbook.

31. Graph $f(x) = \dfrac{1}{x-1}$ and $g(x) = \dfrac{x+1}{x}$. Refer to the graph in the back of the textbook.

33. Graph $f(x) = \sqrt[3]{x} + 4$ and $g(x) = (x-4)^3$. Refer to the graph in the back of the textbook.

35. a. Yes, it passes the horizontal line test.
 b. No, it fails the horizontal line test.
 c. No, it fails the horizontal line test.
 d. Yes, it passes the horizontal line test.

37. a. Yes, it passes the horizontal line test.
 b. No, it fails the horizontal line test.

39. a. Yes, it passes the horizontal line test.
 b. No, it fails the horizontal line test.

41. Solve $F(t) = 5$.
$$\tfrac{2}{3}t + 1 = 5$$
$$\tfrac{2}{3}t = 4$$
$$t = 6$$
Therefore, $F^{-1}(5) = 6$.

43. Solve $m(v) = -3$.
$$6 - \tfrac{2}{v} = -3$$
$$-\tfrac{2}{v} = -9$$
$$v = \tfrac{2}{9}$$
Therefore, $m^{-1}(-3) = \tfrac{2}{9}$.

45. Write the equation for f in the form $y = \dfrac{x+2}{x-1}$. Interchange the variables and solve for y:

342

Chapter 10, Section 10.5

$x = \dfrac{y+2}{y-1}$

$x(y-1) = y+2$

$xy - x = y + 2$

$xy - y = x + 2$

$y(x-1) = x + 2$

$y = \dfrac{x+2}{x-1}$

Therefore, $f^{-1}(x) = \dfrac{x+2}{x-1}$.

47. a. Solve $f(x) = 1$. Observe that $x^3 + x + 1 = 1$ when $x = 0$. Therefore, $f^{-1}(1) = 0$.

b. Solve $f(x) = 3$. Observe that $x^3 + x + 1 = 3$ when $x = 1$. Therefore, $f^{-1}(3) = 1$.

49. a. Since $f(0) = 1$, $f^{-1}(1) = 0$.

b. Since $f(2) = -1$, $f^{-1}(-1) = 2$.

51. a.

x	$f(x)$
-2	$2^{-2} = \dfrac{1}{4}$
-1	$2^{-1} = \dfrac{1}{2}$
0	$2^0 = 1$
1	$2^1 = 2$
2	$2^2 = 4$

b. Refer to the graph in the back of the textbook.

c.

x	$f^{-1}(x)$
$\dfrac{1}{4}$	-2
$\dfrac{1}{2}$	-1
1	0
2	1
4	2

d. Refer to the graph in the back of the textbook.

Section 10.5

1. $\begin{bmatrix} -2 & 1 & | & 0 \\ 3 & -1 & | & 2 \end{bmatrix} \rightarrow \begin{bmatrix} -2 & 1 & | & 0 \\ -9 & 3 & | & -6 \end{bmatrix}$

3. $\begin{bmatrix} 1 & -3 & | & 6 \\ -2 & 4 & | & -1 \end{bmatrix} \rightarrow \begin{bmatrix} 1 & -3 & | & 6 \\ 0 & -2 & | & 11 \end{bmatrix}$

Chapter 10, Section 10.5

5. $\begin{bmatrix} 0 & -3 & 2 & | & -3 \\ 2 & 6 & -1 & | & 4 \\ 1 & 0 & -2 & | & 5 \end{bmatrix} \rightarrow \begin{bmatrix} 1 & 0 & -2 & | & 5 \\ 2 & 6 & -1 & | & 4 \\ 0 & -3 & 2 & | & -3 \end{bmatrix}$

7. $\begin{bmatrix} 1 & 2 & 1 & | & -5 \\ 0 & 4 & -2 & | & 3 \\ 4 & -1 & 6 & | & -8 \end{bmatrix} \rightarrow \begin{bmatrix} 1 & 2 & 1 & | & -5 \\ 0 & 4 & -2 & | & 3 \\ 0 & -9 & 2 & | & 12 \end{bmatrix}$

9. To obtain 0 as the first entry in the second row, add $-2(\text{row } 1)$ to row 2:

$-2(\text{row } 1) + \text{row } 2 \quad \begin{bmatrix} 1 & -3 & | & 2 \\ 2 & 1 & | & 4 \end{bmatrix} \rightarrow \begin{bmatrix} 1 & -3 & | & 2 \\ 0 & 7 & | & 0 \end{bmatrix}$

11. To obtain 0 as the second entry in the second row, add $-\dfrac{1}{2}(\text{row } 1)$ to row 2:

$-\dfrac{1}{2}(\text{row } 1) + \text{row } 2 \quad \begin{bmatrix} 2 & 6 & | & -4 \\ 5 & 3 & | & 1 \end{bmatrix} \rightarrow \begin{bmatrix} 2 & 6 & | & -4 \\ 4 & 0 & | & 3 \end{bmatrix}$

13. To obtain 0 as the first entry in the second and third rows, add $-2(\text{row } 1)$ to row 2 and $-4\,(\text{row } 1)$ to row 3:

$\begin{array}{l} -2(\text{row } 1) + \text{row } 2 \\ -4(\text{row } 1) + \text{row } 3 \end{array} \quad \begin{bmatrix} 1 & -2 & 2 & | & 1 \\ 2 & 3 & -1 & | & 6 \\ 4 & 1 & -3 & | & 3 \end{bmatrix} \rightarrow \begin{bmatrix} 1 & -2 & 2 & | & 1 \\ 0 & 7 & -5 & | & 4 \\ 0 & 9 & -11 & | & -1 \end{bmatrix}$

15. To obtain 0 as the second entry in the second and third rows, add $\dfrac{1}{2}(\text{row } 1)$ to row 2 and $-\dfrac{1}{2}(\text{row } 1)$ to row 3:

$\begin{array}{l} \tfrac{1}{2}(\text{row } 1) + \text{row } 2 \\ -\tfrac{1}{2}(\text{row } 1) + \text{row } 3 \end{array} \quad \begin{bmatrix} -1 & 4 & 3 & | & 2 \\ 2 & -2 & -4 & | & 6 \\ 1 & 2 & 3 & | & -3 \end{bmatrix} \rightarrow \begin{bmatrix} -1 & 4 & 3 & | & 2 \\ \tfrac{3}{2} & 0 & -\tfrac{5}{2} & | & 7 \\ \tfrac{3}{2} & 0 & \tfrac{3}{2} & | & -4 \end{bmatrix}$

This is the equivalent to the matrix in the back of the textbook, which can be obtained by multiplying row 2 by 2 and row 3 by -2.

Chapter 10, Section 10.5

17. To obtain 0 as the first entry in the second and third rows, add 2(row 1) to row 2 and 3(row 1) to row 3:

$$\begin{matrix} \\ 2(\text{row 1}) + \text{row 2} \\ 3(\text{row 1}) + \text{row 3} \end{matrix} \begin{bmatrix} -2 & 1 & -3 & | & -2 \\ 4 & 2 & 0 & | & 2 \\ 6 & -1 & 2 & | & 0 \end{bmatrix} \to \begin{bmatrix} -2 & 1 & -3 & | & -2 \\ 0 & 4 & -6 & | & -2 \\ 0 & 2 & -7 & | & -6 \end{bmatrix}$$

To obtain 0 as the second entry in the third row, add $-\dfrac{1}{2}$(row 2) to row 3:

$$\begin{matrix} \\ \\ -\tfrac{1}{2}(\text{row 2}) + \text{row 3} \end{matrix} \begin{bmatrix} -2 & 1 & -3 & | & -2 \\ 0 & 4 & -6 & | & -2 \\ 0 & 2 & -7 & | & -6 \end{bmatrix} \to \begin{bmatrix} -2 & 1 & -3 & | & -2 \\ 0 & 4 & -6 & | & -2 \\ 0 & 0 & -4 & | & -5 \end{bmatrix}$$

19. The augmented matrix is $\begin{bmatrix} 1 & 3 & | & 11 \\ 2 & -1 & | & 1 \end{bmatrix}$. Perform Gaussian reduction:

$$\begin{matrix} \\ -2(\text{row 1}) + \text{row 2} \end{matrix} \begin{bmatrix} 1 & 3 & | & 11 \\ 2 & -1 & | & 1 \end{bmatrix} \to \begin{bmatrix} 1 & 3 & | & 11 \\ 0 & -7 & | & -21 \end{bmatrix}$$

Which corresponds to the system
$x + 3y = 11$

$-7y = -21.$
Therefore, $y = 3$. Substitute 3 for y to find
$x + 3(3) = 11$

$x = 2$
The solution is the ordered pair (2, 3).

21. The augmented matrix is $\begin{bmatrix} 1 & -4 & | & -6 \\ 3 & 1 & | & -5 \end{bmatrix}$. Perform Gaussian reduction:

$$\begin{matrix} \\ -3(\text{row 1}) + \text{row 2} \end{matrix} \begin{bmatrix} 1 & -4 & | & -6 \\ 3 & 1 & | & -5 \end{bmatrix} \to \begin{bmatrix} 1 & -4 & | & -6 \\ 0 & 13 & | & 13 \end{bmatrix}$$

Which corresponds to the system
$x - 4y = -6$

$13y = 13.$
Therefore, $y = 1$. Substitute 1 for y to find
$x - 4(1) = -6$

$x = -2$
The solution is the ordered pair (−2, 1).

Chapter 10, Section 10.5

23. The augmented matrix is $\begin{bmatrix} 2 & 1 & | & 5 \\ 3 & -5 & | & 14 \end{bmatrix}$. Perform Gaussian reduction:

$-\frac{3}{2}(\text{row 1}) + \text{row 2} \begin{bmatrix} 2 & 1 & | & 5 \\ 3 & -5 & | & 14 \end{bmatrix} \to \begin{bmatrix} 2 & 1 & | & 5 \\ 0 & -\frac{13}{2} & | & \frac{13}{2} \end{bmatrix}$

Which corresponds to the system
$2x + y = 5$
$-\frac{13}{2}y = \frac{13}{2}.$

Therefore, $y = -1$. Substitute -1 for y to find
$2x + (-1) = 5$

$x = 3.$
The solution is the ordered pair $(3, -1)$.

25. The augmented matrix is $\begin{bmatrix} 1 & -1 & | & -8 \\ 1 & 2 & | & 9 \end{bmatrix}$. Perform Gaussian reduction:

$-(\text{row 1}) + \text{row 2} \begin{bmatrix} 1 & -1 & | & -8 \\ 1 & 2 & | & 9 \end{bmatrix} \to \begin{bmatrix} 1 & -1 & | & -8 \\ 0 & 3 & | & 17 \end{bmatrix}$

which corresponds to the system
$x - y = -8$
$3y = 17.$

Therefore, $y = \frac{17}{3}$. Substitute $\frac{17}{3}$ for y to find

$x - \frac{17}{3} = -8$

$x = -\frac{7}{3}.$

The solution is the ordered pair $\left(-\frac{7}{3}, \frac{17}{3}\right)$.

27. The augmented matrix is $\begin{bmatrix} 1 & 3 & -1 & | & 5 \\ 3 & -1 & 2 & | & 5 \\ 1 & 1 & 2 & | & 7 \end{bmatrix}$. Perform Gaussian reduction:

$$\begin{array}{c} \\ -3(\text{row 1}) + \text{row 2} \\ -(\text{row 1}) + \text{row 3} \end{array} \begin{bmatrix} 1 & 3 & -1 & | & 5 \\ 3 & -1 & 2 & | & 5 \\ 1 & 1 & 2 & | & 7 \end{bmatrix} \rightarrow \begin{array}{c} \\ \\ -\frac{1}{5}(\text{row 2}) + \text{row 3} \end{array} \begin{bmatrix} 1 & 3 & -1 & | & 5 \\ 0 & -10 & 5 & | & -10 \\ 0 & -2 & 3 & | & 2 \end{bmatrix}$$

$$\rightarrow \begin{bmatrix} 1 & 3 & -1 & | & 5 \\ 0 & -10 & 5 & | & -10 \\ 0 & 0 & 2 & | & 4 \end{bmatrix}$$

which corresponds to the system
$x + 3y - z = 5$
$-10y + 5z = -10$
$ 2z = 4.$
Therefore, $z = 2$. Substitute 2 for z to find
$-10y + 5(2) = -10$
$y = 2.$
Substitute 2 for z and 2 for y to find
$x + 3(2) - 2 = 5$
$x = 1.$
The solution is the ordered triplet (1, 2, 2).

29. The augmented matrix is $\begin{bmatrix} 2 & -1 & 1 & | & 5 \\ 1 & -2 & -2 & | & 2 \\ 3 & 3 & -1 & | & 4 \end{bmatrix}$. Perform Gaussian reduction:

$$\text{Swap row 1 and row 2} \begin{bmatrix} 2 & -1 & 1 & | & 5 \\ 1 & -2 & -2 & | & 2 \\ 3 & 3 & -1 & | & 4 \end{bmatrix} \rightarrow \begin{array}{c} \\ -2(\text{row 1}) + \text{row 2} \\ -3(\text{row 1}) + \text{row 3} \end{array} \begin{bmatrix} 1 & -2 & -2 & | & 2 \\ 2 & -1 & 1 & | & 5 \\ 3 & 3 & -1 & | & 4 \end{bmatrix}$$

$$\rightarrow \begin{array}{c} \\ \\ -3(\text{row 2}) + \text{row 3} \end{array} \begin{bmatrix} 1 & -2 & -2 & | & 2 \\ 0 & 3 & 5 & | & 1 \\ 0 & 9 & 5 & | & -2 \end{bmatrix} \rightarrow \begin{bmatrix} 1 & -2 & -2 & | & 2 \\ 0 & 3 & 5 & | & 1 \\ 0 & 0 & -10 & | & -5 \end{bmatrix}$$

which corresponds to the system
$x - 2y - 2z = 2$
$ 3y + 5z = 1$
$ -10z = -5.$

347

Chapter 10, Section 10.5

Therefore, $z = \dfrac{1}{2}$. Substitute $\dfrac{1}{2}$ for z to find

$$3y + 5\left(\dfrac{1}{2}\right) = 1$$

$$y = -\dfrac{1}{2}.$$

Substitute $\dfrac{1}{2}$ for z and $-\dfrac{1}{2}$ for y to find

$$x - 2\left(-\dfrac{1}{2}\right) - 2\left(\dfrac{1}{2}\right) = 2$$

$$x = 2.$$

The solution is the ordered triplet $\left(2, -\dfrac{1}{2}, \dfrac{1}{2}\right)$.

31. The augmented matrix is $\begin{bmatrix} 2 & -1 & -1 & | & -4 \\ 1 & 1 & 1 & | & -5 \\ 1 & 3 & -4 & | & 12 \end{bmatrix}$. Perform Gaussian reduction:

Swap row 1 and row 2 $\begin{bmatrix} 2 & -1 & -1 & | & -4 \\ 1 & 1 & 1 & | & -5 \\ 1 & 3 & -4 & | & 12 \end{bmatrix} \rightarrow \begin{matrix} -2(\text{row 1}) + \text{row 2} \\ -(\text{row 1}) + \text{row 3} \end{matrix} \begin{bmatrix} 1 & 1 & 1 & | & -5 \\ 2 & -1 & -1 & | & -4 \\ 1 & 3 & -4 & | & 12 \end{bmatrix}$

$\rightarrow \begin{bmatrix} 1 & 1 & 1 & | & -5 \\ 0 & -3 & -3 & | & 6 \\ 0 & 2 & -5 & | & 17 \end{bmatrix} \rightarrow \begin{bmatrix} 1 & 1 & 1 & | & -5 \\ 0 & -3 & -3 & | & 6 \\ 0 & 0 & -7 & | & 21 \end{bmatrix}$
$\tfrac{2}{3}(\text{row 2}) + \text{row 3}$

which corresponds to the system

$x + y + z = -5$

$-3y - 3z = 6$

$-7z = 21.$

Therefore, $z = -3$. Substitute -3 for z to find

$-3y - 3(-3) = 6$

$y = 1.$

Substitute -3 for z and 1 for y to find

$x + 1 + (-3) = -5$

$x = -3.$

The solution is the ordered triplet $(-3, 1, -3)$.

33. The augmented matrix is $\begin{bmatrix} 2 & -1 & 0 & | & 0 \\ 0 & 3 & 1 & | & 7 \\ 2 & 0 & 3 & | & 1 \end{bmatrix}$. Perform Gaussian reduction:

$$\begin{array}{c} \\ \\ -(\text{row 1}) + \text{row 3} \end{array} \begin{bmatrix} 2 & -1 & 0 & | & 0 \\ 0 & 3 & 1 & | & 7 \\ 2 & 0 & 3 & | & 1 \end{bmatrix} \rightarrow \text{Swap row 2 and row 3} \begin{bmatrix} 2 & -1 & 0 & | & 0 \\ 0 & 3 & 1 & | & 7 \\ 0 & 1 & 3 & | & 1 \end{bmatrix}$$

$$\rightarrow \begin{array}{c} \\ \\ -3(\text{row 2}) + \text{row 3} \end{array} \begin{bmatrix} 2 & -1 & 0 & | & 0 \\ 0 & 1 & 3 & | & 1 \\ 0 & 3 & 1 & | & 7 \end{bmatrix} \rightarrow \begin{bmatrix} 2 & -1 & 0 & | & 0 \\ 0 & 1 & 3 & | & 1 \\ 0 & 0 & -8 & | & 4 \end{bmatrix}$$

which corresponds to the system
$2x - y = 0$
$y + 3z = 1$
$-8z = 4.$

Therefore, $z = -\dfrac{1}{2}$. Substitute $-\dfrac{1}{2}$ for z to find

$y + 3\left(-\dfrac{1}{2}\right) = 1$

$y = \dfrac{5}{2}.$

Substitute $\dfrac{5}{2}$ for y to find

$2x - \dfrac{5}{2} = 0$

$x = \dfrac{5}{4}.$

The solution is the ordered triplet $\left(\dfrac{5}{4}, \dfrac{5}{2}, -\dfrac{1}{2}\right)$.

Chapter 10 Review

1. The graph of $g(x) = |x| + 2$ is a translation of the basic graph of $y = |x|$ shifted upward two units. Refer to the graph in the back of the textbook.

2. The graph of $F(t) = \dfrac{1}{t} - 2$ is a translation of the basic graph of $y = \dfrac{1}{t}$ shifted downward two units. Refer to the graph in the back of the textbook.

Chapter 10 Review

3. The graph of $f(s) = \sqrt{s} + 3$ is a translation of the basic graph of $y = \sqrt{s}$ shifted upward three units. Refer to the graph in the back of the textbook.

4. The graph of $h(r) = (r-2)^3$ is a translation of the basic graph of $y = r^3$ shifted two units to the right. Refer to the graph in the back of the textbook.

5. The graph of $f(x) = (x-2)^2 - 4$ is a translation of the basic graph of $y = x^2$ shifted two units to the right and downward four units. Refer to the graph in the back of the textbook.

6. The graph of $g(u) = \sqrt{u+2} - 3$ is a translation of the basic graph of $y = \sqrt{u}$ shifted two units to the left and downward three units. Refer to the graph in the back of the textbook.

7. The graph of $G(t) = |t+2| - 3$ is a translation of the basic graph of $y = |t|$ shifted two units to the left and downward three units. Refer to the graph in the back of the textbook.

8. The graph of $H(t) = \dfrac{1}{(t-2)^2} + 3$ is a translation of the basic graph of $y = \dfrac{1}{t^2}$ shifted two units to the right and upward three units. Refer to the graph in the back of the textbook.

9. The graph of $h(s) = -2\sqrt{s}$ is an expansion of the basic graph of $y = \sqrt{s}$ by a factor of two combined with a reflection about the x-axis. Refer to the graph in the back of the textbook.

10. The graph of $g(s) = \dfrac{1}{2|s|}$ is a compression of the basic graph of $y = \dfrac{1}{|s|}$ by a factor of $\dfrac{1}{2}$. Refer to the graph in the back of the textbook.

11. The graph is a translation of the basic graph $y = x^2$, shifted two units to the right and downward three units.
$$y = (x-2)^2 - 3$$

12. The graph is a translation of the basic graph $y = |x|$ reflected about the x-axis, shifted upward three units.
$$y = -|x| + 3$$

13. $x^2 + y^2 = 9$
$x^2 + y^2 = 3^2$
This is the equation for a circle of radius 3 centered at (0, 0). Refer to the graph in the back of the textbook.

14. $(x-2)^2 + (y+3)^2 = 16$
$(x-2)^2 + (y+3)^2 = 4^2$
This is the equation for a circle of radius 4 centered at (2, −3). Refer to the graph in the back of the textbook.

15. $x^2 + y^2 - 4x + 2y - 4 = 0$

$(x^2 - 4x + 4) + (y^2 + 2y + 1)$

$ = 4 + 4 + 1$

$(x - 2)^2 + (y + 1)^2 = 3^2$

This is the equation for a circle of radius 3 centered at (2, −1). Refer to the graph in the back of the textbook.

16. $x^2 + y^2 - 6y - 4 = 0$

$x^2 + (y^2 - 6y + 9) = 4 + 9$

$x^2 + (y - 3)^2 = \left(\sqrt{13}\right)^2$

This is the equation for a circle of radius $\sqrt{13}$ centered at (0, 3). Refer to the graph in the back of the textbook.

17. $\dfrac{x^2}{9} + y^2 = 1$

$\dfrac{x^2}{3^2} + \dfrac{y^2}{1^2} = 1$

This is the equation for an ellipse centered at (0, 0). The major axis is horizontal, $a^2 = 9$, and $b^2 = 1$. The vertices lie three units to the left and right of the center at (−3, 0) and (3, 0), and the covertices lie one unit above and below the center at (0, 1) and (0, −1). Refer to the graph in the back of the textbook.

18. $\dfrac{x^2}{4} + \dfrac{y^2}{16} = 1$

$\dfrac{x^2}{2^2} + \dfrac{y^2}{4^2} = 1$

This is the equation for an ellipse centered at (0, 0). The major axis is vertical, $a^2 = 16$, and $b^2 = 4$. The vertices lie four units above and below the center at (0, 4) and (0, −4), and the covertices lie two units to the left and right of the center at (−2, 0) and (2, 0). Refer to the graph in the back of the textbook.

19. $\dfrac{(x-2)^2}{4} + \dfrac{(y+3)^2}{9} = 1$

$\dfrac{(x-2)^2}{2^2} + \dfrac{(y+3)^2}{3^2} = 1$

This is the equation for an ellipse centered at (2, −3). The major axis is vertical, $a^2 = 9$, and $b^2 = 4$. The vertices lie three units above and below the center at (2, 0) and (2, −6) and the covertices lie two units to the left and right of the center at (0, −3) and (4, −3). Refer to the graph in the back of the textbook.

20. $\dfrac{(x+4)^2}{12} + \dfrac{(y-2)^2}{6} = 1$

$\dfrac{(x+4)^2}{\left(\sqrt{12}\right)^2} + \dfrac{(y-2)^2}{\left(\sqrt{6}\right)^2} = 1$

This is the equation for an ellipse centered at (−4, 2). the major axis is horizontal, $a^2 = 12$, and $b^2 = 6$. the vertices lie $\sqrt{12}$ to the left and right of the center at approximately (−7.5, 2) and (−0.5, 2), and the covertices lie $\sqrt{6}$ units above and below the center at approximately (−4, 4.4) and (−4, −0.4). Refer to the graph in the back of the textbook.

Chapter 10 Review

21. $4x^2 + y^2 - 16x + 4y + 4 = 0$

$4(x^2 - 4x + 4) + (y^2 + 4y + 4)$
$= -4 + 16 + 4$

$4(x-2)^2 + (y+2)^2 = 16$

$\dfrac{(x-2)^2}{4} + \dfrac{(y+2)^2}{16} = 1$

$\dfrac{(x-2)^2}{2^2} + \dfrac{(y+2)^2}{4^2} = 1$

This is the equation for an ellipse centered at (2, –2). The major axis is vertical, $a^2 = 16$, and $b^2 = 4$. The vertices lie four units above and below the center at (2, 2) and (2, –6), and the covertices lie two units to the left and right of the center at (0, –2) and (4, –2). Refer to the graph in the back of the textbook.

22. $8x^2 + 5y^2 + 16x - 20y - 12 = 0$

$8(x^2 + 2x + 1) + 5(y^2 - 4y + 4)$
$= 12 + 8 + 20$

$8(x+1)^2 + 5(y-2)^2 = 40$

$\dfrac{(x+1)^2}{5} + \dfrac{(y-2)^2}{8} = 1$

$\dfrac{(x+1)^2}{\left(\sqrt{5}\right)^2} + \dfrac{(y-2)^2}{\left(\sqrt{8}\right)^2} = 1$

This is the equation for an ellipse centered at (–1, 2). The major axis is vertical, $a^2 = 8$, and $b^2 = 5$. The vertices lie $\sqrt{8}$ units above and below the center at approximately (–1, 4.8) and (–1, –0.8), and the covertices lie $\sqrt{5}$ units to the left and right of the center at approximately (–3.2, 2) and (1.2, 2). Refer to the graph in the back of the textbook.

23. $4(y-2) = (x+3)^2$

$(x+3)^2 = 4(1)(y-2)$

This is the equation for a parabola with vertex at (–3, 2) and opening upward. $p = 1$, the focus is at (–3, 3), and the directrix is the horizontal line $y = 1$. Refer to the graph in the back of the textbook.

24. $(x-2)^2 + 4y = 4$

$(x-2)^2 = -4y + 4$

$(x-2)^2 = -4(1)(y-1)$

This is the equation for a parabola with vertex at (2, 1) and opening downward. $p = 1$, the focus is at (2, 0), and the directrix is the horizontal line $y = 2$. Refer to the graph in the back of the textbook.

25. $x^2 - 8x - y + 6 = 0$

$(x^2 - 8x + 16) = y - 6 + 16$

$(x-4)^2 = 4\left(\dfrac{1}{4}\right)(y+10)$

This is the equation for a parabola with vertex at (4, –10) and opening upward. $p = \dfrac{1}{4}$, the focus is at $\left(4, -\dfrac{39}{4}\right)$, and the directrix is the horizontal line $y = -\dfrac{41}{4}$. Refer to the graph in the back of the textbook.

26. $y^2 + 6y + 4x + 1 = 0$

$(y^2 + 6y + 9) = -4x - 1 + 9$

$(y+3)^2 = -4(1)(x-2)$

This is the equation for a parabola with vertex at $(2, -3)$ and opening to the left. $p = 1$, the focus is at $(1, -3)$, and the directrix is the vertical line $x = 3$. Refer to the graph in the back of the textbook.

27. $x^2 + y = 4x - 6$

$x^2 - 4x + 4 = -y - 6 + 4$

$(x-2)^2 = -4\left(\dfrac{1}{4}\right)(y+2)$

This is the equation for a parabola with vertex at $(2, -2)$ and opening downward. $p = \dfrac{1}{4}$, the focus is at $\left(2, -\dfrac{9}{4}\right)$, and the directrix is the horizontal line $y = -\dfrac{7}{4}$. Refer to the graph in the back of the textbook.

28. $y^2 = 2y + 2x + 2$

$(y^2 - 2y + 1) = 2x + 2 + 1$

$(y-1)^2 = 4\left(\dfrac{1}{2}\right)\left(x + \dfrac{3}{2}\right)$

This is the equation for a parabola with vertex at $\left(-\dfrac{3}{2}, 1\right)$ and opening to the right. $p = \dfrac{1}{2}$, the focus is at $\left(-\dfrac{3}{2}, \dfrac{3}{2}\right)$, and the directrix is the vertical line $x = \dfrac{1}{2}$. Refer to the graph in the back of the textbook.

29. $\dfrac{y^2}{6} - \dfrac{x^2}{8} = 1$

$\dfrac{y^2}{\left(\sqrt{6}\right)^2} - \dfrac{x^2}{\left(\sqrt{8}\right)^2} = 1$

This is the equation for a hyperbola centered at $(0, 0)$. The transverse axis lies on the y-axis, $a^2 = 6$, and $b^2 = 8$. The vertices lie $\sqrt{6}$ units above and below the center at approximately $(0, 2.4)$ and $(0, -2.4)$. Refer to the graph in the back of the textbook.

30. $\dfrac{(x-2)^2}{4} - \dfrac{(y+3)^2}{9} = 1$

$\dfrac{(x-2)^2}{2^2} - \dfrac{(y+3)^2}{3^2} = 1$

This is the equation for a hyperbola centered at $(2, -3)$. The transverse axis is parallel to the x-axis, $a^2 = 4$, and $b^2 = 9$. The vertices lie two units to the left and right of the center at $(0, -3)$ and $(4, -3)$. Refer to the graph in the back of the textbook.

Chapter 10 Review

31. $2y^2 - 3x^2 - 16y - 12x + 8 = 0$

$2(y^2 - 8y + 16) - 3(x^2 + 4x + 4)$
$\qquad = -8 + 32 - 12$

$2(y - 4)^2 - 3(x + 2)^2 = 12$

$\dfrac{(y-4)^2}{6} - \dfrac{(x+2)^2}{4} = 1$

$\dfrac{(y-4)^2}{\left(\sqrt{6}\right)^2} - \dfrac{(x+2)^2}{2^2} = 1$

This is the equation for a hyperbola centered at (−2, 4). The transverse axis is parallel to the y-axis, $a^2 = 6$, and $b^2 = 4$. The vertices lie $\sqrt{6}$ units above and below the center at approximately (−2, 6.4) and (−2, 1.6). Refer to the graph in the back of the textbook.

32. $9x^2 - 4y^2 - 72x - 24y + 72 = 0$

$9(x^2 - 8x + 16) - 4(y^2 + 6y + 9)$
$\qquad = -72 + 144 - 36$

$9(x - 4)^2 - 4(y + 3)^2 = 36$

$\dfrac{(x-4)^2}{4} - \dfrac{(y+3)^2}{9} = 1$

$\dfrac{(x-4)^2}{2^2} - \dfrac{(y+3)^2}{3^2} = 1$

This is the equation for a hyperbola centered at (4, −3). The transverse axis is parallel to the x-axis, $a^2 = 4$, and $b^2 = 9$. The vertices lie two units to the left and right of the center at (2, −3) and (6, −3). Refer to the graph in the back of the textbook.

33. $2x^2 - y^2 + 6y - 19 = 0$

$2x^2 - (y^2 - 6y + 9) = 19 - 9$

$2x^2 - (y - 3)^2 = 10$

$\dfrac{x^2}{5} - \dfrac{(y-3)^2}{10} = 1$

$\dfrac{x^2}{\left(\sqrt{5}\right)^2} - \dfrac{(y-3)^2}{\left(\sqrt{10}\right)^2} = 1$

This is the equation for a hyperbola centered at (0, 3). The transverse axis is parallel to the x-axis, $a^2 = 5$, and $b^2 = 9$. The vertices lie $\sqrt{5}$ units to the left and right of the center at approximately (−2.2, 3) and (2.2, 3). Refer to the graph in the back of the textbook.

34. $4y^2 - x^2 + 8x - 28 = 0$

$4y^2 - (x^2 - 8x + 16) = 28 - 16$

$4y^2 - (x - 4)^2 = 12$

$\dfrac{y^2}{3} - \dfrac{(x-4)^2}{12} = 1$

$\dfrac{y^2}{\left(\sqrt{3}\right)^2} - \dfrac{(x-4)^2}{\left(\sqrt{12}\right)^2} = 1$

This is the equation for a hyperbola centered at (4, 0). The transverse axis is parallel to the y-axis, $a^2 = 3$, and $b^2 = 12$. The vertices lie $\sqrt{3}$ units above and below the center at approximately (4, 1.7) and (4, −1.7). Refer to the graph in the back of the textbook.

Chapter 10 Review

35. $(x+4)^2 + (y-3)^2 = (2\sqrt{5})^2$

$(x+4)^2 + (y-3)^2 = 20$

36. The center is between the two endpoints at
$\left(\dfrac{-5+1}{2}, \dfrac{2+6}{2}\right) = (-2, 4)$.
The length of the diameter is
$\sqrt{(1+5)^2 + (6-2)^2} = \sqrt{52}$
$= 2\sqrt{13}$.
Therefore, the radius is $\sqrt{13}$.
$(x+2)^2 + (y-4)^2 = (\sqrt{13})^2$
$(x+2)^2 + (y-4)^2 = 13$

37. $\dfrac{(x+1)^2}{2^2} + \dfrac{(y-4)^2}{4^2} = 1$

$\dfrac{(x+1)^2}{4} + \dfrac{(y-4)^2}{16} = 1$

38. The center is between the vertices at (3, 1). The major axis length is the distance between the vertices, which is 10 units. The minor axis length is the distance between the covertices, which is 4 units. Therefore, $2a = 10$, so $a = 5$, and $2b = 4$, so $b = 2$.
$\dfrac{(x-3)^2}{2^2} + \dfrac{(y-1)^2}{5^2} = 1$

$\dfrac{(x-3)^2}{4} + \dfrac{(y-1)^2}{25} = 1$

39. p is equal to the distance from the focus to the vertex, which is 4. The parabola opens upward since the focus is above the vertex.
$(x-0)^2 = 4(4)(y-0)$
$x^2 = 16y$

40. p is equal to the distance from the focus to the vertex, which is 4. The parabola opens to the left since the focus is to the left of the vertex.
$(y-4)^2 = -4(4)(x-2)$
$(y-4)^2 = -16(x-2)$

41. p is equal to the distance from the vertex to the directrix, which is 2. The parabola opens upward since the directrix is below the vertex.
$(x-2)^2 = 4(2)(y+3)$
$(x-2)^2 = 8(y+3)$

42. The vertex is between the focus and directrix at $(-1, 1)$. p is equal to the distance from the focus to the vertex, which is 3. The parabola opens to the left since the focus is to the left of the vertex.
$(y-1)^2 = -4(3)(x+1)$
$(y-1)^2 = -12(x+1)$

43. Write the equation for f in the form $y = x + 4$. Interchange the variables and solve for y:
$x = y + 4$
$y = x - 4$
Therefore, $f^{-1}(x) = x - 4$.

Chapter 10 Review

44. Write the equation for f in the form $y = \dfrac{x-2}{4}$. Interchange the variables and solve for y:

$x = \dfrac{y-2}{4}$

$4x = y - 2$

$y = 4x + 2$

Therefore, $f^{-1}(x) = 4x + 2$.

45. Write the equation for f in the form $y = x^3 - 1$. Interchange the variables and solve for y:

$x = y^3 - 1$

$y^3 = x + 1$

$y = \sqrt[3]{x+1}$

Therefore, $f^{-1}(x) = \sqrt[3]{x+1}$.

46. Write the equation for f in the form $y = \dfrac{1}{x+2}$. Interchange the variables and solve for y:

$x = \dfrac{1}{y+2}$

$y + 2 = \dfrac{1}{x}$

$y = \dfrac{1}{x} - 2$

Therefore, $f^{-1}(x) = \dfrac{1}{x} - 2$.

47. Write the equation for f in the form $y = \dfrac{1}{x} + 2$. Interchange the variables and solve for y:

$x = \dfrac{1}{y} + 2$

$\dfrac{1}{y} = x - 2$

$y = \dfrac{1}{x-2}$

Therefore, $f^{-1}(x) = \dfrac{1}{x-2}$.

48. Write the equation for f in the form $y = \sqrt[3]{x} - 2$. Interchange the variables and solve for y:

$x = \sqrt[3]{y} - 2$

$\sqrt[3]{y} = x + 2$

$y = (x+2)^3$

Therefore $f^{-1}(x) = (x+2)^3$.

49. Solve $F(t) = 2$.

$\dfrac{3}{4}t + 2 = 2$

$\dfrac{3}{4}t = 0$

$t = 0$

Therefore, $F^{-1}(2) = 0$

50. Solve $G(x) = 3$.

$\dfrac{1}{x} - 4 = 3$

$\dfrac{1}{x} = 7$

$x = \dfrac{1}{7}$

Therefore, $G^{-1}(3) = \dfrac{1}{7}$.

Chapter 10 Review

51. Graph $f(x) = x + 4$ and $f^{-1}(x) = x - 4$. Refer to the graph in the back of the textbook.

52. Graph $f(x) = \dfrac{x-2}{4}$ and $f^{-1}(x) = 4x + 2$. Refer to the graph in the back of the textbook.

53. Graph $f(x) = x^3 - 1$ and $f^{-1}(x) = \sqrt[3]{x+1}$. Refer to the graph in the back of the textbook.

54. Graph $f(x) = \dfrac{1}{x+2}$ and $f^{-1}(x) = \dfrac{1}{x} - 2$. Refer to the graph in the back of the textbook.

55. Graph $f(x) = \dfrac{1}{x} + 2$ and $f^{-1}(x) = \dfrac{1}{x-2}$. Refer to the graph in the back of the textbook.

56. Graph $f(x) = \sqrt[3]{x} - 2$ and $f^{-1}(x) = (x+2)^3$. Refer to the graph in the back of the textbook.

57. The augmented matrix is $\begin{bmatrix} 1 & -2 & | & 5 \\ 2 & 1 & | & 5 \end{bmatrix}$. Perform Gaussian reduction:

$-2(\text{row 1}) + \text{row 2}$ $\begin{bmatrix} 1 & -2 & | & 5 \\ 2 & 1 & | & 5 \end{bmatrix} \rightarrow \begin{bmatrix} 1 & -2 & | & 5 \\ 0 & 5 & | & -5 \end{bmatrix}$

which corresponds to the system
$x - 2y = 5$
$5y = -5$.
Therefore $y = -1$. Substitute -1 for y to find
$x - 2(-1) = 5$
$x = 3$.
The solution is the ordered pair $(3, -1)$.

58. The augmented matrix is $\begin{bmatrix} 4 & -3 & | & 16 \\ 2 & 1 & | & 8 \end{bmatrix}$. Perform Gaussian reduction:

$-\tfrac{1}{2}(\text{row 1}) + \text{row 2}$ $\begin{bmatrix} 4 & -3 & | & 16 \\ 2 & 1 & | & 8 \end{bmatrix} \rightarrow \begin{bmatrix} 4 & -3 & | & 16 \\ 0 & \tfrac{5}{2} & | & 0 \end{bmatrix}$

which corresponds to the system
$4x - 3y = 16$
$\dfrac{5}{2} y = 0$.

Chapter 10 Review

Therefore, $y = 0$. Substitute 0 for y to find
$4x - 3(0) = 16$
$x = 4$.
The solution is the ordered pair $(4, 0)$.

59. The augmented matrix is $\begin{bmatrix} 2 & -1 & | & 7 \\ 3 & 2 & | & 14 \end{bmatrix}$. Perform Gaussian reduction:

$-\frac{3}{2}(\text{row 1}) + \text{row 2} \begin{bmatrix} 2 & -1 & | & 7 \\ 3 & 2 & | & 14 \end{bmatrix} \rightarrow \begin{bmatrix} 2 & -1 & | & 7 \\ 0 & \frac{7}{2} & | & \frac{7}{2} \end{bmatrix}$

which corresponds to the system
$2x - y = 7$
$\frac{7}{2}y = \frac{7}{2}$.

Therefore, $y = 1$. Substitute 1 for y to find
$2x - 1 = 7$
$x = 4$.
The solution is the ordered pair $(4, 1)$.

60. The augmented matrix is $\begin{bmatrix} 2 & -1 & 3 & | & -6 \\ 1 & 2 & -1 & | & 7 \\ 3 & 1 & 1 & | & 2 \end{bmatrix}$. Perform Gaussian reduction:

Swap row 1 and row 2 $\begin{bmatrix} 2 & -1 & 3 & | & -6 \\ 1 & 2 & -1 & | & 7 \\ 3 & 1 & 1 & | & 2 \end{bmatrix} \rightarrow \begin{matrix} -2(\text{row 1}) + \text{row 2} \\ -3(\text{row 1}) + \text{row 3} \end{matrix} \begin{bmatrix} 1 & 2 & -1 & | & 7 \\ 2 & -1 & 3 & | & -6 \\ 3 & 1 & 1 & | & 2 \end{bmatrix}$

$\rightarrow \begin{bmatrix} 1 & 2 & -1 & | & 7 \\ 0 & -5 & 5 & | & -20 \\ 0 & -5 & 4 & | & -19 \end{bmatrix} \rightarrow \begin{bmatrix} 1 & 2 & -1 & | & 7 \\ 0 & -5 & 5 & | & -20 \\ 0 & 0 & -1 & | & 1 \end{bmatrix}$
$-(\text{row 2}) + \text{row 3}$

which corresponds to the system
$x + 2y - z = 7$
$-5y + 5z = -20$
$-z = 1$.

Therefore, $z = -1$. Substitute -1 for z to find
$-5y + 5(-1) = -20$
$y = 3$.

Substitute −1 for z and 3 for y to find
$x + 2(3) − (−1) = 7$
$x = 0$.
The solution is the ordered triplet $(0, 3, −1)$.

61. The augmented matrix is $\begin{bmatrix} 1 & 2 & -1 & | & -3 \\ 2 & -3 & 2 & | & 2 \\ 1 & -1 & 4 & | & 7 \end{bmatrix}$. Perform Gaussian reduction:

$\begin{array}{c} -2(\text{row 1}) + \text{row 2} \\ -(\text{row 1}) + \text{row 3} \end{array} \begin{bmatrix} 1 & 2 & -1 & | & -3 \\ 2 & -3 & 2 & | & 2 \\ 1 & -1 & 4 & | & 7 \end{bmatrix} \rightarrow \begin{array}{c} \\ -\frac{3}{7}(\text{row 2}) + \text{row 3} \end{array} \begin{bmatrix} 1 & 2 & -1 & | & -3 \\ 0 & -7 & 4 & | & 8 \\ 0 & -3 & 5 & | & 10 \end{bmatrix}$

$\rightarrow \begin{bmatrix} 1 & 2 & -1 & | & -3 \\ 0 & -7 & 4 & | & 8 \\ 0 & 0 & \frac{23}{7} & | & \frac{46}{7} \end{bmatrix}$

which corresponds to the system
$x + 2y − z = −3$
$−7y + 4z = 8$
$\frac{23}{7}z = \frac{46}{7}$.

Therefore, $z = 2$. Substitute 2 for z to find
$−7y + 4(2) = 8$
$y = 0$.
Substitute 2 for z and 0 for y to find
$x + 2(0) − 2 = −3$
$x = −1$.
The solution is the ordered triplet $(−1, 0, 2)$.

Chapter 10 Review

62. The augmented matrix is $\begin{bmatrix} 1 & 1 & 1 & | & 1 \\ 2 & -1 & -1 & | & 2 \\ 2 & -1 & 3 & | & 2 \end{bmatrix}$. Perform Gaussian reduction:

$\begin{matrix} -2(\text{row 1}) + \text{row 2} \\ -2(\text{row 1}) + \text{row 3} \end{matrix} \begin{bmatrix} 1 & 1 & 1 & | & 1 \\ 2 & -1 & -1 & | & 2 \\ 2 & -1 & 3 & | & 2 \end{bmatrix} \rightarrow \begin{matrix} \\ -(\text{row 2}) + \text{row 3} \end{matrix} \begin{bmatrix} 1 & 1 & 1 & | & 1 \\ 0 & -3 & -3 & | & 0 \\ 0 & -3 & 1 & | & 0 \end{bmatrix}$

$\rightarrow \begin{bmatrix} 1 & 1 & 1 & | & 1 \\ 0 & -3 & -3 & | & 0 \\ 0 & 0 & 4 & | & 0 \end{bmatrix}$

which corresponds to the system
$x + y + z = 1$
$-3y - 3z = 0$
$\quad\quad\;\; 4z = 0.$
Therefore, $z = 0$. Substitute 0 for z to find
$-3y - 3(0) = 0$
$y = 0.$
Substitute 0 for z and 0 for y to find
$x + 0 + 0 = 1$
$x = 1.$
The solution is the ordered triplet $(1, 0, 0)$.

63. Let d be the number of dimes and q be the number of quarters. Then the system of equations is
$d = q + 25$ or $d - q = 25$
$0.10d + 0.25q = 4.95.$

The augmented matrix is $\begin{bmatrix} 1 & -1 & | & 25 \\ 0.10 & 0.25 & | & 4.95 \end{bmatrix}$. Perform Gaussian reduction:

$-0.10(\text{row 1}) + \text{row 2} \begin{bmatrix} 1 & -1 & | & 25 \\ 0.10 & 0.25 & | & 4.95 \end{bmatrix} \rightarrow \begin{bmatrix} 1 & -1 & | & 25 \\ 0 & 0.35 & | & 2.45 \end{bmatrix}$

which corresponds to the system
$d - q = 25$
$0.35q = 2.45.$
Therefore $q = 7$. Substitute 7 for q to find
$d - 7 = 25$
$d = 32.$
There are 32 dimes.

Chapter 10 Review

64. Let x be the number of first class tickets and y be the number of tourist tickets. Then the system of equations is
$$x + y = 64$$
$$280x + 160y = 12{,}160.$$

The augmented matrix is $\begin{bmatrix} 1 & 1 & | & 64 \\ 280 & 160 & | & 12{,}160 \end{bmatrix}$. Perform Gaussian reduction:

$-280(\text{row 1}) + \text{row 2}$ $\begin{bmatrix} 1 & 1 & | & 64 \\ 280 & 160 & | & 12{,}160 \end{bmatrix} \rightarrow \begin{bmatrix} 1 & 1 & | & 64 \\ 0 & -120 & | & -5760 \end{bmatrix}$

which corresponds to the system
$$x + y = 64$$
$$-120y = -5760.$$
Therefore, $y = 48$. Substitute 48 for y to find
$x + 48 = 64$
$x = 16.$
There are 16 first-class tickets and 48 tourist tickets sold.

65. The equation of the parabola can be expressed in the form $y = ax^2 + bx + c$. Substitute the three points (0, 3), (1, 8), and (3, 30) into the equation to get the system:
$$c = 3$$
$$a + b + c = 8$$
$$9a + 3b + c = 30$$
Since $c = 3$, we get the system:
$$a + b = 5$$
$$9a + 3b = 27$$
Multiply the first equation by -3 and add to the second to get $6a = 12$.

Therefore $a = 2$ and $b = 3$. An equation for the parabola is $y = 2x^2 + 3x + 3$.

66. Let x be the amount of cranberry juice, y be the amount of apricot nectar, and z be the amount of club soda. The system of equations is
$$x + y + z = 1$$
$$1200x + 1600y = 800$$
$$25x + 20y + 5y = 16$$

Chapter 10 Review

The augmented matrix is $\begin{bmatrix} 1 & 1 & 1 & | & 1 \\ 1200 & 1600 & 0 & | & 800 \\ 25 & 20 & 5 & | & 16 \end{bmatrix}$. Perform Gaussian reduction:

$-1200(\text{row 1}) + \text{row 2}$
$-25(\text{row 1}) + \text{row 3}$
$\begin{bmatrix} 1 & 1 & 1 & | & 1 \\ 1200 & 1600 & 0 & | & 800 \\ 25 & 20 & 5 & | & 16 \end{bmatrix}$

$\rightarrow$
$\frac{1}{80}(\text{row 2}) + \text{row 3}$
$\begin{bmatrix} 1 & 1 & 1 & | & 1 \\ 0 & 400 & -1200 & | & -400 \\ 0 & -5 & -20 & | & -9 \end{bmatrix} \rightarrow \begin{bmatrix} 1 & 1 & 1 & | & 1 \\ 0 & 400 & -1200 & | & -400 \\ 0 & 0 & -35 & | & -14 \end{bmatrix}$

which corresponds to the system
$x + y + z = 1$
$400y - 1200z = -400$
$-35z = -14$.

Therefore, $z = 0.4$. Substitute 0.4 for z to find
$400y - 1200(0.4) = -400$
$y = 0.2$.
Substitute 0.4 for z and 0.2 for y to find
$x + 0.2 + 0.4 = 1$
$x = 0.4$.
Therefore, Juan should use 0.4 quarts of cranberry juice, 0.2 quarts of apricot nectar, and 0.4 quarts of club soda.

Appendices

Appendix 1

1. $-\frac{5}{8}$ is negative, so it is not in N or W. It cannot be reduced so it is not in J. Because it can be written as $\frac{-5}{8}$, $-\frac{5}{8} \in Q$ and R.

3. Since 8 is not a perfect square (like 4 or 16), $\sqrt{8}$ is not a rational number. $\sqrt{8} \in H$ and R.

5. $-36 \in J$, Q, and R. It is negative so it is not an element of N or W. $-36 = \frac{-36}{1}$ so it is a rational number.

7. $0 \in W$, J, Q, and R. 0 is not a natural number. Since $0 = \frac{0}{1}$, 0 is a rational number.

9. $13.\overline{289} \in Q$ and R. The decimal representation is repeating, as indicated by the bar, so $13.\overline{289}$ is a rational number.

11. $6.468725... \in H$ and R. The ellipsis (...) indicates that the decimal representation continues without repeating, so $6.468725...$ is an irrational number.

Appendix 2

1. $4y(x - 2y) = 4y(x) + 4y(-2)$
   $= 4xy - 8y^2$

3. $-6x(2x^2 - x + 1)$
   $= -6x(2x^2) - 6x(-x) - 6x(1)$
   $= -12x^3 + 6x^2 - 6x$

5. $a^2b(3a^2 - 2ab - b)$
   $= a^2b(3a^2) + a^2b(-2ab) + a^2b(-b)$
   $= 3a^4b - 2a^3b^2 - a^2b^2$

7. $2x^2y^3(4xy^4 - 2x^2y - 3x^3y^2)$
   $= 2x^2y^3(4xy^4) + 2x^2y^3(-2x^2y)$
   $\quad + 2x^2y^3(-3x^3y^2)$
   $= 8x^3y^7 - 4x^4y^4 - 6x^5y^5$

9. $(n + 2)(n + 8) = n^2 + 8n + 2n + 16$
   $= n^2 + 10n + 16$

11. $(r + 5)(r - 2) = r^2 - 2r + 5r - 10$
    $= r^2 + 3r - 10$

13. $(2z + 1)(z - 3) = 2z^2 - 6z + z - 3$
    $= 2z^2 - 5z - 3$

15. $(4r + 3s)(2r - s)$
    $= 8r^2 - 4rs + 6rs - 3s^2$
    $= 8r^2 + 2rs - 3s^2$

Appendix 2

17. $(2x - 3y)(3x - 2y)$
$= 6x^2 - 4xy - 9xy + 6y^2$
$= 6x^2 - 13xy + 6y^2$

19. $(3t - 4s)(3t + 4s)$
$= 9t^2 + 12st - 12st - 16s^2$
$= 9t^2 - 16s^2$

21. $(2a^2 + b^2)(a^2 - 3b^2)$
$= 2a^4 - 6a^2b^2 + a^2b^2 - 3b^4$
$= 2a^4 - 5a^2b^2 - 3b^4$

23. $4x^2z + 8xz = (4xz)(x) + (4xz)(2)$
$= 4xz(x + 2)$

25. $3n^4 - 6n^3 + 12n^2$
$= (3n^2)(n^2) + (3n^2)(-2n) + 3n^2(4)$
$= 3n^2(n^2 - 2n + 4)$

27. $15r^2s + 18rs^2 - 3r$
$= (3r)(5rs) + (3r)(6s^2) + (3r)(-1)$
$= 3r(5rs + 6s^2 - 1)$

29. $3m^2n^4 - 6m^3n^3 + 14m^3n^2$
$= (m^2n^2)(3n^2) + (m^2n^2)(-6mn)$
$\qquad + (m^2n^2)(14m)$
$= m^2n^2(3n^2 - 6mn + 14m)$

31. $15a^4b^3c^4 - 12a^2b^2c^5 + 6a^2b^3c^4$
$= (3a^2b^2c^4)(5a^2b)$
$\qquad + (3a^2b^2c^4)(-4c)$
$\qquad + (3a^2b^2c^4)(2b)$
$= 3a^2b^2c^4(5a^2b - 4c + 2b)$

33. $a(a + 3) + b(a + 3) = (a + b)(a + 3)$

35. $y(y - 2) - 3x(y - 2)$
$= (y - 3x)(y - 2)$

37. $4(x - 2)^2 - 8x(x - 2)^3 = [4(x - 2)^2][1] - [4(x - 2)^2][2x(x - 2)]$
$= [4(x - 2)^2][1 - 2x(x - 2)]$
$= 4(x - 2)^2(1 - 2x^2 + 4x)$
$= 4(x - 2)^2(-2x^2 + 4x + 1)$

39. $x(x - 5)^2 - x^2(x - 5)^3 = [x(x - 5)^2][1] - [x(x - 5)^2][x(x - 5)]$
$= [x(x - 5)^2][1 - x(x - 5)]$
$= x(x - 5)^2(1 - x^2 + 5x)$
$= x(x - 5)^2(-x^2 + 5x + 1)$

41. $3m - 2n = (-1)(-3m) + (-1)(2n)$
$= -(-3m + 2n)$
$= -(2n - 3m)$

43. $-2x + 2 = (-2)(x) + (-2)(-1)$
$= -2(x - 1)$

45. $-ab - ac = (-a)(b) + (-a)(c)$
$= (-a)(b + c)$
$= -a(b + c)$

47. $2x - y + 3z$
$= (-1)(-2x) + (-1)(y) + (-1)(-3z)$
$= -(-2x + y - 3z)$

49. Since $2 + 3 = 5$ and $2 \cdot 3 = 6$,
$x^2 + 5x + 6 = (x + 2)(x + 3)$

51. Since $4 \cdot 3 = 12$ and $-4 - 3 = -7$,
$y^2 - 7y + 12 = (y - 3)(y - 4)$

53. $x^2 - 6 - x = x^2 - x - 6$
Since $2 \cdot 3 = 6$ and $2 - 3 = -1$,
$x^2 - x - 6 = (x - 3)(x + 2)$

55. $2x^2 + 3x - 2$; 2 factors as $1 \cdot 2$ and -2 factors as $-1 \cdot 2$ or $1 \cdot (-2)$, so the possibilities are:
$(2x - 1)(x + 2)$
$(2x + 1)(x - 2)$
$(x - 1)(2x + 2)$
$(x + 1)(2x - 2)$
From checkintg the middle terms of the product,
$2x^2 + 3x - 2 = (2x - 1)(x + 2)$

57. $7x + 4x^2 - 2 = 4x^2 + 7x - 2$; 4 factors as $1 \cdot 4$ or $2 \cdot 2$; -2 factors as $-1 \cdot 2$ or $1 \cdot (-2)$. The possibilities are:
$(x - 1)(4x + 2), (x + 1)(4x - 2),$
$(4x - 1)(x + 2), (4x + 1)(x - 2),$
$(2x - 1)(2x + 2), (2x + 1)(2x - 2).$
Check the middle terms of the product.
$4x^2 + 7x - 2 = (4x - 1)(x + 2)$

59. $9y^2 - 21y - 8$
$= (\ - \ )(\ + \ )$ or $(\ + \ )(\ - \ )$
$9 = 1 \cdot 9$ or $3 \cdot 3$;
$-8 = (-1) \cdot 8, \ -8 \cdot (1), \ (2)(-4),$
or $(-2)(4)$
$3 \cdot (-8) + 3 \cdot 1 = -24 + 3 = -21$ so
$9y^2 - 21y - 8 = (3y + 1)(3y - 8)$

61. $10u^2 - 3 - u = 10u^2 - u - 3$
$10 = 1 \cdot 10$ or $2 \cdot 5$,
$-3 = 1 \cdot (-3)$ or $(-1) \cdot 3$
$2 \cdot (-3) + 5 \cdot 1 = -6 + 5 = -1$
so $10u^2 - u - 3 = (2u + 1)(5u - 3)$

63. $21x^2 - 43x - 14$;
$21 = 3 \cdot 7,$
$-14 = (-1) \cdot 14, \ 1 \cdot (-14), \ (-2) \cdot 7,$
or $2 \cdot (-7)$
$3 \cdot (2) + 7 \cdot (-7) = 6 - 49 = -43;$
$21x^2 - 43x - 14 = (3x - 7)(7x + 2)$

Appendix 2

65. $5a + 72a^2 - 12 = 72a^2 + 5a - 12$
$72 = 1 \cdot 72,\ 2 \cdot 36,\ 3 \cdot 24,\ 4 \cdot 18,$
$\quad\quad 6 \cdot 12,\ \text{or}\ 8 \cdot 9$
$-12 = 1 \cdot (-12),\ (-1) \cdot 12,\ 2 \cdot (-6),$
$\quad\quad (-2) \cdot 6,\ 3 \cdot (-4),\ \text{or}\ (-3) \cdot 4$
Try $72 = 3 \cdot 24,\ -12 = 6 \cdot (-2)$
$3 \cdot 6 + 24(-2) = 18 - 48 = -30$
$3 \cdot (-2) + 24(6) = -6 + 144 = 138$
Since the sum of the middle terms is $5a$, try to factor using terms that are close together such as
$72 = 8 \cdot 9,\ -12 = (3)(-4)\ \text{or}\ (4)(-3).$
$8 \cdot 3 + 9(-4) = 24 - 36 = -12$
$8 \cdot (-4) + 9(3) = -32 + 27 = -5$
Switch the signs on the -4 and 3.
$72a^2 + 5a - 12 = (9a + 4)(8a - 3)$

67. $12 - 53x + 30x^2 = 30x^2 - 53x + 12$
$30 = 1 \cdot 30,\ 2 \cdot 15,\ 3 \cdot 10,\ \text{or}\ 5 \cdot 6.$
$12 = (-1) \cdot (-12),\ (-2) \cdot (-6),$
$\quad\quad \text{or}\ (-3) \cdot (-4)$
Since the coefficient of x is negative, the answer will have the form: $(\ -\ )(\ -\ )$. We need terms that add to 53 (or -53).
Try $30 = 3 \cdot 10$ and $12 = (-2) \cdot (-6)$
$3 \cdot (-2) + 10(-6) = -6 - 60 = -66$
$3 \cdot (-6) + 10(-2) = -18 - 20 = -38$
Try the combinations.
$30x^2 - 53x + 12 = (2x - 3)(15x - 4)$

69. $-30t - 44 + 54t^2 = 2(27t^2 - 15t - 22);$
$27 = 1 \cdot 27\ \text{or}\ 3 \cdot 9$
$-22 = (1) \cdot (-22),\ (-1) \cdot (22),\ (2) \cdot (-11), \text{or}\ (-2) \cdot (11)$
$54t^2 - 30t - 44 = 2(3t + 2)(9t - 11)$
since $3(-11) + 9(2) = -33 + 18 = -15$

71. $3x^2 - 7ax + 2a^2$

$3x^2 = x \cdot 3x$ and $2a^2 = (-a) \cdot (-2a)$
so the possibilities are:
$(x - a)(3x - 2a)$ or $(x - 2a)(3x - a);$
$3x^2 - 7ax + 2a^2 = (x - 2a)(3x - a)$

Appendix 2

73. $15x^2 - 4xy - 4y^2$

$15x^2 = x \cdot 15x$ or $3x \cdot 5x$

$-4y^2 = -y \cdot 4y,\ y \cdot -4y,$ or $2y \cdot -2y$

$-4xy = -2y \cdot 5x + 2y \cdot 3x = -10xy + 6xy$

so $15x^2 - 4xy - 4y^2 = (3x - 2y)(5x + 2y)$

75. $18u^2 + 20v^2 - 39uv = 18u^2 - 39uv + 20v^2$

$18u^2 = u \cdot 18u,\ 2u \cdot 9u,$ or $3u \cdot 6u$

$20v^2 = -v \cdot -20v,\ -2v \cdot -10v,$ or $-4v \cdot -5v$

$-39uv = 6u(-4v) + 3u(-5v) = -24uv - 15uv$

so $18u^2 + 20v^2 - 39v = (3u - 4v)(6u - 5v)$

77. $12a^2 - 14b^2 - 13ab = 12a^2 - 13ab - 14b^2$

$12a^2 = a \cdot 12a,\ 2a \cdot 6a,$ or $3a \cdot 4a$

$-14b^2 = b \cdot -14b,\ -b \cdot 14b,\ 2b \cdot -7b,$ or $-2b \cdot 7b$

$-13ab = 3a \cdot -7b + 4a \cdot 2b = -21ab + 8ab$

so $12a^2 - 14b^2 - 13ab = (3a + 2b)(4a - 7b)$

79. $10a^2b^2 - 19ab + 6$

$10a^2b^2 = ab \cdot 10ab$ or $2ab \cdot 5ab$

$6 = -1 \cdot -b$ or $-2 \cdot -3$

$-19ab = 2ab \cdot -2 + 5ab \cdot -3 = -4ab - 15ab$

so $10a^2b^2 - 19ab + 6 = (5ab - 2)(2ab - 3)$

81. $56x^2y^2 - 2xy - 4 = 2(28x^2y^2 - xy - 2)$

$28x^2y^2 = xy \cdot 28xy,\ 2xy \cdot 14xy,$ or $4xy \cdot 7xy$

$-2 = -1 \cdot 2$ or $-2 \cdot 1$

$-xy = 4xy \cdot -2 + 7xy \cdot 1 = -8xy + 7xy$

so $56x^2y^2 - 2xy - 4 = 2(4xy + 1)(7xy - 2)$

Appendix 2

83. $22a^2z^2 - 21 - 19az = 22a^2z^2 - 19az - 21$

$22a^2z^2 = az \cdot 22az$ or $2az \cdot 11az$

$-21 = 1 \cdot -21, \; -1 \cdot 21, \; 3 \cdot -7, \text{ or } -3 \cdot 7$

$-19az = 11az \cdot -3 + 2az \cdot 7 = -33az + 14az$

so $22a^2z^2 - 21 - 19az = (2az - 3)(11az + 7)$

85. $(x+3)^2 = (x+3)(x+3)$
$= x^2 + 3x + 3x + 9$
$= x^2 + 6x + 9$

87. $(2y-5)^2 = (2y-5)(2y-5)$
$= 4y^2 - 10y - 10y + 25$
$= 4y^2 - 20y + 25$

89. $(x+3)(x-3) = x^2 - 3x + 3x - 9$
$= x^2 - 9$
which is $x^2 - 3^2$

91. $(3t - 4s)(3t + 4s)$
$= 9t^2 + 12st - 12st - 16s^2$
$= 9t^2 - 16s^2$
which is $(3t)^2 - (4s)^2$

93. $(5a - 2b)^2$
$= (5a - 2b)(5a - 2b)$
$= 25a^2 - 10ab - 10ab + 4b^2$
$= 25a^2 - 20ab + 4b^2$

95. $(8xz + 3)^2$
$= (8xz + 3)(8xz + 3)$
$= 64x^2z^2 + 24xz + 24xz + 9$
$= 64x^2z^2 + 48xz + 9$

97. $x^2 - 25 = (x)^2 - (5)^2$
$= (x+5)(x-5)$

99. $x^2 - 24x + 144$
$= (x)^2 - 2(12)x + (12)^2$
$= (x - 12)^2$

101. $x^2 - 4y^2 = (x)^2 - (2y)^2$
$= (x + 2y)(x - 2y)$

103. $4x^2 + 12x + 9$
$= (2x)^2 + 2(3)(2x) + (3)^2$
$= (2x + 3)^2$

105. $9u^2 - 30uv + 25v^2$
$= (3u)^2 - 2(5v)(3u) + (5v)^2$
$= (3u - 5v)^2$

107. $4a^2 - 25b^2 = (2a)^2 - (5b)^2$
$= (2a + 5b)(2a - 5b)$

Appendix 3

109. $x^2y^2 - 81 = (xy)^2 - (9)^2$
$= (xy+9)(xy-9)$

111. $9x^2y^2 + 6xy + 1$
$= (3xy)^2 + 2(1)(3xy) + (1)^2$
$= (3xy+1)^2$

113. $16x^2y^2 - 1 = (4xy)^2 - (1)^2$
$= (4xy+1)(4xy-1)$

115. $(x+2)^2 - y^2$
$= (x+2)^2 - (y)^2$
$= (x+2+y)(x+2-y)$

Appendix 3

1. $\sqrt{-4} = \sqrt{-1} \cdot \sqrt{4} = i \cdot 2 = 2i$

3. $\sqrt{-32} = \sqrt{-1} \cdot \sqrt{32}$
$= i\sqrt{16} \cdot \sqrt{2}$
$= i \cdot 4 \cdot \sqrt{2} = 4i\sqrt{2}$

5. $3\sqrt{-8} = 3\sqrt{-1} \cdot \sqrt{8}$
$= 3i\sqrt{4} \cdot \sqrt{2}$
$= 3i(2\sqrt{2}) = 6i\sqrt{2}$

7. $3\sqrt{-24} = 3\sqrt{-1}\sqrt{4}\sqrt{6}$
$= 3 \cdot i \cdot 2\sqrt{6} = 6i\sqrt{6}$

9. $5\sqrt{-64} = 5\sqrt{-1}\sqrt{64}$
$= 5 \cdot i \cdot 8 = 40i$

11. $-2\sqrt{-12} = -2\sqrt{-1}\sqrt{12}$
$= -2 \cdot i \cdot \sqrt{4} \cdot \sqrt{3}$
$= -2 \cdot i \cdot 2 \cdot \sqrt{3} = -4i\sqrt{3}$

13. $4 + 2\sqrt{-1} = 4 + 2i$

15. $3\sqrt{-50} + 2 = 3\sqrt{-1}\sqrt{25}\sqrt{2} + 2$
$= 3 \cdot i \cdot 5\sqrt{2} + 2$
$= 15i\sqrt{2} + 2$
$= 2 + 15i\sqrt{2}$

17. $\sqrt{4} + \sqrt{-4} = 2 + \sqrt{-1}\sqrt{4}$
$= 2 + i \cdot 2 = 2 + 2i$

19. $(2+4i) + (3+i) = (2+3) + (4i+i)$
$= 5 + 5i$

21. $(4-i) - (6-2i)$
$= (4-6) + [-i-(-2i)]$
$= -2 + (-i + 2i)$
$= -2 + i$

23. $3 - (4+2i) = (3+0i) - (4+2i)$
$= (3-4) + (0i-2i)$
$= -1 - 2i$

25. $(2-i)(3+2i) = 6 + 4i - 3i - 2i^2$
$= 6 + i - 2(-1)$
$= 6 + i + 2$
$= 8 + i$

27. $(3+2i)(5+i) = 15 + 3i + 10i + 2i^2$
$= 15 + 13i + 2(-1)$
$= 15 + 13i - 2$
$= 13 + 13i$

29. $(6-3i)(4-i) = 24 - 6i - 12i + 3i^2$
$= 24 - 18i + 3(-1)$
$= 24 - 18i - 3$
$= 21 - 18i$

Appendix 3

31. $(2-i)^2 = (2-i)(2-i)$
$ = 4 - 2i - 2i + i^2$
$ = 4 - 4i - 1$
$ = 3 - 4i$

33. $(2-i)(2+i) = 4 + 2i - 2i - i^2$
$ = 4 - (-1)$
$ = 4 + 1$
$ = 5$
which is $(2)^2 - (i)^2$

35. $\dfrac{1}{3i} = \dfrac{1(i)}{3i(i)} = \dfrac{i}{3i^2}$
$\phantom{\dfrac{1}{3i}} = \dfrac{i}{3(-1)} = -\dfrac{1}{3}i$

37. $\dfrac{3-i}{5i} = \dfrac{(3-i)(i)}{(5i)(i)} = \dfrac{3i - i^2}{5i^2}$
$\phantom{\dfrac{3-i}{5i}} = \dfrac{3i - (-1)}{5(-1)} = \dfrac{1 + 3i}{-5}$
$\phantom{\dfrac{3-i}{5i}} = \dfrac{-1 - 3i}{5} = -\dfrac{1}{5} - \dfrac{3}{5}i$

39. $\dfrac{2}{1-i} = \dfrac{2(1+i)}{(1-i)(1+i)} = \dfrac{2 + 2i}{(1)^2 - (i)^2}$
$\phantom{\dfrac{2}{1-i}} = \dfrac{2 + 2i}{1 - (-1)} = \dfrac{2 + 2i}{2} = 1 + i$

41. $\dfrac{2+i}{1+3i} = \dfrac{(2+i)(1-3i)}{(1+3i)(1-3i)}$
$\phantom{\dfrac{2+i}{1+3i}} = \dfrac{2 - 6i + i - 3i^2}{(1)^2 - (3i)^2}$
$\phantom{\dfrac{2+i}{1+3i}} = \dfrac{2 - 5i - 3(-1)}{1 - 9i^2}$
$\phantom{\dfrac{2+i}{1+3i}} = \dfrac{2 - 5i + 3}{1 + 9} = \dfrac{5 - 5i}{10}$
$\phantom{\dfrac{2+i}{1+3i}} = \dfrac{1 - i}{2} = \dfrac{1}{2} - \dfrac{1}{2}i$

43. $\dfrac{2-3i}{3-2i} = \dfrac{(2-3i)(3+2i)}{(3-2i)(3+2i)}$
$\phantom{\dfrac{2-3i}{3-2i}} = \dfrac{6 + 4i - 9i - 6i^2}{(3)^2 - (2i)^2}$
$\phantom{\dfrac{2-3i}{3-2i}} = \dfrac{6 - 5i - 6(-1)}{9 - 4i^2} = \dfrac{6 - 5i + 6}{9 - 4(-1)}$
$\phantom{\dfrac{2-3i}{3-2i}} = \dfrac{12 - 5i}{9 + 4} = \dfrac{12}{13} - \dfrac{5i}{13}$

45. $\dfrac{3+2i}{5-3i} = \dfrac{(3+2i)(5+3i)}{(5-3i)(5+3i)}$
$\phantom{\dfrac{3+2i}{5-3i}} = \dfrac{15 + 9i + 10i + 6i^2}{(5)^2 - (3i)^2}$
$\phantom{\dfrac{3+2i}{5-3i}} = \dfrac{15 + 19i + 6(-1)}{25 - 9(-1)} = \dfrac{9 + 19i}{34}$
$\phantom{\dfrac{3+2i}{5-3i}} = \dfrac{9}{34} + \dfrac{19}{34}i$

47. $\sqrt{-4}(1 - \sqrt{-4}) = i\sqrt{4}(1 - i\sqrt{4})$
$\phantom{\sqrt{-4}(1 - \sqrt{-4})} = 2i(1 - 2i)$
$\phantom{\sqrt{-4}(1 - \sqrt{-4})} = 2i - 4i^2$
$\phantom{\sqrt{-4}(1 - \sqrt{-4})} = 2i - 4(-1)$
$\phantom{\sqrt{-4}(1 - \sqrt{-4})} = 2i + 4 = 4 + 2i$

49. $(2 + \sqrt{-9})(3 - \sqrt{-9})$
$= (2 + i\sqrt{9})(3 - i\sqrt{9})$
$= (2 + 3i)(3 - 3i)$
$= 6 - 6i + 9i - 9i^2$
$= 6 + 3i - 9(-1)$
$= 6 + 3i + 9 = 15 + 3i$

51. $\dfrac{3}{\sqrt{-4}} = \dfrac{3}{i\sqrt{4}} = \dfrac{3}{2i}$

$\qquad = \dfrac{3(i)}{2i(i)} = \dfrac{3i}{2i^2}$

$\qquad = -\dfrac{3}{2}i$

53. $\dfrac{2-\sqrt{-1}}{2+\sqrt{-1}} = \dfrac{2-i\sqrt{1}}{2+i\sqrt{1}} = \dfrac{2-i}{2+i}$

$\qquad = \dfrac{(2-i)(2-i)}{(2+i)(2-i)}$

$\qquad = \dfrac{4-2i-2i+i^2}{4-i^2}$

$\qquad = \dfrac{4-4i-1}{4-(-1)}$

$\qquad = \dfrac{3-4i}{5} = \dfrac{3}{5} - \dfrac{4}{5}i$

55. $\sqrt{x-5}$ is real when $x - 5 \geq 0$ or $x \geq 5$. $\sqrt{x-5}$ is imaginary when $x - 5 < 0$ or $x < 5$.

57. a. $i^6 = i^2 \cdot i^4 = (-1)(1) = -1$

  b. $i^{12} = (i^4)^3 = (1)^3 = 1$

  c. $i^{15} = i(i^2)(i^{12}) = i(-1)(1) = -i$

  d. $i^{102} = i^{100+2} = i^{100} \cdot i^2$

  $\qquad = (i^4)^{25} \cdot i^2 = (1)^{25} \cdot (-1)$

  $\qquad = 1 \cdot (-1) = -1$

59. $(1+i)^2 + 2(1+i) + 3$

$\qquad = 1 + 2(1)(i) + i^2 + 2 + 2i + 3$

$\qquad = 1 + 2i + (-1) + 5 + 2i$

$\qquad = 5 + 4i$